What is far and yet near. Paradise.

Sei Shōnagon 2017, entry 160.

In Memoriam Walter †

← Heini Mathys studied the art of bonsai and kept photographic records of it.

↑ A sea of fog over the Rossberg, Uto (summit of the Uetliberg), and Glärnisch, photographed by Leo Wehrli in 1912. This was the imposing natural backdrop to Evariste Mertens's "Schuler-Ganzoni Garden", designed in 1905.

Borrowed Sceneries

The Influence of Japanese Garden Art on Swiss Landscape Architecture

Rahel Hartmann Schweizer

Birkhäuser Basel

Japan as prima materia

A dilapidated house with a garden overgrown with trailing weeds and knee-deep in wormwood, flooded with moonlight from a brilliant rising moon. Also the moonlight that seeps in through the cracks in the decaying boards of such a house.
The sound of the wind, as long as it isn't fierce.
A place with a pond, in the fifth month when the rains are falling, is a very moving thing. It's deeply affecting to sit for hours on end staring out at the garden, a sea of monochrome soft green with the pond's water as deep green as the sweet flag and reeds that crowd it, and the heavy rain clouds hanging above! In general, I like garden ponds at any time. Indeed all places with ponds are at all times moving and delightful, and of course this is too on winter mornings when the water is frozen over.
Rather than a carefully tended pond, I find delightful the sort that have been left neglected, where patches of reflected moonlight gleam whitely on the water here and there between the swathes of green.
All moonlight is moving, wherever it may be.[1]

Sei Shōnagon

Numerous parks and gardens designed in Switzerland in the 20th century have drawn inspiration from Japan – in fact, many seem almost inconceivable without Japanese influences. Evidence of this can be found in the Archive of Swiss Landscape Architecture (ASLA) in Rapperswil and in the archives of the Teaching and Research Institute "History and Theory of Architecture" (gta) at ETH Zurich. Hansjörg Gadient, Professor of Landscape Architecture at the OST University of Applied Sciences and Director of the ASLA suspected that the bequests in the archives could harbour a wealth of evidence by which to trace this influence. When he approached me three years ago with the idea of literally awakening them from their slumber – the oldest designs date back to the 1890s – and examining them for references to Japan, my curiosity was sparked.

To begin with, I discovered initial indications – a freehand sketch of an alpine motif by Adolf Zürcher [→ p. 277], the model for a green space design by Ernst Cramer for the SUVA in Bellikon [→ p. 300], or an over five-metre-long plan of the lakeside path in Zurich by Willi Neukom [→ pp. 100–101] – but with each new foray into the archives, my sense of discovery was spurred on by new, unexpected finds. The more I researched, the more our initial hunch began to consolidate into a concrete hypothesis. After trawling through thousands of plans, drawings, and photographs, the yield proved to be unexpectedly sizeable, with the legacies of ten designer offices emerging as the most productive sources: Gustav Ammann, Albert Baumann, Ernst Baumann, Ernst Cramer, Hans Graf, Fritz and Fredy Klauser, Walter Leder, Evariste Mertens, the Mertens brothers, Mertens & Nussbaumer, Willi Neukom and Adolf Zürcher.[2]

10 estates – 1000 objects

A first collection of around 1000 objects that exhibited Japanese features was chosen. To distil these to a more manageable number, a set of selection criteria was needed. These took into account the respective biographical background, the period of contact with Japanese models, the way they were conveyed – books, exhibitions, travels, professional colleagues – and how they were perceived by the author in their own writings. A further criterion was to focus on objects in which Japanese influences were formative for a project's overall composition, for example, because it employed a number of Japanese design elements, or that its design was shaped overwhelmingly by Japanese principles such as the interpenetration of inside and outside, the use of borrowed or miniaturised landscapes, the creation of painterly tableaus, the beauty of imperfection, or the phenomenon of emptiness. Finally, the quality of the archived material and its reproducibility were also considered.

The geographical distribution is largely a product of the place of residence and radius of activity of the respective landscape architects, which resulted in a focus on the cantons of Aargau, Bern, Thurgau, Zug, and Zurich – followed by the neighbouring cantons of Solothurn, Lucerne, Schwyz, St. Gallen and, occasionally, Vaud and Ticino, as well as neighbouring countries. The time span of the realised projects ranges from the "Eugster Garden" designed by Gustav Ammann in Altstätten in 1929 to Hans Graf's "Schmid Garden" in Steffisburg in 1997.

This publication marks the first comprehensive examination of this phenomenon based on the work of ten landscape architecture offices. Up to now, influences – if mentioned at all – have only been touched upon in individual cases in the study of a person's body of work – as in the case of Willi Neukom, Ernst Cramer, or Gustav Ammann[3] – rather than identified as generative factors. Until now, no publication has systematically traced the impact of traditional Japanese garden design on Swiss landscape architecture.

Parallels

Switzerland's close contacts with Japan, which began with the first trade treaty between Japan and a Western non-major power in 1864, were in part a product of parallels between the Alpine Nation and the Land of the Rising Sun. These were also decisive for the reception of Japanese garden design: **1.** Their connotation as a place of paradise, **2.** Their respective garden design training, **3.** The relationship between garden and house, **4.** Their small-scale spatial structure, **5.** The influence from outside.

1 Shōnagon 2017, p. 316, entry 323.
2 The surviving estates of women are rare. So far, only Verena Dubach and Ursula Schmocker-Willi have bequeathed their estates to the Archives of Swiss Landscape Architecture (ASLA). Regrettably, they also only considered a modest portion of their works worthy of preservation. Interview with Hansjörg Gadient, Professor of Planning and Design of Urban Open Spaces, Head of the Archive for Swiss Landscape Architecture ASLA, University of Applied Sciences of Eastern Switzerland, 15 Feb. 2021. Only Verena Dubach's works contain isolated Japanese inspirations, which are discussed in the chapter "Inspiration, imitation, integration".
3 Johannes Stoffler devoted himself most extensively to the topic. See Stoffler 2006a, pp. 147–152, pp. 170, 178f, 220, 250f.

1. Like Japan, Switzerland was perceived internationally as a paradisiacal sphere. Each of the two countries also regarded the other as a natural idyll.

2. In Japan, the art of garden design was passed on to successive generations almost as if it were a secret skill. In Switzerland, too, there was initially very little formal training in aspects of garden design. Such subjects had only been taught since 1887 at the École cantonale d'horticulture de Châtelaine (Châtelaine School of Horticulture) in Geneva, and even then design aspects were only a small part of the overall curriculum. In German-speaking Switzerland, there were no training opportunities at all until the opening of Oeschberg Cantonal School for Horticulture and Fruit and Vegetable Cultivation in Koppigen in 1920.[4] Most young landscape architects therefore sought to hone their creative skills abroad. Of the landscape architects featured here, most travelled to Germany or England, and the reception of Japan in these two countries is therefore also examined in more detail in the chapter "From Zukai teien zōhō to Sakuteiki".

3. Garden design has consistently enjoyed a prominent status as an art form in Japan and has defined its relationship to architecture. In Switzerland, as in Germany, avant-garde architects exerted a strong influence on garden design. Some of them – Hermann Muthesius (1861–1927) and Peter Behrens (1868–1940), for example, considered it indispensable that a house and garden be planned together and felt suitably qualified to also take on the design of the garden.[5] Similarly, other garden designers saw the house and garden as a whole, conceived by the same mind, but accorded primacy to the architecture: "[...] the house plays the dominant role, and the garden must by necessity respond to the forms and lines of the house".[6]

At the same time, there was the reverse tendency: "Many garden architects were interested in reclaiming professional ground from architects," arguing for the need for "dedicated garden concepts along with the requisite knowledge of plants"[7].

4. During the Heian period (平安時代, 794–1185), the political centre of Heian-kyō (now Kyōto) became increasingly urbanised. To find unspoiled nature, courtiers had to look beyond the city limits, and brought it back into the "confines of the city gardens". Using trees, grasses, stones, and streams, they created "idealised natural landscapes".[8]

In the 1930s, a change came about in Switzerland due in part to the world economic crisis. The wealthy class of landowners began to decline, and garden designers began to receive commissions from middle-class single-family homeowners for their comparatively small home gardens.[9]

5. Garden design in Japan developed from the adoption of foreign, that is, Chinese – Taoist and Buddhist – influences. These were absorbed, distilled, reinterpreted, and, over the centuries, transformed into what finally consolidated into the Japanese garden, an art form of its own (see "From the plenty of paradisiacal islands to the beauty of emptiness").

In Switzerland, the most historically significant gardens were the product of "impulses from outside"[10], that frequently had their roots in "English, American and Japanese landscapes, gardens from Sweden, and German garden design"[11]. Gustav Ammann compared the task facing Swiss garden designers with what "the Japanese people had learned about a simple way of life through the appearance of the Zen sect of Buddhism" in the 13th century: imbuing one's work with the "value of elegance and the refinement of simplicity" was to be the new goal of Swiss garden design.[12]

A back and forth between landscape and architectural garden, nature, and geometry

Two steps forward, one step back aptly describes the process by which Swiss garden design developed, embedded as it was in the European context, in response to Japanese influences.

From the middle of the 19th century, the English landscape garden began to replace the absolutist "geometric mastery of nature"[13] that for decades had shaped the design of gardens. With time, however, its enlightened principles of "empathy with nature through its imitation" began increasingly to degenerate into "superficial stylistic concepts"[14] and ornamental decoration, which Heini Mathys once disparagingly called "carpet bed gardening".[15] Fuelled by the Arts and Crafts movement and Art Nouveau, the architectural garden stepped into the breach at the turn of the 19th and 20th centuries "as a reaction to naturalism, which had become staid and lacklustre". These new gardens were characterised by orderly layouts, axial spaces, and decorative Art Nouveau elements.

The opening of Oeschberg Cantonal Horticultural School in Koppigen, Bern, in 1920 coincided with this period. Albert Baumann, who shaped teaching at the school for many years, designed a park that both reflected and supported the teaching. It epitomises and embodies the character of symmetrical compositions and axiality. In addition, however, Baumann incorporated a Japanese garden within it.

The founding of the Bund Schweizerischer Gartengestalter (Federation of Swiss Garden Designers, BSG) in 1925 took place in the wake of a new wave of reforms and was supported by garden designers as well as architects. The impetus was the Weißenhofsiedlung built in Stuttgart in 1927. Le Corbusier, in his book *Towards a New Architecture*, and Leberecht Migge, in his article "Towards a New Garden", both argued for a design approach based around living with nature and called for more spatial interplay between house and garden (see "Inspiration, imitation, integration").[16]

The Zürcher Gartenbauausstellung (Zurich Horticultural Exhibition, Züga) in 1933 was the first large-scale project to showcase this trend towards a new aesthetic of "organic design".[17] And it contained not one but two Japanese-inspired gardens, by Otto Spross and by

4 Stoffler 2006a, p. 7.
5 Muthesius also published a book on this subject. See Muthesius 1907; Bucher 1996, p. 25.
6 Encke subsumed his assessments under the term "residential garden". Encke 1910, p. 8.
7 Wolschke-Bulmahn 2009, p. 174.
8 Itoh 1999, p. 72.
9 See Osoegawa 2020c, p. 24.
10 Mathys 1972, pp. 1–2.
11 Ammann 1936.
12 Ammann 1943b, [p. 287].
13 Bucher 1998, p. 4.
14 Bucher 1998, p. 6.
15 Mathys 1972, p. 6.
16 See Le Corbusier 1926; Migge 1927b. The German title of *Vers une architecture* is *Kommende Baukunst*. In response Migge penned an article entitled *Der kommende Garten*.
17 Mathys 1972, pp. 8, 10.

***Usnea longissima* or *Dolichousnea longissima* is the principal species of *nagasagarigoke* (hanging mosses) in Japan and is traditionally associated with jewel stones.**

Paul Schädlich. Also in 1933, an exhibition on "Japanese Architecture and Gardens" at the Kunstgewerbemuseum (Museum of Decorative Arts) in Zurich fuelled a bourgeoning interest in Japanese garden design in Switzerland – precisely at a time when the architectural garden, which developed as a reaction to the landscape garden, was being abandoned in favour of the dwelling garden[18], in order to make room for "a 'natural' design style [through] the dissolution of the formal […]".[19]

"Jewel Tower" and "Jewel Tree"

In 1937, Ammann declared that a "new awareness of landscape" had become a defining factor in garden design. This new appreciation of landscape was reflected in Ammann's own use of plants, which was influenced by the English concept of "wild gardening" propagated by William Robinson, as well as German "wilderness garden art", as advocated by Karl Foerster.[20] These in turn reflected a "steadily growing interest in Japanese garden culture and Far Eastern plants of a picturesque character".[21]

Foerster, who was well received in Switzerland, invented a nomenclature for his cultivars whose poetic epithets bear a striking resemblance to some Japanese names, such as "Jewel Tower" for the Larkspur *Delphinium cultorum* or "Silk Jewel" for the Turkish poppy, *Papaver orientale*.[22] In Japan, *Magnolia compressa* is considered a "jewel tree" and the hanging *mosses Usneaceae Lichenophyta* "jewel stones" [→ p. 7].[23]

At the Landesausstellung 1939 (National Exhibition), a new Japanese inspired unity of the house and garden had become evident. No longer was the garden wrapped around the house, instead "a modest house was placed within the informal, natural surroundings of the garden"[24].

A more significant waypoint in the adoption and assimilation of Japanese garden design, was the G59 Swiss Horticultural Exhibition: it was here that Japanese influences served in a sense as prima materia from which, as once in Japan, an own quintessence had emerged.

In 1964, a hundred years after the first trade relations had been agreed, a dozen Swiss garden designers travelled to Tōkyō for the 9th Congress of the International Federation of Landscape Architects (IFLA). Of the landscape architects discussed in this book, Ernst Baumann, Ernst Cramer, Walter Leder, Hans Nussbaumer, and Willi Neukom were part of this group, as was Richard Arioli in the function of a reporter. As part of the accompanying field trips, they visited several of the most famous Japanese gardens – including scenic, picturesque settings such as the park of the Katsura residence (桂離宮) and reductive, symbolically charged abstract compositions such as the garden of the Zen temple Ryōan-ji (龍安寺) in Kyōto. The latter would go on to inspire a tendency towards stylisation and an appreciation of emptiness.

Misunderstood reduction, however, soon resulted in a proliferation of low-maintenance but desolate "gardens". In the mid-1970s, the natural garden movement entered the resulting cultural vacuum. Its most prominent theoretical proponent was Louis G. Le Roy, with his book "*Natuur uitschakelen. Natuur inschakelen*" *(Switch on nature. Switch off nature)* [25], published in 1973, and its most well-known example the University of Irchel designed by Atelier Stern+Partner in 1982,[26] the first large park to be built in the natural garden style. The Grün 80, held in Münchenstein, brought together both poles, with biodiverse plant communities and formal compartmentation – each underpinned by Japanese influences – acting in unison.

Since the early 1990s, naturalistic approaches and architectural design elements have finally been on an equal footing.[27] The landscape architects featured in this book cover the entire spectrum of approaches.

Water, islands, borrowed sceneries, dry landscapes

In some cases, the "Japanese garden" is expressly cited as an influence, while other gardens feature certain characteristic elements of Japanese origin: bodies of water, islands, stone settings, stepping stones, water basins, bridges, and plants with an interesting character or personality. The use of water was the first conspicuous sign, and several designers published articles on this element alone.[28] The most active of these was Gustav Ammann. His design for the "Eugster Garden" in Altstätten exemplifies how Japanese influences inform the changing form of the water basins: "Until the beginning of the twenties, such basins were only ever neatly rectangular in shape. Flowing forms more akin to the nature of water only became more popular over time."[29]

18 Bucher 1998, p. 6.
19 Stoffler 2006a, p. ix.
20 Dümpelmann 2001, p. 71.
21 Stoffler 2016a, p. 13.
22 Cf. Groeningen Van 1996, pp. 592, 598.
23 Cf. Rüttermann 1999, pp. 82, 83.
24 Ammann 1943b, p. 285.
25 See Roy 1983.
26 Cf. Bucher 1998. Christian Stern formed the studio in 1974 together with Klaus Holzhausen, Edmund Badeja, Gerwin Engel and Hans-Ulrich Weber. Schwerin 2015, p. 71.
27 In the journal *Anthos*, books on each topic were reviewed parallel to one another in the same issue. Cf. anon. 1991.
28 See Mertens 1881c; Zülli 1932; Ammann 1947; Leder 1952; Ammann 1953a.
29 p. 1945; Gustav Ammann Belegbuch IV, NSL 2-8-S-4.

Ammann himself summarised this change as a transformation that moved from the "earlier formal expression of the water basin" as an architectural frame "as with a mirror" to the dissolution of the edges to interweave them with the surrounding planting so that the transparent element connects with the vegetation. He also outlined a contour as being "like an island in a river's current, except that here flowing water appears as land and the island as water, as it were a negative representation of the actual state".[30] Willi Neukom was particularly fond of the principle of inversion and experimented with it frequently, honing it to a degree of perfection.

Ammann likewise incorporated means of crossing a water's surface into his repertoire, using Japanese-inspired slabs or bridging elements[31], as well as shallow waterside areas imitating the impression of a shoreline at the edge of the pond in the garden of the Katsura residence. Borrowing from nature was also a theme: "The irregular water lily pond stands in place of the former marshy meadow in the midst of the flowering perennials garden."[32] A further popular Japanese reference was the mutual interpenetration of house and garden – "much like in a Japanese house and garden, both of which now have an inescapable influence in Europe".[33] Rock gardens were likewise a strong connecting element, with the Japanese influence acting as a means of converting the alpinum native to Switzerland into a dry landscape.

From Ammann to Zürcher

Gustav Ammann (1885–1955), despite at first proposing the architectural garden in opposition to the landscape garden, propagated around 1927 "the dissolution of the formal in favour of a 'natural' way of designing the dwelling garden"[34]. He regarded Japan's gardens as exhibiting a cultural superiority from which "not only the tendency towards the non-architectural, but also the deepening of spiritual values in the garden" could be learned.[35]

Albert Baumann (1891–1976), who was director of horticulture at Oeschberg Cantonal Horticultural School from 1920–1957, made a name for himself as a garden architect in the field of reform garden architecture. At first glance, one would not credit him with an affinity for Japanese culture; however, with the exception of Heini Mathys, none of the other landscape architects whose estates are in the ASLA possessed such a large number of books on traditional Japanese garden design.

Ernst Baumann (1907–1992) possessed a sensitivity comparable to Zen art. Photographs of individual parts of his gardens, when juxtaposed with photographs of Japanese gardens, often bear a striking similarity.

Ernst Cramer (1898–1980) was responsible for the famous "Poet's Garden" at the G59. Although not generally associated with Japan, this sculptural creation can be read as embodying a certain Japanese spirit. In addition to designing gardens with an organic layout that are clearly Japanese in style, he also created two gardens that were expressly conceived as "Japanese gardens".

Hans Graf (1919–2014) often used the motif of the miniature garden and even studied the art of bonsai. For him, influences from the Far East served increasingly as a basis for naturalistic garden designs, which grew in relevance with the need for more ecological approaches to garden design and biodiversity.

Fritz and Fredy Klauser (1885–1950, 1921–2007) differed markedly in their drawing styles. Fritz Klauser's plans are orderly, and his geometric-ornamental garden layouts sometimes conceal their underlying Japanese motifs. Fredy Klauser's attitude is closer to the Japanese, exhibiting a fascination with the timeless, the simple, and unattractive.

Walter Leder (1892–1985) was the quiet advocate of Japanese garden design and perhaps penetrated its soul most deeply of all. He maintained intensive contacts with the Japanese colleagues he met at the IFLA Congress in Tōkyō in 1964, from where he returned with Japan at heart.

Evariste Mertens (1846–1907) created his work long before Japanese influences became more widely known in Switzerland, but at a time of lively exchange with the "Land of the Rising Sun", which began with its forced opening in 1853. His comments on garden design draw on Japanese principles. "Let us strive to gather together what the forest, mountain and valley offer us to observe every day, and let us seek to fuse them intelligently into a whole; […]."[36] His recommendations for the use of water and stones are likewise reminiscent of East Asian practices: "Waterfalls, which can be executed in infinite variations, contribute greatly to enlivening a garden and should be included whenever possible."[37] "Rock sections should, wherever possible, be made exclusively of externally weathered blocks […]."[38]

Walter and Oskar Mertens (1885–1943, 1887–1976), who continued their father's firm as the Mertens Brothers, designed a "rock garden" in 1933, taking their cue from their father's advice. In their creations, and especially their handling of water, one sees recurring evidence of Japanese borrowings.

Hans Nussbaumer (1913–1992) continued this design approach in the team Mertens & Nussbaumer.

Willi Neukom (1917–1983) made lavish and liberal use of Japanese elements – whether for private gardens, public parks, or cemeteries. For a long time, however, most were aware only of his designs for the shoreline in Zurich[39] and the grounds of the ETH Zurich[40].

Adolf Zürcher (1934–2011) was equally at home with organic-biomorphic and geometric-reductionist scenarios. His depictions of trees and stones in his plans have a distinctly Japanese feel. Several of the drawings in his estate clearly reveal the influence of East Asian models.

Common to all the landscape architects featured in this book is that their work – which spans the oscillating development of Swiss garden design from architectural-geometric stringency to landscape-organic fluidity – bears particular witness to the catalytic function of the Japanese model for Swiss landscape architecture.

30 Ammann 1953 gtaA, p. 1.
31 Ammann 1953 gtaA, p. 1.
32 Zülli 1932, p. 5.
33 Ammann 1934, p. 115.
34 Stoffler 2006a, p. ix.
35 Stoffler 2006a, p. 151.
36 Mertens 1881b, p. 90.
37 Mertens 1881c, p. 203.
38 Mertens 1881c, p. 204.
39 Sigel/Jong 2010, pp. 41–48.
40 Stoffler 2016b, pp. 24–28.

From
the plenty of
paradisiacal
islands

to
the beauty
of emptiness

← p. 9: The lush pond in the Mertens brothers' Enzenbühl Cemetery in Zollikon is fed by a waterfall that tumbles over rocks beneath the bridge.

pp. 10–15: The Mertens brothers designed a miniature landscape in the Japanese style for the rock garden at Villa Tannhof (now Landgut Reinhart), Ebertswil in Hausen am Albis.

p. 16: The red Japanese maple in the former Hegner Garden (now Mirani) in Meilen, designed by Ernst Cramer.

It should be reiterated that when ideas about gardens were borrowed, virtually no one correctly understood what they were borrowing. Misinterpretations and reinterpretations – whether by ignorance or accident – combined with native Japanese garden ideas to generate still different forms.[1] Teiji Itoh

A look at the reception of Japanese culture in the West reveals an incoherent understanding of the Land of the Rising Sun. This has to do on the one hand with a somewhat selective perception of Nippon as a haven of unspoilt nature and traditional craftsmanship, and on the other with the self-referential interpretation of forms of artistic and architectural expression with own stylistic principles. The Nabis group of artists, for example, saw their conception of art in Japanese woodblock prints, and the Bauhaus members their architectural maxims in the traditional wooden architecture of Japan. Garden designers, too, eagerly embraced the lushness of the plant world, initially likening the Japanese with the Chinese and the English landscape garden, before later interpreting the range of planting in Japanese gardens solely in terms of a reduced colour palette.

At the same time, the image of Japan was itself distorted by the fact that Japanese culture – religion, philosophy, art, architecture, poetry, and garden design – did not develop "endemically", immune to foreign influences, but has rather imported, imitated, and sublimated these over the years.

The assimilation of foreign influences in early Shintoist Japan went through three main phases. The first contact with the mainland can be traced back to the 6th century, when Chinese Buddhist monks arrived via Korea, bringing Buddhism, the logographic writing script, the classics of literature, Taoism, and Confucianism as well as architecture and garden design to the country. In the 10th century – to continue with this abbreviated overview – the spirit and aesthetics of Zen (禅) spread into the country, again mediated by Chinese monks, and in the early Japanese Middle Ages (Muromachi period 室町時代, 1333–1573), the monochrome ink paintings of the Chinese Song period (960–1279) began to influence Japanese garden art. Through a process of imitation, adaption, and refinement, the Japanese garden finally emerged. Three styles – Shinden (寝殿造り), Shoin (書院造り), and Sukiya (数寄屋造り) – correspond roughly to these three phases.

"Although deemed to be the foundational religion of Japan, the Shinto view of cosmos is a complex worldview that contains influences from various parts of Asia and even Oceania."[2] According to this belief system, the physical universe is divided into five parts. The first is the High Plane of Heaven, Takamanohara (高天原), from which the creation of the world began, inhabited by the heavenly kami (神). Earth is the second part, and beneath it is the kingdom of the dead and evil spirits Yomi no Kuni (黄泉の国). The oceans make up the fourth part, referred to as Watatsumi no Kuni (綿津見), and are the domain of all kinds of creatures, from ordinary fish to dragon kings. The fifth part lies in a mythical sphere beyond the sea – Tokoyo no Kuni (常世の国)[3] – "a utopian land whose inhabitants neither age nor die".[4]

The Japanese garden can only be understood through its historical development. This applies in particular to the still fundamental art of placing stones. The *Sakuteiki* (作庭記, *Records of Garden Making*), which opens with the heading "Ishi wo taten koto" (石を立てん事, "The art of setting stones") qualifies stones as being "indispensable in the creation of a garden"[5], and equates stone setting with the act of garden design in general.[6]

Written during the Heian period (平安時代, 794–1185/1192) in the mid to late 11th century by Tachibana no Toshitsuna (橘俊綱, 1028–1094), it is the oldest example of documenting knowledge hitherto passed down only orally or "otherwise kept secret".[7] Originally, the *Sakuteiki* had neither a title nor illustrations. The oldest surviving version is a set of manuscript scrolls dating from the Kamakura period (鎌倉時代, 1185–1333).[8] A supplementary scroll was added in 1289, and the subsequent two-volume work was known as *Zensai Hishou* or *Senzai Hisshō* (前栽秘抄, *Secret Text on Gardens*).[9] In the Edo period (江戸時代, 1603–1868), the book acquired its current title *Sakuteiki*.[10] As the only written record of that era, it is an inexhaustible source for garden designs influenced by geomancy and Buddhism in early Japan.

1 Itoh 1998, p. 74.
2 Dutton 2010, p. 244.
3 Cf. Dutton 2010, p. 244.
4 Hall/Brown/Jansen 1988–1999, p. 329.
5 Takei/Keane 2004, p. 153.
6 Cf. Takei/Keane 2004, p. 8.
7 Takei/Keane 2004, p. 28.
8 Natsumi 2008, p. 5.
9 Sfakiotaki 2005, p. 46.
10 Natsumi 2008, p. 5.

The treatise was written for the aristocracy, who were the main patrons in the Heian period.

The setting of stones has its roots in the belief in sacred places – mountains or islands – as dwellings of immortals, sages, or the blessed, or in an eternal, paradisiacal land. The ancient Japanese garden developed around the central element of a pond with islands – symbolising the sea, where the mystical land of the blessed or immortals, Tokoyo (常世), lies, as well as the three sacred islands Hōrai (蓬莱), Hōjō (方丈), and Eishū (瀛州).[11]

The Japanese historian Minoru Senda translates "toko" as "solid foundation" and "yo" as "world" so that "Tokoyo" can be understood as the "world that exists for eternity". He associates the land where immortals could live for eternity with Riso-kyo (理想郷), the Japanese term for utopia,[12] and traces the infiltration of the idea of immortality as the formative element of Chinese Taoism in Japan to the 3rd century AD, long before the infiltration of Buddhism.

The earliest descriptions of Tokoyo, influenced by the Chinese concept of Taoist immortals, Shenxian (神仙), as "a place across the sea where ageless Taoist immortals live",[13] can be found in the annals *Kojiki* (古事記, *The Records of Ancient Matters*) and *Nihonshoki* (日本書紀, *The Chronicles of Japan*) compiled in 712 and 720.

The latter mentions in an entry for the year 478 a certain Urashima Tarō (浦島太郎) who, together with his wife, reached the mystical sacred islands and spoke with the immortal beings.[14] Tokoyo is described as a fairy tale-like realm shrouded in smoke and mists, dotted with lush flowers, with towers of silver and palaces of gold, furnished with silken curtains and patterned wallpaper. The immortals played music that blended harmoniously with the singing of flocks of birds. Similar to the western idea of the land of milk and honey, it had a never-ending supply of food.[15] Tokoyo is also identified with Mount Hōrai (Hōraisan, 蓬莱山) by the way it is written – Hōraisan Tokoyo no Kuni (蓬莱山常世の国).[16]

Senda also relates the ideas of a land of immortality to the Ise shrine of the sun goddess Amaterasu-ō-mikami (天照大御神, Tenshō Daijin). According to the *Nihonshoki*, the founder of the Japanese imperial family is said to have chosen this spot because waves from Tokoyo were constantly lapping against it. "This implies a relationship between the founding of Ise shrine and the concept of Shenxian. [...] Thus we can think of Ise as representing a paradise or land of the immortals."[17]

The adaptation of the idea of immortality by the Japanese is documented in a text from the Heian period. Around 1100, the scribe Ōe no Masafusa (大江 匡房, 1041–1111) compiled a collection of originally 37 tales of Japanese immortals. Entitled *Honchō shinsenden* (本朝神仙伝) or just *Shinsenden*, the work reveals the influence of its Taoist Chinese predecessor, the *Biographies of Divine Immortals* (神仙傳, *Shenxian zhuan*) written by Ge Hong (葛洪, 261–341) in the 4th century AD.[18]

It also mentions the three islands of the immortals. In the account of number 19, words are put into the mouth of Hijiri, the disciple of Yosho,[19] in which he claims there is not one among the three mountains of the immortals, Penglai (蓬莱, Hōrai), Fāngzhàng (方丈, Hōjō), and Yíngzhōu (瀛州, Eishu, Eijū), and the five sacred peaks of China that he had not passed.[20]

The work, however, is not solely a compendium of exclusively Taoist sagas. It also includes personalities such as crown prince Shōtoku Taishi (聖徳太子, 574–621), as well as Buddhist intellectuals such as the monks Kūkai (空海, 774–835), founder of the esoteric Shingon sect, and Ennin (圓仁 / 円仁, 794–864), whose diaries achieved fame,[21] as well as the legendary Urashima Tarō mentioned in the *Nihonshoki*.

Lièzǐ (列子, c. 450 BC) also mentions two other mountains besides Penglai, Fāngzhàng, and Yíngzhōu in his famous work *True Scripture of Emptiness*[22]: Dàiyú (岱輿) and Yuánjiāo (員嶠).[23] The immortals enjoyed an existence in harmony and flew about their peaks on the backs of giant cranes. But since the islands were not anchored but drifted freely in the ocean, the Celestial Ruler – according to art historian François Berthier – decreed that turtles should act as pillars to stabilise them [→ pp. 26, 27].[24]

The legend reached Japan in the 5th century AD and not only "found its way into the world of Japanese myths, fairy tales and legends"[25] but also into garden art: dry landscape gardens (Kare-san-sui-teien, 枯山水庭園) began featuring Hōrai Mountain and Turtle and Crane Island as symbolical figures of the Taoist islands of the immortals.[26]

The Lièzǐ also explains why Penglai, Fāngzhàng, and Yíngzhōu appear in Japanese garden creations, while Dàiyú and Yuánjiāo play no role. The latter are said to have broken free of their roots in the form of the aforementioned gigantic turtles and drifted to the North Pole where they sunk.[27]

The most frequently cited mountain of the immortals – alongside the older Kūnlún Shān (崑崙山) – is Penglai, also known as Hōrai. Kūnlún is the western paradise where Xīwángmǔ (西王母), the "Queen Mother of the West", resides as one of the oldest Chinese deities. Penglai is the eastern sphere of eternal life. Together they represent the two branches of the classical Chinese myth system, which merged around the middle of the time of the Warring States (戰國時代,475–221 BC).[28]

In the *Shānhǎijīng* (山海經, *Classics of Mountains and Seas*), an approximate location of Penglai is given in the Gulf of Bohai, that is, in the Sea of Japan.[29] As a result, Penglai and Hōrai were identified with Japan itself.

The myth of the hemispheres of immortality was so powerful that in China from the 4th century BC onwards, several rulers sent expeditions in search of the islands of Penglai, Fāngzhàng, and Yíngzhōu. According to legend, the regent of the first unified Chinese empire, Qín Shǐhuángdì (秦始皇帝, 259–210 BC) feared death so greatly that he sent the scholar, magician, and alchemist Xú Fú (徐福, jap. Jofuku) on a quest to find the elixir of life he believed to be on the islands. The story passed down in the *Shǐjì* (史記), the *Records of the*

11 Itoh 1998, p. 75.
12 Senda 1997, p. 37 (Jap. p. 45).
13 Ibid.
14 Cf. Kleine 2004, p. 69.
15 Cf. Kleine 2004, p. 68.
16 Cf. Kleine 2004, p. 69.
17 Senda 1997, p. 41 (Jap. p. 47).
18 Cf. Kleine 2004, p. 65.
19 He has no actual proper name but is only generally referred to as Hijiri, holy man, blessed man and "disciple of Yosho".
20 Cf. Kleine/Kohn 1999, p. 169.
21 See Reischauer 1955.
22 See Liä Dsi 1921.
23 Cf. "中有五山焉: 一曰岱輿, 二曰員嶠, 三曰方壺, 四曰瀛洲, 五曰蓬莱. Zhōng yǒu wǔshān yān: Yī yuē dài yú, èr yuē yuán jiào, sān yuē fāng hú, sì yuē yíng zhōu, wǔ yuē pénglái. There are five mountains: the first is Daiyu, the second is Yuanqiao, the third is Fanghu, the fourth is Yingzhou, and the fifth is Penglai." Translation after: Wilhelm 1980, p. 98.
24 Cf. Berthier/Parkes 2000, p. 13.
25 Pauly 2014, pp. 20–21.
26 Cf. Berthier/Parkes, 2000, p. 13.
27 Cf. Wilhelm 1980, pp. 98–99.
28 Cf. Gu 1979, pp. 31–57.
29 Cf. Wu 2021, pp. 11, 15.

Chikutō Nakabayashi (中林 竹洞, 1776–1853),
Landscape with Waterfall, 1841, ink on paper,
157.3 × 86.8 cm.

Lan Ying (藍瑛, 1585–1664),
Red Friend, hanging scroll, ink and
colour on paper, 148.9 × 47.3 cm

Sesshū Tōyō (雪舟 等楊, 1420–1506),
View of Ama-no-hashidate, 1501–1506, ink and
light colour on paper, 90 × 178.2 cm.

Ike Gyokuran (池 玉瀾, 1727–1784),
Old Plum Tree, six-panel screen, ink and
light colour on paper, 135 × 257 cm.

Grand Historian, written by Sīmǎ Qiān (司馬遷), says that Xú Fú set off a second time from China never to return again.[30]

The saga gained popularity in Japan because an interpretation of the *Records of the Three Kingdoms, Sānguó Zhì* (三國志) and the *Book of the Later Han, Hòu Hànshū* (後漢書) suggested that Xú Fú landed in Japan, and, adopting the name Jimmu, founded the Japanese imperial house and, crowning himself Jimmu-tennō (神武天皇, 711–585 BC). In particular, the fact that Fan Ye (范曄), who compiled the *Book of the Later Han* in the 5th century from numerous older documents, placed the storey of Xú Fú in the *Treatise on Wa, Wō chuán* (倭傳, *Chronicles of Japan*), fuelled this interpretation. The sinologist Joshua A. Fogel points out, however, that it does not unequivocally state that Xú Fú ended up in Wō, the then common synonym for Japan.[31]

After Qín Shǐhuángdì, it was Hàn Wǔdì (漢武帝, 156–87 BC) who was obsessed more than anyone with the idea of immortality. After many futile attempts to enter the islands, which, no sooner had they been espied seemed to dissolve into mist, they thought it might be possible instead to motivate the immortals to visit the people. Consequently, Hàn Wǔdì set about creating a garden designed to match the residences of the immortals as faithfully as possible. Unable to resist the lure of the replicas of the three islands of Penglai, Fāngzhàng, and Yíngzhōu, it was hoped the sacred immortals would take them over as residences. Placated in this manner, they might then be persuaded to reveal the secret of the elixir's recipe.[32] The park became, as it were, a beacon for the imitation of mythical worlds and the replication of natural landscapes in garden design.

Suí Yángdì (隋煬帝, 560–618), second emperor of the Sui dynasty, emulated the Han ruler. He too created a landscape garden of incredible dimensions with a circumference of 120 kilometres. Out of a 10-kilometre-long lake rose the three "Islands of the Immortals" – imitations of those that Hàn Wǔdì had built – each crowned with pavilions. Other navigable ponds connected by waterways represented the five lakes and four seas of the universe. The banks meandering through the landscape were said to offer a thousand views of unparalleled beauty.[33]

The complex was not just a microcosm of Suí Yángdì's empire, but also of the entire world whose riches he had gathered. Given its gigantic dimensions, it was in effect a reversal of the idea of miniaturisation, extended to virtually cosmic dimensions.

Hàn Wǔdì's "fairy tale islands" served not only as a model for the second Sui emperor, but also set the style for Chinese gardens over the centuries, from which the Japanese concept of "symbolic rock islands in garden lakes" developed.[34]

Indigenous, Shinto, Taoist, Buddhist – the Islands of the Immortals and the Pure Land

The transitions between indigenous and Buddhist beliefs were just as fluid as between the indigenous and Taoist. "The syncretism of Buddhism and Shintō was already underway in the Middle Ages."[35] To the Japanese creation myth came the Buddhist view of the universe, according to which the world consists of nine mountains – with the world mountain Shumisen (須弥山 *Mount Sumeru*) at the centre – and eight oceans. While scholarly circles may have differentiated between the faiths, most lay people saw little substantial difference between Amitābha, the Pure Land of the Buddha, Tokoyo no kuni, and Hōraisan. "All these places were regarded as paradisiacal spheres beyond the reach of common mortals."[36]

Christoph Kleine, professor of religion and philosophy in East Asia, supports this thesis with a quote "from the account of the immortal Yosho": "I have a grandfather in my old village who is a close friend of this old mountain. Please tell [him] that I can see the Buddhas and hear the dharma, ascend to heaven and enter the earth; I can change [my appearance] supernaturally as I wish."[37] Kleine interprets this as evidence of the mixing of the ideas of immortality and birth in one of the Buddha lands.

This argument builds in turn on that of Minoru Senda, who already considered indigenous, Taoist, and Buddhist worldviews to be interlinked: the idea of a Riso-kyo beyond the sea can also be found in Buddhism, specifically in the belief in Putuoluo, where Bodhisattva Guānyīn resides.[38]

The syncretism of historical Japanese religion was also reflected in garden design through a synthesis of the stone arrangements of the Buddhist and Taoist world views. The nine mountains and eight oceans of Hindu-Buddhist cosmology were symbolised by a single stone called Kusen hakkai ishi (九山八海石) placed in a pond. Particularly famous is the Stone of the Nine Mountains and Eight Oceans in the Kyōkochi Pond of the complex known as Kinkaku-ji (金閣寺, Temple of the Golden Pavilion), but actually called Rokuon-ji (鹿苑寺, Deer Garden Temple) in Kyōto. Sanzan-gogaku (三山五岳, Three Mountains and Five Peaks) manages to include the three legendary mountains of the immortals, Penglai, Fāngzhàng, and Yíngzhōu, as well as the five sacred peaks of Taoism: Tài Shān (泰山), Heng Shān (衡山), Song Shān, (嵩山), Huà Shān (華山), and Héng Shān (恆山).[39]

Sometimes they were combined with the four sacred mountains of Buddhism – Wutai Shan, Shanxi; Putuo Shan, Zhejiang; Emei Shan, Sichuan; Jiuhua Shan, Anhui – and supplemented by Fuji-san [→ p. 29], which is revered as sacred in Shintō.[40]

The influx of Chinese garden design influences is first documented in the *Kiki* (記紀),[41] and specifically in the arrival of the Korean craftsman and horticulturist Michinoko no Takumi (路子工) from the kingdom of Paekche in 612. He boasted of a talent for "making figures of hills and mountains",[42] which caused Empress Suiko (推古天皇,

30 Cf. Pauly 2014, p. 20.
31 Cf. Fogel 2013, p. 24.
32 Cf. Keswick 2004, p. 50.
33 Cf. Kuck 1940/2016, p. 11.
34 Keswick 2004, p. 51.
35 Pauly 2014, p. 12.
36 Kleine 2004, pp. 90–91.
37 Kleine 2004, p. 91.
38 Cf. Senda 1997, p. 42 (Jap. p. 47).
39 Tài Shān (泰山), Great Eastern Peak in Shandong Province; Heng Shan (終南山) Great Southern Peak in Hunan Province; Song Shan, (嵩山), Great Middle Peak in Henan Province; Huà Shān (華山), Great Western Peak in Shaanxi Province; Héng Shān (恆山), Great Northern Peak in Shanxi Province.
40 Cf. Wang 2015, p. 38.
41 Collective term for *Kojiki* and *Nihon shoki*.
42 Gyori/Fekete/Herczeg 2018, p. 47.

Suzuki Kiitsu (鈴木 其一, 1796–1858), Island of the Immortals, hanging scroll, gold on indigo dyed silk, 107.32 × 35.56 cm.

554–628), the 33rd and first female Tennō of Japan to commission him to erect a replica of the mystical Mount Meru and to build a "Bridge of Wu"[43] in the garden of her palace.[44]

Also designed during Suiko's reign are the gardens that Soga no Umako (蘇我馬子, 551–626) conceived for himself. The *Nihonshoki* dates their design to 625. Having studied gardens on the Chinese mainland, the aristocrat and advisor to several Japanese emperors during the Yamato period (大和時代, 250–710) designed a garden for his estate on the Asuka River with a pond with islands as its centrepiece.[45] After the fall of the Soga clan, which also erected the first Buddhist temple, Hōkō-ji (方広寺), built between 585 and 609, Umako's house became the residence of Tenmu (天武天皇, ca. 631–686), the 40th Tennō of Japan.[46]

Tenmu relocated the court back to Asuka-kyō (飛鳥京), where he ruled from 672 until 686. A key figure in the formation of the Japanese self-image, he lay the foundation for the writing of the mytho-historical chronicles of the land that later appeared under the names *Kojiki* (712) and *Nihonshoki* (720). The Tenmu dynasty (650–800) also promoted Sinicisation and strengthened Japanese identity. Following the Chinese model, the state system was centralised, historical events were recorded, and Buddhism was promoted through the construction of temples – with Tōdai-ji (東大寺, 752) and its main hall Daibutsuden (大仏殿) in the capital Nara representing the splendour of the dynasty. By assigning the Ise shrine the status of the supreme imperial ancestral shrine, the ruling dynasty cemented its claim to power in the indigenous Japanese self-image.[47]

Amalgamation

The transfer of the imperial city from Asuka-kyō to Fujiwara-kyō (藤原京) south of Kyōto in 694 was accompanied by the construction of numerous Buddhist temples, but the Taoist-based concept of Shenxian remained present.[48] Accordingly, Fujiwara-kyō (694–710), as the first capital of Japan, was not only laid out in accordance with traditional Chinese geomancy and modelled on the Chinese capital Chang'an (Xi'an) from the Tang-period (618–907) but was also equipped with the three islands of the immortals. The same applies to the city of Heijō-kyō (平城京), which served as a residence during most of the Nara period (710–794).[49]

When the capital was moved once again in 794 by the 50th Tennō Kanmu (桓武天皇, 737–806), he had Heian-kyō (平安京) recreated as an almost original replica of the Chinese metropolis Chang'an. He prepared the ground for the "Golden Age" of the Heian period.[50]

This amalgamation can also be seen in the development of garden designs that on the one hand gradually began to fuse secular – agricultural buildings – and religious installations – Shintō shrines – and on the other indigenous, Taoist and Buddhist features.

Niwa (庭) is translated today as garden but corresponds to Chinese Teien (庭園), which originally meant

43 The Wu-style bridge, Kure hashi (呉橋), named after the southern Chinese kingdom of Wu, was probably a typical Chinese ornamented arch bridge.
44 Cf. Deane n.d., p. 3.
45 Cf. Deane n.d., p. 3.
46 Cf. Itoh 1998, p. 75.
47 Cf. Inoue/Watt 2014, pp. 36–46.
48 Cf. Senda 1997, p. 41 (Jap. p. 47).
49 710–740 and 745–784. From 784–794, Nagaoka-kyō (長岡京) in Yamashiro Province was "Imperial Resident City of Nagaoka".
50 Cf. Senda 1997, p. 41 (Jap. p. 47).
51 Cf. Itoh 1998, p. 73.
52 Cf. ibid.
53 Itoh 1998, p. 75.
54 Cf. Itoh 1998, p. 78.
55 Pauly 2014, pp. 22–23.
56 Cf. Itoh 1998, p. 26.
57 Cf. Itoh 1998, p. 76.

Kano Tansui (1839–1907), Island of the Immortals with flanking Chinese landscapes (Eight Views of the Xiao and Xiang Rivers), 19th century, hanging scroll, ink and colour on silk, 95.6 × 34.3 cm.

an area in front of the house spread with fine-grained clay for farming activities.[51] In religious contexts, Niwa acquired the meaning of a sanctified open space, cleansed and purified for the arrival of the kami. This was adapted in the form of white gravel surfaces to the iconography of Shintō shrines and Buddhist temples as religious sites, as well as to the residences of emperors and nobles as centres of secular power, as initially both realms were intertwined.[52]

Coupled with the replication of the indigenous Tokoyo as "a very distant place out in the sea", a paradisiacal sphere "removed from realm of actuality", and "mysterious in many respects",[53] or of the Taoist realm of the immortals Hōrai, Hōjō, and Eishū, or the Buddhist Pure Land of the West, in the form of a pond out of which artificially created islands rose, the prototypical Japanese garden emerged, from which all later developments evolved.[54] The idea of the eternal land of Tokoyo thus merged with the imported Chinese concept of the realm of the immortals and the Buddhist-influenced sphere to which the faithful are guided after the death of Buddha Amida – the Pure Land of the West.

This connection offered the nobility, for whom the paradisiacal connotation of a Pure Land in the beyond aroused the desire for an earthly equivalent, a suitable pretext for the creation of a Garden of Eden in the land of the living. Buddhism supplied seductive descriptions of the Pure Lands: "The colourful descriptions of the Pure Lands, wreathed in clouds of incense and conveyed to them during readings of Sūtras in the temples of Buddhism, which were magnificently decorated compared to the modest Shintō shrines, had an overwhelming effect on the people who until then had only the vague notions of an afterlife that had been passed down to them."[55]

The design of gardens of paradise called Jōdo teien (浄土庭園, Pure Land Garden) was fuelled by the emergence of the Jōdo mandalas (浄土), geometrically constructed cult images of the Pure Land of varying complexity – concentrated on a palace surrounded by the Pure Land or ornamented, symbolising the universe illustrated with landscapes, embellished with writing, and populated with Buddhas.

The Japanese conception of nature, which emerged from their original animistic world view[56], in turn justified the use of real landscapes that the nobles encountered on their travels as models. These led them to explore remote areas of the country, never-before-seen stretches of land whose unspoilt beauty could be associated with paradisiacal ideas. The Seto Sea (瀬戸内海, Seto-naikai), an inland sea between the islands of Honshū, Shikoku, and Kyūshū, which served as a waterway for travellers to western Japan, Korea, and China, was one such landscape.[57]

Anticipating images of the afterlife in this world

An area that exemplifies the interaction of Buddhist and indigenous concepts is the region around Uji south of Kyōto. Rulers built their mansions here since the early Heian period, and the Uji area is also the setting for key scenes in *The Tale of Prince Genji, Genji monogatari* (源氏物語), by Murasaki Shikibu (紫式部, documented 990–1010). Its protagonist, Hikaru Genji (光源氏), is traditionally associated with the poet and statesman Minamoto no Tōru (源融, 822–895), son of Emperor Saga. Two of his creations are prime examples of complexes with spiritual connotations on the one hand and naturalistic ones on the other.

The Seikai mandala can be read as a template for the Byōdō-in (平等院). The estate, originally built by Minamoto no Tōru as the Uji-dono country house, was rededicated by Fujiwara no Yorimichi (藤原 頼通, 990–1074), son of Fujiwara no Michinaga (藤原 道長, 966–1028), as a Buddhist Temple of Equality in 1052. Built in the geometric Shinden style, it reflects the cosmic order, the coming and passing, and the cycle of the seasons. The lake that spreads out in front of the temple has a small island representing Mount Hōrai, and the island and temple are connected by a bridge, symbolising the path to paradise. The architecture and landscape of the Byōdō-in symbolise principles of Taoist and Buddhist philosophy. Teiji Itoh states, however, that for Yorimichi the religious background was only a foil for "the creation of a 'world [on this side]' where he could indulge in aesthetic pursuits"[58].

Shrouded in legend is the Kawara-no-in estate, which Minamoto no Tōru is said to have built not far from the Kamo River "on a plot of land four chô in area", recreating the coastal landscape of Shiogama (塩竈市), "whose coast is dotted with islands covered with pines of unusual shapes"[59]. The garden was mentioned in the annals, establishing Tōru's fame as its creator. Researchers believe that the Shosei-en (渉成園) complex reconstructed by Higashi Honganji (東本願寺) and Kobori Enshū (小堀遠州) owes its origins to the park of Minamoto no Tōru.

The abandoned villa that features in the "Yugao" chapter of the *Genji monogatari* has been interpreted as this residence.[60] However, the garden there is described as being covered with "tremendously" old trees and is "indescribably overgrown". "It was just a large, lonely autumnal landscape. The pond was overgrown with marsh grasses. One could truly be afraid."[61]

Ivo Smits, professor of Japanese art and culture at Leiden University, doubts that Shiogama figured as a garden design toponym. As a "poetic place", Shiogama came to be associated with sadness and loneliness, which the poetry of the 10th century almost exclusively associated with the garden of Kawara-no-in. It was only after the 10th century that the original Shiogama established itself as an independent literary toponym. "It is far from obvious, then, that Tōru chose Shiogama because it was so famous: during his lifetime in ninth-century Japan, it was not."[62]

The Japanologist Judit Árokay also emphasises the originally melancholic association of Shiogama, which "initially evoked the sense of mourning of being separated from the capital, and often from a lover still there". This association began to wane in the course of the 11th century, "and the image took on a life of its own as a landscape. From then on, new motifs became associated with this place, especially fog and the moon."[63]

In the garden of his Rokujō-in villa, Onakatomi no Sukechika (大中臣輔親, 954–1038) imitated the natural dam of Ama-no-hashidate (天橋立). The headland on the coast of Tango, also called the "Sky Bridge"[64], belongs together with Matsushima (松島町), originally the name of a group of islands in Shiogama Bay,[65] and Miyajima Island (宮島, Shrine Island), in the Seto Inland Sea off the coast of Honshū, to the so-called Nihon sankei (日本三景), the three most beautiful landscapes in Japan, as defined by the Confucian scholar Hayashi Razan (林 羅山, 1583–1657).

While these seascapes were initially still tied to the belief-based ideas of the land of Tokoyo, the focus gradually shifted towards an aesthetic primacy. The ponds still bore the symbolism of the sea, but the importance of the religious motives diminished in favour of aesthetic aspects. The notion of striving for supernatural bliss and eternal life was gradually supplanted by a desire to recall favourite or famous sceneries.[66]

The imitation of a famous natural scene was defined as a Shukkei (縮景, shrunken scenery) and a garden designed in this manner as a Shukkeien (縮 景 園), while a landscape worthy of imitation was a Meisho (名所).[67]

The earliest mention of this design approach is found in the *Sakuteiki*. Its author Tachibana no Toshitsuna (1028–1094) was the son of Fujiwara no Yorimichi, the guiding spirit of the Byōdō-in, and belonged to one of the Four Great Clans Shisei no uchi (四性ノ内) – Tachibana, Fujiwara, Genji, and Heike – that had extensive estates and would potentially benefit from reading it. Addressing them, Tachibana noted in his family album: "Consider the famous landscapes of our country and attain an understanding of their interesting spots. Recreate the essence of these sceneries in the gardens, but proceed freely, not strictly according to the model."[68] At the same time, the author expressly noted that gardens designed by man could not surpass the beauty of nature.[69]

The esoteric branches of Buddhism gave further impetus to the idea of garden designs borrowed from nature and conceived as earthly paradises. The Buddhist sect Shingon-shū (真言宗, School of the True Word) founded in 807 by the monk Kūkai (空海, 774–835), later celebrated as Kōbō Daishi (弘法大師, The Grand Master who Propagated the Dharma), enthralled many both aesthetically through its ornate ceremonies and ideologically through its teaching that Buddhahood can be attained in this life.

Similarly, the Tendai-shū (天台宗, School of the Lotus Sūtra), which was based on the Lotus Sūtra and was spread in Japan by Saichō (最澄, Highest Clarity, 767–822) – posthumously revered as Dengyō Daishi (伝教大師,

58 Itoh 1998, p. 29.
59 Bernard 1977, p. 1038–1039.
60 Cf. Harper/Shirane 2015, p. 132; Bargen 2015, p. 230.
61 Shikibu 1966, vol. 1, p. 107.
62 Smits 2022, p. 6.
63 Árokay 2003, p. 483.
64 Looking at Ama-no-hashidate from Kasamatsu Park, bent over backwards through spread legs, one has the impression that the sandbank is floating between heaven and earth.
65 "Interestingly, the popular motif Matsushima also derives from Shiogama Bay: a small group of islands in the bay was originally called this. Matsushima then took on a life of its own and became associated with other images as an independent poetic motif, [...]." Cf. Árokay 2003, p. 484.
66 Cf. Itoh 1998, p. 76.
67 縮 景 means close-up, 園 garden.
68 Takei/Keane 2004, p. 119.
69 Cf. Takei/Keane 2004, p. 146.

Ogata Kōrin (尾形 光琳, 1658–1716), Landscape with Mount Fuji seen from the sea, hanging scroll, ink on paper, 34.5 × 49.4 cm.

Gakutei Harunobu (岳亭 春信, 1786–1868),
Murasaki Shikibu (A young woman with writing utensils in a pavilion looking at a river), 19th century, ink, strong colour and gold on paper, 38.5 × 54.8 cm.

Grand Master of the Propagation of the Doctrine) – promised paradise on earth through the attainment of knowledge.

The foundation of the doctrine, the Lotus Sūtra,[70] gained prominence due to its "popular appeal", as it promised salvation to all people "regardless of their offences or merits, status or gender". Especially the latter – the prospect of salvation for women as well – "helped the 'lotus sūtra' gain great popularity in Japan during the late Heian period in the 10th–12th centuries, when especially imperial ladies-in-waiting and educated women of the aristocracy acquired greater self-confidence and began to exert a stronger influence on cultural life."[71]

World literature as references for a rich use of plants

One of the first to name the Lotus Sūtra[72] was Sei Shōnagon (清少納言, c. 966 – c. 1025), the author of *The Pillow Book, Makura no Sōshi* (枕草子).[73] Along with the *Tales of Prince Genji* by her colleague Murasaki Shikibu, it became one of the most famous works of the Heian period, and of Japanese poetry in general, and eventually also of world literature. As a moral portrait of the Heian period, it testifies to the poetic sensibility of the day and conveys a picture of the understanding of nature at the time, including invaluable evidence of the extremely diverse range of plants in the gardens of the Heian period compared to later eras. Sei Shōnagon mentions over a hundred different plants (see plant catalogues).

"The prolific use of flowers and meadow grasses distinguishes the Heian-period gardens from modern Japanese gardens, in which those kinds of plants are almost non-existent",[74] remarked Jirō Takei and Marc P. Keane, who have translated the *Sakuteiki*. They also emphasise the more natural use of plants compared to later centuries, including the creation of meadow areas.[75] The garden was "naturalistic, somewhat uncontrolled, and abundant"[76] with "pampas grass (*Miscanthus sinensis*), Chinese bellflower (*Plytacodon grandiflorum*), and plantain lily, all of which are used to create the feeling of a meadow, for instance along streams in the garden."[77]

In detail, the *Sakuteiki* contains instructions on the creation and design of ponds, streams, and waterfalls as well as islands. Descriptions of the arrangement of stones and the planting of trees are interspersed with geomantic principles and Buddhist teachings. (References to Buddha Amida or the Pure Land are absent in the *Sakuteiki*, although gardens were created at that time according to ideas people had about the Pure Land,[78] as is mentioned, for example, by Sei Shōnagon)[79].

The *Sakuteiki* defines gardening styles according to naturally occurring landscape phenomena, such as ocean style, broad river style, mountain torrent style, wetland style, and reed style.[80] Similarly, island styles are specified as mountain isle, meadow isle, rocky shore isle, cloud type, mist type, cove beach type, slender stream, tide land, pine bark, and so on[81], while waterfalls, which may trickle, shoot out, be set back on one side, be visible in profile, or offset right to left, are classified as cloth fall, twin fall, thread fall, stepped fall, leaping fall, and off-sided fall.[82] Stones are described as standing and reclining, fleeing and chasing, and are arranged in "multilayer shape, crown shape, writing desk shape, bucket shape, and so on".[83]

As valuable as the *Sakuteiki* is as a source for the study of early Japanese garden design, it was first and foremost a manual of a technical nature[84], not an iconographic interpretation. Justifications for certain instructions mostly cite geomantic requirements and make little recourse to any aesthetic commentary. Exceptions are made, for example, to choose "an especially distinctive stone"[85], to "focus on the distant view"[86], to create a "most pleasing sight",[87] a design with "a splendid effect",[88] a "murmuring stream"[89], or to ensure that one allows "the gently burbling water to be easily seen from the halls nearby",[90] while "always keeping the overall design of the garden in mind".[91] Finally, it is surprising that the Taoist symbols for longevity, the turtle and the crane, are not recommended as forms of islands but of ponds.[92]

The *Sakuteiki* is modern in that "for the most part, the same rules and techniques, or variants of this explicit gardening information, are still employed today by traditional gardeners and garden designers."[93] That is, it contains most of the elements that would come to define garden design in the following centuries. Apart from a complex system of taboos, Takei and Keane identify three lines of tradition, that of the image of nature, that of ancient Chinese geomancy, and that of Buddhism.

The *Sakuteiki* marks, in a sense, a pivotal point between the Chinese-Taoist-influenced style of design and the establishment of own Japanese styles. This also applies to the phenomenon of incorporating the external landscape as "borrowed scenery", which has been intensively received in the West. The borrowed landscape has its roots in China. The Chinese concept of Jiejing (借景), borrowing scenery, was popular in China during the Song Dynasty (宋朝, 960–1279), which valued the inclusion of settings in private gardens. Jiejing means the incorporation of images of scenery outside the garden into its design, which reveal themselves as one strolls through the garden, like images on an ink scroll. The landscape should pervade the garden and, in turn, the garden should reveal something of itself to the outside world.

Called shakkei in Japan, the principle can be translated as "to vividly capture". Although it is not expressly mentioned in the *Sakuteiki*, Takei and Keane have no doubt that there was a good awareness at that time of the visual appeal of the interaction between inside and outside. In the capital Heian-kyō, the nobles in their residences would have had an unobstructed, spectacular view of "the long, serpentine ridge of mountains on the eastern or western flanks, the layered ranks of mountains that climb ever higher toward the north" and "especially Mount Hiei, which sits prominently in the

70 jap. Myōhō renge kyō 妙法蓮華経 or Hoke-kyō 法華経.
71 Brinker 2008, p. 36.
72 "Of course the lotus sūtra" in: "Sūtras", entry 195, in: Shōnagon 2017, p. 228.
73 Stein, Michael, "Afterword", in: Shōnagon 2017, p. 331.
74 Takei/Keane 2004, p. 39.
75 Cf. Takei/Keane 2004, p. 42.
76 Ibid.
77 Ibid.
78 Cf. Takei/Keane 2004, p. 74.
79 Cf. entry 31, in: Shōnagon 2017, p. 42, note 76.
80 Cf. Takei/Keane 2004, p. 122.
81 Cf. Takei/Keane 2004, p. 127.
82 Cf. Takei/Keane 2004, p. 132.
83 Takei/Keane 2004, p. 139.
84 Cf. Takei/Keane 2004, p. 32.
85 Takei/Keane 2004, p. 139.
86 Takei/Keane 2004, p. 135.
87 Takei/Keane 2004, p. 138.
88 Takei/Keane 2004, p. 135.
89 Takei/Keane 2004, p. 134.
90 Takei/Keane 2004, p. 138.
91 Takei/Keane 2004, p. 138.
92 "Ponds should be designed in the shape of a turtle and a crane." Takei/Keane 2004, p. 145.
93 Takei/Keane 2004, p. 32.

Tosa Mitsunobu, (土佐 光信, 1434?–1525?), "The Maidens" (Otome, 乙女), Illustration to Chapter 21 of the Tale of Genji (Genji monogatari), 1509–1510, album leaf, ink, colour, and gold on paper, 24.2 × 18 cm.

Tosa Mitsunobu, "Red Plum Blossom" (Kōbai, 紅梅),
Illustration to Chapter 43 of the Tale of Genji (Genji monogatari),
album leaf, ink, colour, and gold on paper, 24.2 × 18 cm.

Tosa Mitsunobu, "Spring Shoots" (Wakana: Ge, 若菜下),
Illustration to Chapter 35 of the Tale of Genji (Genji monogatari),
album leaf, ink, colour, and gold on paper, 24.2 × 18 cm.

Tosa Mitsunobu, "The Eastern Cottage" (Azumaya, 東屋),
Illustration to Chapter 50 of the Tale of Genji (Genji monogatari),
album leaf, ink, colour, and gold on paper, 24.3 × 18 cm.

Tosa Mitsunobu, “The Bluebell” (Asagao, 朝顔),
Illustration to Chapter 20 of the Tale of Genji (Genji monogatari),
album leaf, ink, colour, and gold on paper, 24.2 × 17.9 cm.

Tosa Mitsunobu, “The Mayfly” (Kagerō, 蜻蛉),
Illustration to Chapter 52 of the Tale of Genji (Genji monogatari),
album leaf, ink, colour, and gold on paper, 24.2 × 18.4 cm.

northeast", "from any of the garden pavilions, from a boat on the pond or from a vantage point on one of the small hillocks in the garden"[94].

The concepts of shukkei and shakkei were sometimes equated in their reception. Tetsurō Yoshida wrote that the garden of the imperial Shugakuin Palace on the northern edge of Kyōto was "a particularly good example of the incorporation of the natural landscape not only into the garden itself but also as a background".[95]

Vividly capture

The garden that was created in the grounds of Emperor Saga's (嵯峨 天皇, 786–842) Shinden Mansion after his retreat from rule in 823 in western Kyōto is the oldest to have survived the test of time. In 880, Saga-no-in was rededicated as the Daikaku-ji temple (大覚寺). It contains the most complete surviving garden lake from the Heian period, the Osawa Pond, which was formed by damming the natural course of the river that poured into the plain from the Nakoso Waterfall.[96] The pond, with two small islands, was of a size that could be navigated by boat to enjoy the garden, which is why it is called the Chisen shuyu-shiki-teien (池泉周遊式庭園, pond spring boating garden). It was modelled on Lake Dongting (洞庭湖, Dòngtíng Hú) in the Chinese province of Hunan. The distant place of longing, miniaturised in the manner of Shukkei, is an adaptation of the nearby Nakoso Waterfall, which was recreated in the form of a stone arrangement.

The pond in the garden of Tenryū-ji (天龍寺, 1339), which was set out five hundred years after Daikaku-ji in Arashiyama in western Kyōto, is too small to be navigable. Designed as Chisen-kaiyū-shiki-teien (池泉回遊式庭園, pond spring stroll garden), it is therefore best experienced on foot. With views of the two hills of Kameyama and Arashiyama, it is considered the earliest example of the art of borrowing the outside landscape as shakkei. Tenryū-ji is also a prime example of the connection between classical Chinese iconography, Japanese myth-making, and Zen Buddhist interpretation. The seven stones "floating" as islands in the pond refer to the mythical Hōrai as Hōrai-ishigumi (蓬莱石組, Hōrai stone composition).

A dry waterfall, Karedaki (枯滝), connotated as Ryumon-no-taki (龍門瀑, Dragon Gate Waterfall), derives originally from the Chinese sphere, but here it is given a Zen Buddhist reinterpretation. The carp stone at its foot, symbolising a carp fighting against the current, is an allusion to the Chinese legend according to which a carp, after leaping up the gorge of the Yellow River at Lung Men (Dragon Gate), is transformed into a dragon. In China, the motif was a metaphor for successfully passing the Confucian civil service examinations; in Japan, it was interpreted as a symbol for Zen enlightenment achieved through meditative exercises.[97]

Some parts of the design of the Tenryū-ji have been attributed to Musō Soseki (夢窓 疎石, 1275–1351). The Zen scholar and poet was not only well-versed in Chinese culture and one of the first to champion gardens as a means of meditation, but also inspired by the monochrome ink painting of the Song dynasty.

Likewise associated with the famous Zen priest – although not conclusively determined – the origins of the Saihō-ji (西芳寺) in Kyōto also date back to the Kenkyū era (建久, 1190–1198), more than a hundred years before Soseki visited it in 1334 and transformed it into a Zen temple. In addition to its characteristic expanses of moss, from which it derives its name Koke-dera, a stone cascade once again figures as a garden design paradox.

Matsunosuke Tatsui handed down the importance of this design style, which was reflected in the grounds of Kinkaku-ji (金閣寺, Gold Pavilion Temple, 1397), actually Rokuon-ji (鹿苑寺, Deer Garden Temple), and Ginkaku-ji (銀閣寺, Silver Pavilion Temple, 1482–1490), actually Jishō-ji (慈照寺).[98] Both gardens were founded on the threshold of the Kamakura period as it transitioned into the Muromachi period. The warrior caste that took the sceptre and moved the capital to Kamakura denounced the lavish opulence of the gardens, giving preference instead to the reduced aesthetics of Zen. In the Muromachi period, the nobility revived under Tennō Go-Daigo (後醍醐天皇) and established the already developing spread of Zen philosophy, the appreciation of Song painting, and the introduction of the Shoin style of architecture. In a departure from the Shinden style, asymmetrical compositions became more widespread in architecture. The gardens were characterised by smaller-scale structures and a rejection of decorative aspects in favour of the use of artistic means – such as contrasts of light and dark and staggered depth – to enhance the effect of space.

Amazingly, the *Sakuteiki* already recommends a way of achieving this optical illusion: the technique of "dividing the view",[99] which allows the horizon to be "glimpsed just slightly through the gaps between the hills"[100], has the effect of "making a pond in a small garden look as large as possible".[101]

Naturalistic abstraction and abstracted nature

The dry garden is also already mentioned in the *Sakuteiki*: "There is also a way to create gardens without ponds or streams. This is called the Dry Garden Style."[102] Here, kare-san-sui (枯山水, dry landscape or 涸山水, parched landscape) meant a dry area of a landscape garden that, like the grounds of Tenryū-ji and Saihō-ji, was best experienced by moving through it. In the Muromachi era, however, its meaning shifted. The term was instead associated with a rock garden that was not to be entered but rather meditatively viewed from the interior or veranda.[103] The influence of Song painting unfolded in such a way that Matsunosuke Tatsui interpreted the gardens as "three-dimensional monochrome landscapes [...] drawn with rocks"[104].

The Kare-san-sui of the Daisen-in (大仙院), a sub-temple of the Daitoku-ji (大徳寺, 1315) in northern Kyōto,

94 Takei/Keane 2004, p. 23.
95 Yoshida 1957, p. 18.
96 Cf. Natsumi 2008, p. 6.
97 Cf. ibid.
98 Cf. Tatsui 1949, pp. 14–15.
99 Takei/Keane 2004, p. 173, note 102.
100 Takei/Keane 2004, p. 144.
101 Takei/Keane 2004, p. 173, note 102.
102 Takei/Keane 2004, 2004, p. 122.
103 Cf. Walker 2022, pp. 156–181.
104 Tatsui 1949, p. 16.

dates from around 1509–1513. It is variously attributed to the painter Sōami (相阿弥, c. 1455–1525) because the landscape depiction "The Eight Views of Xiaoxiang" on the sliding walls inside the hall are said to be by him, and to the founder of Daisen-in, the Zen priest Kogaku Sōkō (古岳宗亘, 1464–1548).[105] As an L-shaped courtyard bounded by walls and hedges, the garden wraps around the main hall, with its main focus in the corner area. It is dominated by a dry waterfall which, passing through washed-out gorges between towering crags, plunges into the plain, where it now creates a gravel bed as a river meandering between stone islands. It flows under a bridge made of a single stone slab that leads to a stone symbolising Crane Island. Finally, it flows leisurely as a stream before dividing. On one side, bounded by a pierced wall[106], it pours into the sea, where the fictitious water, drawn in raked gravel, washes around a stone shaped like a sailing junk[107] and another associated with a turtle struggling against the tide.[108] On the long side of the Hōjō Hall (北条), the waves flow around another group of stones and "feed" a stone water basin before spreading out over a wider area, also adorned with several islands [→ pp. 52, 53].[109] Despite its symbolism, the garden of the Daisen-in still has a naturalistic quality. The stone formations represent imposing rocky crags, spectacular cliffs, thundering waterfalls.

In contrast, the garden of the Ryōan-ji (龍安寺, Temple of the Dragon at Peace, from 1488) in northwest Kyōto is highly abstracted. Laid out in the 15th century, its current manifestation dates from the Edo period (江戸時代, 1603–1868). Sometimes associated with Sōami, his authorship is now considered doubtful. The dimensions of the garden are just 9×23 metres. It is enclosed by walls and a wooden veranda that affords a view onto the meticulously raked sand surface and is populated by 15 stones composed in groups of two, three, and five. In contrast to the Daisen-in, the stones are comparatively plain, as if no unusual shape or colouring should distract from their relationship within the individual arrangements or the relationship of the groups to each other and to the "empty" space surrounding them. The latter appears all the more stylised because while the sand forms small wave crests and troughs, their strictly linear orientation makes them seem completely abstract, disrupting the association with a body of water. As a concession to the naturalistic conception, the sand is raked in concentric circles around the stones, which also rest on a bed of moss reminiscent of a coastline. Nevertheless, the garden of the Ryōan-ji is unmistakably minimalist and the apotheosis of the beauty of empty space, also known as the remaining white, Yohaku no bi (余白の美), and with it that facet of the mysterious and the hidden associated with yūgen (幽玄). As a garden design, it is an oxymoron, a built Zen-Kōan, an "enlightening enigma",[110] in which opposites collide and are also brought into harmony. Nevertheless, the Ryōan-ji does not float, as it were, in empty space, but drifts in the shifting sands of traditional Japanese garden styles, since it is placed in a large paradise garden.[111]

Pilgrimage on the back of the turtle and the wings of the crane

An example that epitomises the expression of religious syncretism in garden design is the most famous garden creation attributed to Sesshū Tōyō (雪舟 等楊, c. 1420–1506) at Jōei-ji (常栄寺庭園, Jōeiji teien) in Yamaguchi. In the Sesshutei named after him, he combined a pond in the shape of the Chinese character for "heart" (心字池, Shinjiike, heart character pond), symbolically fed by a dry waterfall, with a meadow in which he placed 13 stone formations. These in turn embody the real mountains Wutai Shan, Zhongnan Shan, Song Shan, Huà Shān, and Fuji-san as well as the mythical elevations Héng Shān and Hōrai, the legendary islands of the turtle (亀島, Kamejima) and the crane (鶴島, Tsurujima), as well as historical figures: Buddha (Butsuiwa), the sixteen Arhats (十六羅漢岩, Jūroku rakan iwa, Stones of the Sixteen Arhats), and the Chinese Chan master Bǎizhàng Huáihái (百丈懷海, Hyakujō Ekai, 720–814).[112]

In contrast to the highly abstracted stone images in the more renowned contemporary Kare-san-sui in Ryōan-ji, the rocks at Sesshutei represent actual mountains and historical figures and are integrated into a naturalistic setting – not an artificial-looking gravel bed. The way it is meant to be experienced – from a seated position on the veranda of the temple – is, however, the same as in Ryōan-ji. It is therefore referred to as Chisen-kanshō-shiki-teien (池泉観賞式庭園, pond spring appreciation garden). David Slawson defines this landscape typology as a scroll garden: "[...] the garden is revealed all at once in a single frame."[113] At Sesshutei one can enact "a pilgrimage through China, ancient Buddhist and Taoist mythologies and histories and the sacred territory of Mt. Fuji"[114] from a seated position on the temple veranda.

An introspective atmosphere of seclusion from the world outside also prevails in the Roji (露地, dew-covered ground) or Rojiniwa (露地庭) tea gardens, whose development reached their peak in the Momoyama era (安土桃山時代, 1573–1603) and found its most dedicated advocate in Sen no Rikyū (利休, 1522–1591).[115] Unlike the Chisen-kanshō-shiki-teien, the garden cannot be viewed from the house. Rather, it is designed to be walked through in the sense of a stroll garden. Unlike the Chisen-kaiyū-shiki, which focuses on a pond, it focuses on the Chashitsu (茶室) teahouse, which is hidden from view and must be walked to in a procession along a winding path. In its seclusion, the Roji is also divorced from the surrounding landscape. In that sense, it is the opposite of a Shakkei garden (借景), which incorporates a distant view into the composition.[116]

The "sublime synthesis" of Villa Katsura

A strolling garden also surrounds Villa Katsura. Walking along the paths, the garden spaces unfold sequentially in the manner of a *promenade architecturale*, revealing constantly changing perspectives and surprising views.

105 Cf. Kuck 1940/2016, pp. 147–148.
106 In the photograph of Harada Sunao that Kuck depicted, this wall did not yet exist. Cf. Kuck 1940/2016, n.d. (between p. 140 and p. 141).
107 Cf. Kuck 1940/2016, pp. 143–149.
108 See Walker 2022, p. 160.
109 Cf. Kuck 1940/2016, pp. 143–149.
110 Berger 2014, p. vii.
111 Cf. Walker 2022, p. 188.
112 Cf. Fowler 2018, p. 148.
113 Slawson 1991, p. 81.
114 Fowler 2018, p. 149.
115 Cf. Sfakiotaki 2005, p. 105.
116 Cf. Sfakiotaki 2005, p. 20.

It is an orchestration of vistas in the technique of Miegakure (見え隠れ, hiding and revealing), which creates a varied series of events while simultaneously deceiving one's perception of the actual dimensions and stimulating the imagination.[117]

Extending across 56,000 m^2, the Katsura Rikyū (桂離宮, 1620–1645) estate has been regarded since the beginning of the 20th century as a "sublime synthesis"[118] of Japanese garden aesthetics. Although built in the 17th century, its origins date to the Heian period. Perceived as a picturesque place of longing, predestined for watching the moon, it formed the fictional setting of the site described in the 18th chapter of the *Genji monogatari*, "Matsukaze" (松風, Wind in the Pines), where Prince Genji built his home.[119]

Inspired by his favourite book, Prince Hachijō no miya Toshihito (八条宮 智仁親王, 1579-1629) began construction in the 1620s on what was then known as the Katsura teahouse (Katsura no chaya 桂の茶屋) or Katsura annexe villa (Katsura betsugyô 桂別業). Twenty years later, his son, Prince Toshitada (智忠親王, 1620–1662), inherited the property, which in the 20th century was named Katsura Rikyū (Imperial Detached Palace);[120] he renovated it and added teahouses and pavilions. Its design is attributed to the tea master and garden designer Kobori Enshū,[121] who knew how to depict the artistic milieu of Kyōto in the first half of the 17th century. He fused the intersecting influences of the refined tastes of aristocratic society in the Katsura villa, the religious asceticism of Zen Buddhism at the Hassō no seki tea pavilion (八窓席) of the Konchi-in temple (金地院), and the artistic impressionism of monochrome ink painting in the gardens of the Nijō-jō imperial villa (元離宮二条城).[122]

The residential buildings – Koshoin (古書院, old Shoin), Chūshoin (中書院], and Shinshoin 新書院, new Shoin) – are built in the Sukiya style[123] and are compelling in their rustic simplicity. Teahouses scattered around the garden – Gepparō (月波楼, Moonwave Teahouse), Shōkin-tei (松琴亭, Pine Harp Teahouse), Shōka-tei (賞花亭, Flower Appreciation Pavilion), Shōi-ken (笑意軒, Humour Teahouse) – can be discovered along the paths that wind around the centrally positioned pond and lead to intimate places to admire the moon, across bridges onto islands, or to the headland that, as a shukkei, reproduces the longed-for landscape of Ama-no-hashidate in miniature.

117 See Fowler 2013, p. 106; Sfakiotaki 2005, pp. 21, 30.
118 Natsumi, 2008, p. 5.
119 Cf. Shikibu 1966, vol. 1, pp. 525–547.
120 Cf. Fiévé 2012a, p. 21.
121 See Tatsui 1949, p. 50. Benoît Jacquet refers to research by Ozawa Keijirō (小澤圭次郎, 1842–1932). Jacquet 2013, p. 99. By contrast, Fiévé writes that no source unequivocally attests to Enshū's authorship. Fiévé 2012b, p. 47.
122 Cf. Fiévé 2012b, pp. 30–31.
123 The Daimyō (大名) Toyotomi Hideyoshi (豊臣 秀吉, 1537–1598) is considered the initiator of Sukiya architecture. In 1587, he hired the tea master Sen no Rikyū (千 利休, 1522–1591) to build a teahouse around his Jurakudai castle (聚楽第) in Kyōto. The building, known as the colourful Shoin, is considered the first example of Sukiya architecture to have perfected the stylistic features of the Shoin style – the softening of symmetry and blurring of the relationship between inside and outside. Cf. Itoh 1972, p. 12.

From Zukai teien zōhō to Sakuteiki

Projection, feedback, cross-pollination[1]

← pp. 41–42: The Eugster Garden in Altstätten, designed by Gustav Ammann, is now maintained by Karl Zünd Park AG.

pp. 43–46: Willi Neukom's lakeside path in Zurich is a popular recreational area.

pp. 47–48: Originally designed by Willi Neukom for the G59, the Nymph Pond on the Zürichhorn with its iconic stepping stones exudes Japanese flair.

> To the Japanese mind, the dew drop glittering on a lotus leaf in the morning sun does not merely reflect, but concentrates and contains the sky and surrounding world on its shimmering surface. Similarly the whole garden, down to the flashing dewdrop, is not a thing isolated from the world but a condensation of all nature.[2] Teiji Itoh

The first, albeit comparatively cursory, descriptions of Japanese gardens to surface in the German-speaking world were provided by the German physician Engelbert Kaempfer (1651–1716), who was the first European field botanist to visit Japan and set a milestone in the study of the island nation's flora.[3] Kaempfer also felt compelled to write nothing less than a complete account of the history of Japan, a work that was published posthumously and initially only in English: *The History of Japan*. In it, he described the grounds surrounding the "Tsuganin"[4], a temple in Kyōto, as a "small pleasure garden, as it were in miniature, curiously laid out after the Japanese manner, and with as much regularity as the narrowness of the place would admit". "The walks were finely and neatly gravell'd with a whitish sand. Many scarce uncommon plants and trees, brought by art to a great degree of perfection, and curiously twisted, with some uncommon stones adorn'd the beds. But what was most pleasing to the eyes, was a row of small hills artfully made in imitation of nature, with the most beautiful plants and flowers growing thereon, and a shallow brook running across with an agreeable murmuring noise, over which were laid four small stone bridges for ornament, as well as for easier communication with all parts of the garden."[5]

Although Kaempfer was aware of key concepts of the art of Japanese gardens, the emerging influence of the distant and unknown country in East Asia was initially most evident as a fascination with the island's flora. The first person to import Japanese plants was the physician and naturalist Philipp Franz Balthasar von Siebold (1796–1866), who in 1830, despite Japan's isolation, succeeded in exporting herbarium specimens of 2,000 species and a further 500 living plants. His botanical research culminated in the *Flora japonica*[6], which he edited together with the botanist Joseph Gerhard Zuccarini (1797–1848), also from Bavaria, as a replica of the work of the same name published in 1784 by Carl Peter Thunberg (1743–1828)[7]. His work is still honoured today in the numerous author citations appended to botanical plant names. In the case of the Japanese swamp iris[8], he shares the honour with his colleague Kaempfer.

The Japanese flora embarked on a veritable triumphal march through Europe's gardens, expanding the palette of plants enormously. The opulent flowers of Japanese perennials and the rich colours of Japanese deciduous shrubs were enthusiastically received. And they created the impression of a paradisiacal lushness.

Siebold's, Thunberg's, and Kaempfer's influences continued unabated until the end of the 19th century. Remarkably, in the very year of the forced opening of Japan, 1853, a new edition of Kaempfer's historical work was published – albeit in an abridged version.[9]

Relationship pairing with India, China, and Japan

The early reception of Japanese culture, architecture, and garden art in the West, which began in 1853, was characterised by interpretations and projections that were commonly subsumed under the heading "the Oriental"[10], and which sparked debates on the question of authenticity. Filtered by the perception of Western eyes, relationships with other cultures were fantasised. The revival of the Gothic was linked to an enthusiasm for India that began to emerge in the early 19th century: "There is a surprising and striking resemblance between Indian architecture and Gothic architecture. [...]. The extravagant taste of the Orientals, which eschews all restraint and strives for the monumental, can be seen unmistakably in the spirit of Gothic architecture, and is surpassed only by the colossal dimensions of Indian architecture."[11] The exploration of China in the early 17th century was associated with a longing for antiquity, whose fusion with the euphoric reception of China reached its peak in the English landscape garden towards the end of the 18th century.[12] Its most authoritative representative,

1 The expression is borrowed from: Moos 2004, p. 106.
2 Itoh 1999, p. 35. "Dew pearls on a lotus leaf symbolise the Pure Land, the Buddhist paradise." Shōnagon 2017, p. 43, note 76.
3 See Kaempfer 1712, pp. 765–912.
4 "Chion-in" (知恩院) is the present-day spelling of the name of the temple located in Higashiyama-ku (東山区), the eastern mountain district of Kyōtō. "Schreib- bzw. Druckfehlerverzeichnis", in: Kaempfer 2013, p. 161.
5 Kaempfer 1727, p. 550.
6 See Siebold/Zuccarini 1835.
7 Thunberg stayed in Japan as an employee of the Dutch East India Company 1775–1776. See Thunberg 1784; Thunberg 1793–1795.
8 *Iris ensata* = *Iris kaempferi* Sieb. ex Lem. ("Lem" stands for "Lemgoviae", i.e., Lemgo, Kaempfer's birthplace).
9 See Kaempfer 1853.
10 Although focused on Chinese garden art, William Chambers subsumed the content of his dissertation under the term "oriental". See Chambers 1772.
11 Schelling von 1859, p. 585.
12 "Both Chinese and English horticulturists drew particular inspiration from their respective classical pasts." Keswick 2004, p. 24.

the Scottish architect William Chambers (1723–1796), not only encouraged this new natural style in his treatises on Chinese architecture, art, and gardens, but also argued that the natural garden style had its roots in China. Therefore, European garden artists should increasingly orient themselves towards China."[13] He even coined his own style for it, the "Anglo-Chinese garden" or "Jardin anglo-Chinois". Peter Meyer, who as editor of the magazines *Schweizerische Bauzeitung* and *Das Werk* had studied Japan in depth and linked the "revival of the Middle Ages", which had been in the offing from the 1840s onwards, with "knowledge of the East, especially Japan and its domestic culture". The common characteristic, he argues, is "master craftsmanship" – a quality that in turn has gained renewed relevance in the present, as reflected in the "most modern slogans of material authenticity, correct processing methods and absence of ornamentation".[14]

The first non-ephemeral Japanese garden in Europe, built in 1902 in the park of Halbau Castle in the Polish part of Upper Lusatia, illustrates this hybridisation in an exemplary manner. Fritz Hanisch described it "seen from the terrace of the castle", from where it presented itself in a "lowland with a large expanse of water".[15] "Rocks, stone slabs as path surfaces, brooks, arched wooden bridges with low heavy wooden railings, stone lanterns, gates and temples" are what give it "its characteristic appearance", according to the garden engineer. The garden was incorporated, much like in a botanical garden, between "'wilderness', rosarium and perennials garden"[16]. At the same time, it vied for attention with the Chinese garden – distinguished by its pagoda and oval and circular moon gates [→p. 191].

As with earlier gardens created in the Indian and Chinese style, their Japanese counterparts were criticised on the grounds of authenticity – especially as their creators claimed to have created replicas that were true to the original.[17]

Japan's self-representation – between archaic utopia and modern economic miracle

Among the settings in which the aspect of authenticity was most keenly debated were the World's Fairs. Art and decorative arts from Japan first appeared at a larger scale at the 1862 International Exhibition in London. The World's Fair in Vienna in 1873, the Centennial Exhibition in Philadelphia in 1876, as well as the World's Columbian Exposition in Chicago in 1893 were similarly important for the communication of Japanese garden and architectural design. Now, though, Japan appeared in its own right and could use the exhibitions to present the arts and traditions that so fascinated the Western world in such a way as to convey a certain self-image. In Vienna, the country showed itself with a Shintō shrine and a rustic-looking garden decorated with stone lanterns.[18] In Philadelphia, the first Japanese garden was built on North American soil.[19] In Chicago, Nippon recreated a Buddhist temple – the 11th-century Phoenix

Hall Hō-ō-dō of the Byōdō-in in Uji near Kyōto[20] – and teahouse[21], surrounded by a garden that was recognisably Japanese, not least through its use once more of stone lanterns.

Japan was perceived by the West "as a quaint and exotic land just emerging from centuries of sleep"[22], a romanticisation that for the most part failed to recognise Japan's economic and geopolitical ambitions after the country's opening: "Yet by the middle of the 1890s, the era when America's great fairs began in earnest, Japan was fast becoming a Western-style colonial power. As the Great Empire of Japan, the small island nation was rapidly industrializing at home and creating its own dominion in Asia."[23]

As a result, a dual mode of presentation was adopted that was tradition-conscious in parts and progressive in others. In the exhibition halls, the island state demonstrated economic progress, social change, and military potential, while in the pavilions that it had built in the traditional manner and placed in lush gardens, it conveyed the image of a country steeped in history that "epitomized Asian tranquility, tradition and beauty"[24], "a kind of pre-modern utopia"[25].

The catalogues commissioned by the Japanese government matched this. The authors of the *Histoire de l'art du Japon* produced in 1900 for the Paris Exposition Universelle – the first official history of Japanese art – spanned the arc from its mythological origins with Tennō Jimmu. They defined the Japanese people as

13 Bucher 2015, p. 63.
14 Meyer 1937b, p. 136.
15 Hanisch 1910, pp. 108, 110, 111–112.
16 Ibid.
17 See Tachibana/Daniels/Watkins 2004, pp. 370–371.
18 See illustration in: Schöppler 2020, p. 99.
19 See Brown K 2008, p. 13.
20 See illustration of the original in: Japanese temples 1910, p. 61, plate 52. Ill. of the replica, in: Brown K 2008, p. 14.
21 See illustration in: Schöppler 2020, p. 110.
22 Brown K 2008, p. 14.
23 Ibid.
24 Brown K 2008, p. 15.
25 Brown K 2008, p. 14

← In the Kare-san-sui at Daisen-in, a dry waterfall pours into a stream of gravel. This flows into the sea, where the fictitious water washes around a stone whose shape resembles a junk.

→ In the south garden at Daisen-in, the landscape flows into the extremely abstract Zen Garden with raked gravel and two sand cones.

↓ The dry waterfall in the Kare-san-sui at Daisen-in – one of the most attractive Hōrai-ishigumi (蓬莱石組, Hōrai rock composition).

↓ Northern Zen Garden designed by Shigemori Mirei in 1961 in the subtemple of Daitoku-ji, Zuihō-in (瑞峯院), built in 1535.

→ Heini Mathys' estate includes a portfolio of "Estampes Japonaises" – Japanese woodblock prints – from the collection of Henri Vever.

Harunobu. — *Jeune femme et sa suivante*

gravure et d'en faire rapidement un art parfait par le goût des couleurs choisies, la délicatesse et le fondu des diverses planches colorées, en même temps que par la grâce du dessin et de la composition.

Elève de Shigénaga, Harunobu, né à Yédo vers 1730, débuta par des gravures en noir et à deux tons, portraits d'acteurs ou même quelques danseurs de Nô, drames sacrés fort goûtés de ses contemporains. Mais il abandonne vite ces sujets pour leur préférer, comme Masanobu, l'étude de la vie qui l'entoure. Ce faisant, il acquiert un style bien à lui.

Harunobu introduit, en somme, dans l'estampe,

descendants of a single family, direct descendants of the Tenshō Daijin, and a homogeneous "race"[26]. Despite this mythical glorification, they associated the flourishing of the fine arts with the foreign influence of Buddhism and Confucianism.[27]

While the *Histoire* covered almost all artistic activities – painting, sculpture, architecture, ceramics, metalwork, textiles, lacquerware – the Japanese Ministry of the Interior focused on *Japanese Temples and their Treasures* at the Japan-British Exhibition in London in 1910.[28] Once again the authors anchored Nippon's culture in antiquity, portraying architectural monuments from their origins in pre-Buddhist times, the Izumo Ōyashiro shrine (出雲大社), to the Edo period and the Zen Buddhist Zuigan-ji (瑞巌寺) in Matsushima, founded in 828 and restored in 1609. The catalogue "was also nominally considered the second 'official' Japanese art history, following the *Histoire de l'art du Japon*, which was produced for the Paris Exposition Universelle in 1900"[29]. The catalogue, which included some hand-coloured drawings embossed with gold paint, was more opulent than the comparatively sober *Histoire*, which was illustrated almost exclusively with black and white photographs.

Garden of Peace, Garden of the Floating Isle

The exhibition that was decisive for its reception in Switzerland – as one of the most renowned Swiss garden designers, Gustav Ammann, is known to have visited the show[30] – was the Japan-British Exhibition. The binational show, which took place in London, was the last international exhibition during the Meiji era (1868–1912) and was de facto a "Japan Fair".[31] Nippon featured two gardens based on master plans by Keijirō Ozawa and Kinkichirō Honda. Ozawa designed the "Garden of Peace" in the hill-garden style *tsukiyama or tsukiyama senzui* (築,山), Honda the "Garden of the Floating Isle" in the flat garden style *hiraniwa*.

Ozawa had made a name for himself with the "rescue and restoration of traditional gardens in Japan" and with the design of a Japanese garden in 1889 in the public Hibiya Park. This park also included an English landscape garden and a French baroque garden, and was one of the first Western-style public parks planned by the city of Tōkyō.[32]

Honda was impressed by European garden designs, and he liked to combine them with traditional Japanese landscape ideas featuring ponds and waterfalls. For him the Grande Cascade of the Bois de Boulogne in Paris represented an ideal model for the idea of a "crossover" between a European-English hybrid park and a traditional Japanese garden.[33]

Both of the gardens at the Japan-British Exhibition had a winding serpentine pond running in the middle over the length of the garden.[34] Ozawa's approximately one-hectare "Garden of Peace" was – as the Dutch landscape architect Wybe Kuitert recounts – equipped with an "island, bridges of various designs, a small Shinto shrine with gate, a Buddhist statue, and a five-storied stone pagoda"[35]. In contrast to the traditional design of Tsukiyama, which with mountains, hills, lakes, cliffs, forests, waterfalls, and rivers conveys "an impression of that grandeur which is the natural accompaniment of so much of the scenery of Japan"[36], "immense painted screens were set up to form the background to the design"[37] so that the garden was to be viewed from only one direction.

Kinkichirō Honda designed the second area, the "Garden of the Floating Isle", on an area of around 7,300 m^2 as a *hiraniwa* (平庭), that is, in the flat garden style – "although interpreted quite liberally".[38] The focal point was a hill made of rocks from which a waterfall tumbled, with pavilions set around it.[39]

26 Histoire de l'art du Japon 1900, p. 15.
27 Cf. Histoire de l'art du Japon 1900, p. 11.
28 See Japanese temples 1910.
29 Hayashi 2016, p. 36.
30 Ammann 1910 gtaA.
31 Cf. Hayashi 2016, p. 36.
32 Cf. Tachibana/Daniels/Watkins 2004 p. 368, note p. 393; Fiévé 2012a, p. 98, note 19.
33 Cf. Schöppler 2020, p. 278.
34 Cf. Kuitert 2002, p. 223.
35 Kuitert 2002, p. 224.
36 Shiota 1916, pp. 12–13.
37 Kuitert 2002, p. 223.
38 Kuitert 2002, p. 224.
39 Cf. Kuitert 2002, p. 224.

The garden was more lush than the traditional "flat garden style" as described by the Japanese-American landscape architect Takeo Shiota: the "principal features" of the garden laid out on flat ground were some rocks or trees "notable for [their] picturesqueness and beauty". "It is always so arranged in part as to suggest the sea coast and an island. Gardens of this type are treated in a more idealistic and poetic way than are those of any of the other forms."[40]

The "Garden of Peace" met with a positive response, despite, or perhaps precisely because its opulent design deviated from and lacked the refinement of the traditional Japanese garden, with "aesthetics triumphing over form or function."[41] The "Garden of the Floating Isle" seemed to mesh together "the historical heritage, with modern images of Japan and including modern ideas around landscape gardening"[42]. This was underlined by Honda's book *Zukai teien zōhō* (図解庭造法, *Illustrated Horticultural Methods*)[43], reissued for the 1909 show, which included illustrations of modern Japanese garden styles. Some of these also showed geometrically designed grounds with Western elements that deviated from tradition. Plate 40, for example, documents a focally placed hexagonal planting bed featuring rock and tree arrangements, plate 41 a baroque-style fountain in a circular pond, and plate 12 a garden whose view is obstructed by a staggered box hedge of tall rectangular blocks.

While Honda's "signature" impressed the British architect Josiah Conder (1852–1920), who adopted illustrations from Honda's 1890 first edition *Zukai teien zōhō*[44] for his own 1893 book *Landscape Gardening in Japan* – it was slated by English writer Reginald Farrer (1880–1920): "The 'Japanese' gardens at the White City [Exhibition of 1910] were quite the most cruel and insolent practical jokes that I have ever seen levelled at our native innocence – vast puddings of unrelated pebbles, shapeless, ridiculous, peppered with toys and bronzes and haphazard shrubs. How their creators must have laughed in the building ..."[45]

Authentic copy versus adapted analogy

Josiah Conder was the leading pioneer of the reception of Japanese garden culture in the Anglophone world. Called to Japan in 1877 at the age of 25, he followed in the footsteps of Edward Sylvester Morse (1838–1925). The North American zoologist had paved the way for the reception of Japanese art, culture, architecture, and garden design in 1885 with his book *Japanese Homes and their Surroundings*.[46] The architect Conder came to Tōkyō Imperial University as a teacher, founded an architectural practice in 1888, and remained in Japan until his death. In 1893, he published *Landscape Gardening in Japan*, probably the most cited standard work on Japanese gardens in a Western language.[47] It came out a year after Lafcadio Hearn's (1850–1904) essay "In a Japanese Garden", and *Glimpses of Unfamiliar Japan*, which followed soon after.[48] Morse, Conder, and Hearn were the three primary proponents who decisively shaped the reception of the Japanese garden in the West, with Hearn and Conder taking contrasting approaches.

Hearn cultivated an essayistic style and refrained from any scientific pretensions. He avoided listing the elements of the Japanese garden, let alone categorising them. Rather, he tried to fathom its essence, enveloping it in an exotic haze, often with a tendency to mystifying it that underlined quite effectively his conviction that for foreigners the secrets of such gardens were impenetrable.

In contrast, Conder built his treatise scientifically. It was divided into three parts. In the first, he devoted himself to the history of Japanese garden art; in the second, he meticulously documented all the elements from which it derives; and in the third, he illustrated and discussed examples of Japanese gardens. Nevertheless, he was not afraid to acknowledge the simplification of such methodical accuracy, noting that in the books he relied on, Japanese garden art was shrouded in "an accumulation of abstruse terms, secret meanings, fancies and superstitions, which render it highly complicated."[59] These complications are the result of the tradition of transmitting knowledge about landscape gardening orally or in "secret", vaguely formulated texts that are only accessible to the initiated.[50]

Despite this restriction, *Landscape Gardening in Japan* became the leading manual for the construction of Japanese gardens in the West and its author the guiding spirit behind the gardens at Newstead Abbey, Nottinghamshire, Cowden Castle, Clackmannanshire, and Ingleborough Hall, Yorkshire. In this last case, Japanese specialists were employed to ensure the creation of a garden that was as authentic as possible. Conder even provided a model for the German landscape architect Reinhold Hoemann (1870–1961) and the "Japanisches Theehaus" [51] [Japanese teahouse] he created for the Internationale Kunst-, kunsthistorische und grosse Gartenbau-Ausstellung Düsseldorf (International Art and, art historical Great Horticulture Exhibition) in 1904, where he is said to have created a replica of the "Fukiage Garden" that Conder had shown in *Landscape Gardening in Japan*.[52] Photographs of the two gardens, however, reveal several differences between the two.

Conder's antagonist Reginald Farrer, who had already planted a rock garden as a teenager on a cliff at Ingleborough, the second highest mountain in the Yorkshire Dales, subsequently studied the genre in detail, publishing articles about it[53] and founding a nursery specialising in rock garden plants in Clapham, Yorkshire, in 1901. While he was fascinated by Japan[54], he rejected any suggestion of imitation, understanding his creations at the family estate Ingleborough Hall in the parish of Clapham as hybrid compositions. His attitude was resolute. He accused the West of imitating soullessly, "line by line", so that the copy was a "dead rendering" that exuded nothing of the spirit "that had inspired the original". The East, on the other hand, copied by studying the original and grasping its principles in order to translate it into a new form and formulate a perpetual renewal of life.[55]

40 Shiota 1916, p. 13.
41 Cf. Schöppler 2020, p. 268.
42 Schöppler 2020, p. 265.
43 See Honda 1890–1926.
44 See Honda 1890–1926. The 1909 and 1926 editions of 図解庭造法 are translated synonymously as *Zukai Nihon teizōhō*, *Zukai teien zōhō*, *Zukai teizōhō* or *Zukai niwa zōhō*. "Niwa" meant a flat, open space during the Nara period and acquired the meaning "garden" in the Middle Ages. "Teien" was originally used in the Edo period and later gained currency through the *Meiji teienki* (明治 低延期, Chronicle of Meiji Gardens) edited by Keijirō Ozawa; https://www.nabunken.go.jp/org/bunka/jgd/pages/Garden_teien.html (retrieved 9 Jan. 2023).
45 Tachibana/Daniels/Watkins 2004 p. 386.
46 See Morse 1972 (1885).
47 See Conder 1893/1894.
48 See Hearn 1892; Hearn 1894.
49 Conder 1886, p. 124.
50 Cf. Conder 1886, p. 125.
51 See International Horticultural Exhibition 1904, n. pag. [p. 12].
52 Cf. Gröning 1997, p. 242, ill. p. 243.
53 See Farrer 1913.
54 See Farrer 1904.
55 Cf. Farrer 1912, pp. 330–331.

→ The "Japangarten" created by Ludwig Weinbrenner at the Vienna Horticultural Exhibition in 1913 in an article entitled "Wiener Ausstellung" in the journal *Die Gartenkunst*.

↓ The Japanese-style theme garden by Friedrich Henkel at the Mannheim Jubilee Exhibition in 1907 in an article entitled "Von der Mannheimer Gartenbauausstellung 1907" in the journal *Die Gartenkunst*.

Teiji Itoh's more conciliatory description of Japanese garden art as the quintessence of a series of "misinterpretations and reinterpretations"[56] offers a more fitting appraisal of the beginnings of the reception of Japanese garden design in the West in general and to the adoption of its elements in Swiss landscape architecture in particular. These misinterpretations arose, for example, as a product of the Chinese reminiscences in Japanese garden art, which resulted in a somewhat fuzzy differentiation between Chinese and Japanese motifs in the West. A further reason was that the gardens of the Meiji era were initially used as references, some of which already exhibited European influences. Only later did a more differentiated view of the principles of garden design begin to emerge.

Cross-pollination between Japan and the West

The fascination with oriental exoticism that arose through the World's Fairs probably distorted Western perception, but its mechanisms were used by Japan itself to curate its own identity, as analysed the German Japanologist Irmela Hijiya-Kirschnereit.[57] The US landscape designer and garden conservator Elizabeth Barlow Rogers explained the process of feedback that followed the opening of Japan's borders and the influx of Western culture into the country.[58] Before the Meiji Restoration (1868–1898), all Japanese gardens were planned with a religious background, although perhaps only in form and verging on the superstitious in many cases. "After the Restoration, the religious background was abandoned, and efforts were made to base garden construction on aesthetics pure and simple."[59] Accordingly, this is how the gardens were perceived and recreated in Europe – focusing primarily on the aesthetic effect.

The American writer and garden designer Loraine Kuck, whose treatise on Japanese garden design remains an indispensable reference to this day, is of a similar opinion. From 1868 onwards, a remarkable period of change began in Japan accompanied by the adoption of foreign ideas. She compares this cross-pollination to the Japanese assimilation of Chinese cultural expressions and civilisational achievements thirteen hundred years earlier.[60]

As Barlow Rogers elaborates in a slightly more nuanced fashion: "The Japanese garden continued to flourish during the Edo period (1603–1868) as a deeply traditional yet continually innovative art form. But the impetus for change from within diminished after the appearance of American naval commander Commodore [Matthew] Perry in Tokyo Bay in 1853."[61]

At the same time, the political leadership was enamoured with the idea of restoring the old Japan with its centralist Tennō monarchy and re-establishing the religio-political order "as the essence of the Shintō, the way of the *kami*"[62]. Accordingly, the Meiji period (1867–1912) was "probably characterised by similar dynamics and contradictions"[63] as the seventh century, when under Tenmu Tennō, the state and administration had been reformed according to the Chinese model, and at the same time, national identity was reinforced through the "enthronement" of the shrine of the sun deity Amaterasu in Ise as the most important ancestral shrine of the Tennō house.

Japan as a foil for self-reflection

After this phase of reciprocal inspiration with the import of Japanese garden design and the export of Western garden traditions, a period of emancipation began on both sides. In Japan, a "liberation from Western perception", which will be discussed below, began with reformative self-confidence after the death of Meiji-Tennō Mutsuhito and the end of the Meiji oligarchy, which – accompanied by the emergence of liberal ideas – heralded the begin of the "Taishō democracy". In the West, the focus shifted towards using Japanese ideas and practices as inspirations for one's own creations, rather than aiming to create authentic Japanese gardens. Thomas Mawson (1861–1933), for example, one of the best-known English garden designers, who became known for several Japanese-inspired creations, suggested in his book *The Art and Craft of Garden Making* that one "borrow ideas from their art and their simulation of nature" [64]. Moreover, the Japanese model increasingly mutated into a vehicle for revolutionising one's own architecturally influenced layouts, which were perceived as antiquated, and for establishing ideas of a more naturalistic design. To a certain extent, the Japanese garden was also instrumentalised – as a lance against wilderness gardening and as a source of argumentation for the nationalist-ideologically motivated demand to reduce the plant palette. The image of Japan now served as "a mirror for self-reflection through the foreign", as concisely expressed by the Japanologist Claudia Delank.[65]

The range of interpretations in the most relevant German-speaking area of Switzerland can be seen in the garden periodicals of the day. In particular, *Die Gartenkunst*, *Die Gartenschönheit*, and *Die Gartenwelt* devoted considerable space to Japanese garden design on several occasions. The editorial contributions were complemented by numerous plant reviews and advertisements for Japanese plants – above all the "Japanese ornamental maple"[66], that is, numerous varieties of *Acer palmatum*, as well as Japanese azaleas and *Iris kaempferi* – as well as stone lanterns. While the texts strive to achieve a deeper discussion, they are also rife with latent misunderstandings and exhibit a tendency towards nationalistic appropriations.

While, by the turn of the century, the Japanese garden was widely held to have the character of a work of art, at the same time, little differentiation was drawn between it and the Chinese garden. In fact, the two were largely regarded as the same: "The garden of the Chinese and Japanese is a work of art in the fullest

56 Itoh 1999, p. 70.
57 Hijiya-Kirschnereit 1988, p. 13.
58 Barlow 2008, p. 2.
59 Harada 1956, p. 114.
60 Cf. Kuck 1940/2016, p. 265.
61 Barlow 2008, p. 2.
62 https://www.univie.ac.at/rel_jap/an/Geschichte/Staatsshinto (retrieved 29 Jan. 2022).
63 https://www.univie.ac.at/rel_jap/an/Geschichte/Kami_Kulte (retrieved 29 Jan. 2022).
64 Mawson 1926, p. 211.
65 Delank 1996, p. 25.
66 Unger 1930, p. 76

Heini Mathys was just as taken with
bizarrely shaped trees as Willi Neukom.

Photograph of the Katsura Garden by Willi Neukom showing a bridge that evokes the natural land bridge Ama-no-hashidate.

Richard Arioli took this photograph of the
"Temple of the Golden Pavilion", Kinkaku-ji.

The garden of the Hōshun-in at Daitoku-ji
(大徳寺, 1315) “washes” beneath the buildings.

67 Fintelmann/Weiß 1903, p. 13.
68 Cf. Hoemann 1908, p. 151.
69 Hoemann 1908, p. 153.
70 Boehn 1910, p. 11.
71 Encke 1910, p. 11.
72 Camillo Karl Schneider (1876–1951) was a botanist specialising in dendrology and was active as a landscape architect and writer. Gustav Ammann was on friendly terms with him. Cf. Stoffler 2006a, p. 3.
73 Schneider 1910, pp. 1–2.
74 See Schulze 1907.
75 Cf. Jubiläums-Ausstellung Mannheim 1907, pp. 44–45.
76 Cf. Schneider 1910, p. 1.
77 "I owe this book, the first volume of which appeared in Tokyo in 1893, in which year the supplementary volume also followed, to the kindness of Mr Friedrich Henkel." Schneider 1910, p. 3.
78 Henkel 1910, p. 32.
79 Schneider 1910, p. 2.
80 See Unger 1909, p. 594. Twenty years later, Unger published an eight-part series of articles – Unger 1929/1930 – on his business activities as a partner of Louis Böhmer, who had worked as a horticulturalist for the Japanese government from 1871–1882 before founding a trade specialising in Japanese plants. In 1889, Unger became his assistant and in 1892 a partner in the firm, which he continued under the name L. Böhmer & Co. after Böhmer's retirement in 1894 to run until 1908.
81 Unger 1909, p. 594.
82 See Unger 1929.
83 See Unger 1926.
84 Cf. Foerster 1934, p. 344.
85 Foerster 1935, p. 263; Kühn 2012, p. 23.
86 Foerster 1936, p. 9.
87 Müller-Beeck 1926, pp. 61–64, after Conder 1893, pp. 133–139 (Tsukiyama niwa), pp. 140–141 (Hira niwa), p. 143 (Cha niwa), pp. 150–152 (Roji niwa).
88 Baumann A n.d. [1930–1932], p. 2.

sense of the word, for it is devoted exclusively to aesthetic purposes"[67].

Five years later, Japan was revered as a model for simplicity in garden art, as it also prevails in nature – especially on the seashore, in high mountain areas, in heather moorland, in the steppe, and in the upland forests.[68] Referring to the floral art of the Japanese, Reinhold Hoemann implored his contemporaries to limit the assortments of flowers and woody plants they used: "So you growers, become simpler, stay more natural, get rid of the many inferior varieties and novelties."[69]

Otto Ernst Leo von Boehn felt similarly, arguing that "the unity of woody plants and perennials from a single country in a cohesive picture" was "the new quality of the Japanese garden". "That is the beauty of the Japanese garden, that the different forms fit together because they are native to the same region. The union of plants that belong together physiognomically and geographically always produces good effects."[70] By contrast, Friedrich August Ernst Encke defined the Japanese garden as "a little garden in which Japanese and East Asian plants are united in rich abundance and wild confusion"[71].

Camillo Karl Schneider dealt intensively with the subject,[72] writing a noteworthy and detailed article in 1910 on "Über japanische Gartenkunst" ("On Japanese Garden Art") spread over two issues of the journal *Die Gartenkunst*. The "occasion" was the Japanese teahouse garden at the Internationale Luftschiffahrt-Ausstellung (International Airship Exhibition, I.L.A.), which took place in Frankfurt from 10 July to 17 October 1909 over a period of 100 days. It was designed and executed by Minori Yasuda, "a young Japanese student, currently studying painting in Munich"[73].

The material was supplied by Friedrich Henkel's "Großgärtnerei" in Darmstadt. The plant nursery had already presented itself three years earlier at the Jubiläums-Ausstellung, Mannheim (Mannheim Jubilee Exhibition) in 1907 with a Japanese-style garden[74], whose art was inspired by the "assured style and tact of the Japanese"[75].

It was also Henkel who provided Schneider, who himself had no genuine view on Japanese gardens but had for years been carrying around all kinds of thoughts about Japanese garden art[76], with Josiah Conder's *Landscape Gardening in Japan*[77] and, after writing a short review of it, was in the process of "translating the whole work into German"[78]. Schneider was just as taken with it as Henkel and enthused about the "wonderful pictorial plates" that accompanied Conder's elaborations. He was no less enthusiastic about Yasuda's intervention on the exhibition site, since "for the first time in Germany, a Japanese artist has created a garden that truly conveys the Japanese spirit"[79].

Alfred Unger, on the other hand, was downright scathing in his criticism, saying that having lived in Japan for two decades, he was entitled to make an informed judgement.[80] While conceding that the garden was a well-executed "public attraction", it was "to a good Japanese garden as a wooden rocking horse is to a noble Arabian horse".[81] He did, however, agree with Schneider on the value of Conder's work as a reference.

Reprise: The long arm of Kaempfer and Conder

Two years later, in 1911, Unger created his own garden at his home, which he planted exclusively with Japanese plants from the islands of "Nippon and Hondo" to commemorate his stay in Japan, which he would later report on in 1929.[82] Unger, who was a plant dealer, also recalled Kaempfer's and Siebold's achievements in introducing and spreading Japanese wild plants in the West.[83] The traditional inclusion of plants living in the "wilderness" in Japan and their horticultural domestication became the inspiration for Karl Foerster's (1874–1970) "Wildnisgartenkunst".[84] This approach draws on the idea of "wild gardening", which the gardener William Robinson (1838–1935) contrasted with the Victorian carpet beds. Instead of the formal, framed borders planted with annual flowers, he proposed using mixed, colourful perennials to form a transition to the landscape.

Such "wild" planting also took root in Switzerland, in particular through Gustav Ammann who was greatly inspired by the work of Foerster, and the natural garden movement of the 1970s in turn drew on him (see "Gustav Ammann – Japan as a module").

Foerster was also an ardent advocate of the rock garden. He saw it as having originated in Chinese and Japanese culture with England acting as the role of mediator. "Centuries after its birth in the Far East, rock gardening was transplanted from China and Japan to England."[85] Japan, in particular, was the domain of the setting of wild stones and stone slabs, which "for more than a thousand years" formed the basic framework of Japanese natural garden design.[86] In the decades that followed, rock gardens became an increasingly widespread hobby – also among Swiss landscape architects (see "the Mertens – Japanese masked").

In the context of this "genealogical" relationship to England, Conder's categorisation experienced a comeback. In the "illustrated journal" *Der Erdball*, published between 1926 and 1931 by Georg Buscha "for the entire field of regional studies and ethnology", an article by Fernando George Mueller-Beeck (1854–1928) traced the Buddhist and Shintō or indigenous roots. Under the title "Die japanische Gartenkunst unter dem Einfluss der Lehren des Buddhismus und Shintoismus" (Japanese Garden Art under the Influence of the Teachings of Buddhism and Shintoism), he also proposed a rudimentary classification of the different garden styles, as detailed by Conder: "1. Hill gardens = Tsukiyama niwa, 2. Flat gardens = Hira niwa, 3. Tea gardens = Cha-niwa, 4. Passage gardens = Roji-niwa."[87] Albert Baumann had a list of reference literature that also includes Müller-Beeck's essay.[88]

Schneider also references Conder in his own, similar categorisation of the different styles – "hill, flat, passage and tea gardens" – but goes on to additionally define them in terms of their execution into rough, intermediary, and finished.[89] Their stylistic differentiations, especially of the hill gardens[90], were adapted from Conder's terms "Rocky Ocean Style", "Wide River Style, "Mountain Torrent Style", "Lake Wave Scene", "Reed Marsh Scene" and translated into German.[91] Although he was apparently unaware of the original source, his terms correspond almost exactly to those listed in the *Sakuteiki* (see "From the plenty of paradisiacal islands to the beauty of emptiness").

Reappraising the originals

A step towards a genuine, unadulterated understanding of the Japanese garden, without appropriating it for own intentions, was taken by Helmut Hentrich's and Marta Heisinger's contributions to the journal *Die Form*[92]. Hentrich grasped one of the central Japanese principles that architects such as Adolf Loos, Hermann Muthesius, Ludwig Hilbersheimer, Hugo Häring, Josef Frank, and Bruno Taut also saw in Japan: "The building is not only integrated completely into the landscape, the mountains, etc., and the garden or park not only encloses the house on all sides; rather, the house and garden interpenetrate each other".[93] Implicitly, Hentrich propagated the exemplary nature of this fusion – in contrast to both the supporters of the garden attached to a house and the advocates of the natural garden. The former were convinced that house and garden should form a common whole, but accorded primacy to the architecture: "[...] the house plays the dominant role, and the garden must by necessity respond to the forms and lines of the house"[94]. The latter defined humans and plants as equals and rejected the idea of the garden as an extension of the dwelling. [95]

The visual imagery employed by Hentrich and Heisinger also differed markedly from the photographs published twenty years earlier in *Die Gartenkunst*. Rather than showing "export products", they showcased Japanese gardens in their home country. Most of the illustrations were "taken from a Japanese portfolio presented to the BDA Association of German Architects by Japanese architects on the occasion of the exhibition of German architecture in Tokyo in 1932"[96].

The at the time unspecified "portfolio" was a collection of images edited by Motoi Hashimoto and Naoya Shiga called *Juraku Tōyō kokenchiku teien gashū (Juraku. Pictorial Collection of Japanese Architecture and Gardens)*.[97] An English translation appeared in 1935 under the name *Gardens of Japan*.[98] A veritable treasure trove, it contained numerous photographs of traditional, historically evolved gardens and excluded European-influenced grounds. It also allowed Hentrich and Heisinger to include two photographs depicting the imperial Katsura residence in Kyōto[99], which was hailed in European architectural circles as the *pièce de résistance* of Japanese garden design.

The work of the architect Tetsurō Yoshida (1894–1956), at the time the most important mediator between East Asia and the German-speaking world, was a sensation. Most of the German-language book publications published up until then – as well-founded as they were – focused on arts and crafts, paying only cursory attention to architecture and garden design.[100] With his book *Das japanische Wohnhaus*[101], published in 1935, Yoshida provided "the first systematic and overall description of the Japanese house and its history, with details and many interior photographs"[102] – also in the context of their gardens. The *Deutsche Bauzeitung* immediately heralded the work as: "A book we have been waiting for."[103]

In Germany, soon after taking over the editorship of the journal *Die Gartenkunst*, Michael Mappes published an outline of the history of Japanese gardens in 1936 under the title "Die Kunst des landschaftlichen Gartens in Japan" (The Art of the Landscape Garden in Japan) along with photos of some of the most characteristic gardens. He showed photographs of Kiyosumi Park (清澄庭園, Kiyosumi teien), a prime example of a Japanese landscape and promenade garden, of the Daisen-in complex and of the garden of the Katsura Palace. From the Ryōan-ji stone garden in Kyōto, he reproduced the landscape plan and a stone setting scheme that illustrated the relationship between the stones with their "pictorial balance", the "resolution of the group", and the "perspective depth of the view of a group"[104]. Mappes also referred to Tamura Tsuyoshi (田村剛, 1890–1979), together with Tatsui Matsunosuke (1884–1961)[105], whom Yoshida named as an authority[106] and one of the leading garden scholars of the time.[107]

Liberation from Western perception: Katsura and Ryōan-ji as the pièces de résistance

The featuring of Katsura and Ryōan-ji brought the two *pièces de résistance* of garden architecture, which for around two decades had been the focus of Japanese historiography, back into the centre of attention. Toshio Watanabe associates the "revival" of the two gardens with the end of the Meiji era in 1912, when a shift in their reception at home was becoming apparent. In his eyes, it went hand in hand with the "denigration of Edo period Daimyō gardens"[108] and the rehabilitation of the Zen gardens. Katsura and Ryōan-ji were heralded as the pinnacles of two poles – beauty and spiritualisation. Katsura represented "a highly sophisticated aesthetic sensibility", Ryōan-ji "the deeply thoughtful Zen culture"[109].

This shift in consciousness in Japan was linked with an emancipation from external perception. In particular, it sought to break the powerful influence of Conder. These developments, in turn, resulted in a similarly changing reception in the West.

Akira Satō (1903–2003) criticised Conder's reliance on the 1735 *Tsukiyama Teizōden* (築山庭造伝 *Transmission of Making Mountains and Creating Gardens*) and other manuals published in the Edo period – especially the

89 Schneider translated Conder's expressions "rough", "intermediate" and "finished". Conder 1893, pp. 134, 138, 139, 141, 143.
90 Cf. Schneider 1924, pp. 129–136.
91 Conder 1893, pp. 133–134.
92 See Hentrich 1933, p. 193. Heisinger 1933a. The articles appeared shortly before the 1934/35 takeover and subsequent liquidation by the National Socialists of the journal launched by Walter Curt Behrendt in 1925 for the Deutscher Werkbund and edited by him until the end of 1926.
93 Hentrich 1933, p. 195.
94 Encke 1910, p. 8.
95 Wolschke-Bulmahn 2009, p. 177.
96 Hentrich 1933, p. 193.
97 See Hashimoto/Shiga 1927–1928; Hashimoto/Shiga 1930–1933. The anthology comprised three slipcases of twelve volumes each, each in turn containing twelve plates.
98 See Shiga/Hashimoto 1935.
99 See Hentrich 1933, pp. 201, 203.
100 In the period in question, Oskar Münsterberg in particular subsumed almost every artistic activity under Japanese Art History – painting, woodcut, sculpture, lacquer work, jewellery, ceramics, textiles (carpets and dresses), dance and theatre (masks), weapons (helmets, armour, swords, arrow- and spearheads, quivers) and everyday objects (bronze vessels), also addressing the Chinese influence, while Ernest Francisco Fenollosa concentrated on painting and discussed both China and Japan. See Münsterberg 1904–1907; Fenollosa 1913.
101 See Yoshida 1935a. He was inspired to do so by Hugo Häring (1882–1958) and Ludwig Hilberseimer (1885–1967). Yoshida thanks him for his encouragement in the preface; Yoshida 1935, Preface, n.pag. [p. V].
102 Speidel 2008, p. 84.
103 Yoshida 1935b.
104 Mappes 1936.
105 See Tamura 1935.
106 Cf. Yoshida 1935a, Preface, n.pag. [P. VI].
107 Cf. Katahira 2010, p. 122.
108 Watanabe 2016, p. 1.
109 Ibid.

Photograph by Richard Arioli of the monk's stone in the Jōju-in of the Kiyomizu-dera.

Vistas like this in the Sento Palace Garden inspired
the Swiss landscape architects, especially Ernst Baumann.

Two illustrations from the Manual of the Mustard Seed Garden, *Jieziyuan huapu*, on the art of Chinese ink painting published by Wang Gai and Li Liufang in 1679 and named after the "mustard seed-sized" estate of the editor Li Yu in Nanjing. These drawings were in Heini Mathys estate.

guide to illustrated horticultural methods *Zukai teien zōhō* published by Kinkichirō Honda in 1890 – as unsuitable references. This, he argued, caused Conder to overemphasise the Shin, Gyō, and Sō levels of a composition, leading to an over-formalisation of garden design. In addition, Conder accorded too great importance to the taboos based on the Yin and Yang teachings and Taoism.[110] It was time to revolutionise the perception of Japanese gardens and reveal their true character.[111]

Likewise criticised was Conder's focus on gardens in the capital and relative neglect of those in Kyōto – namely the Ryōan-ji and Katsura gardens. This was not an eccentricity, but an interpretation shared by some Japanese experts at the time. Before 1920, neither Western nor Japanese experts accorded the gardens of Katsura and Ryōan-ji the canonical status bestowed upon them by the subsequent generation after 1920.[112]

Moreover, Conder had certainly mentioned Katsura: "Enshiu, whose real name was Kobori Totomi-no-Kami, was a retainer of the Shogun Iyeyasu, and he became a teacher of accomplished arts to the Regent's son. His most famous work in gardening was the garden of the Katsura Rikiû, a detached palace near Kioto, formerly a villa of the Imperial Prince of Hachijo. The site had the one natural advantage of the river Katsura running through it, but for the rest, the beauties of the landscape design were entirely due to artificial contrivance." The fact that Conder experienced it in a neglected state may have been the reason for not elevating it to the horticultural Olympus it was later to become: "This garden still remains though now much neglected, and it is said that for more than a hundred and fifty years after the death of Enshiu, not a single stone had been removed."[113]

The same was true for Ryōan-ji. The image of Kare-san-sui in Usui Kosaburō's two-volume survey of Japanese gardens, published in 1910, shows the grounds in moderately neglected condition: the gravel surface had not been raked, lacked the ground pattern imitating a flowing body of water, and was interspersed with emerging weeds.[114]

Tsuyoshi Tamura (田村剛, 1890–1979), who taught at Tōkyō Imperial University and was a pioneer in the field of landscape architecture[115], also encountered the garden in a condition unlike that known to contemporary observers. The area depicted in a pre-1918 photograph he published in 1925 lacked accurate maintenance: "[T]he white sand that appears from under the snow in the foreground of the picture shows no evidence of the familiar flowing water design raked into the sand. Not only that, [...] there appear to be some sort of tracks or marks on the sand. Since this picture was taken during the winter, perhaps these are not weeds, but footprints?"[116]

The volatility with which interpretations can change is exemplified in the gardens of Katsura and Ryōan-ji. In the case of Katsura, interpretation shifted from relative neglect to being spoken of in the highest tones, in Ryōan-ji's case from its prosaic interpretation as an allegory of an animal fable to the praising of its geometric-abstract dimension. The attribution of who was responsible for their "rediscovery" also had an aleatory trait – both as architectural and garden design highlights.

In Europe, Manfred Speidel, an architect and authority on Japan who gained his doctorate in Japan and also published Bruno Taut's writings, credited Bruno Taut with having tracked down the Katsura residence.[117] The art historian Corinne Elsesser even attested that "in Japan itself [...] his 'discovery' of Katsura set research into this building in motion".[118] And, Christian Tschumi, a landscape architect and specialist on researching the modernisation of Japanese gardens, found: "Who would know about the Katsura Villa today if Bruno Taut had not promoted it with great enthusiasm among his Japanese hosts?"[119]

110 See Satō 1933.
111 Cf. Satō 1933, pp. 98–101.
112 Cf. Watanabe, 2016, p. 17.
113 Conder 1893, p. 18.
114 See Kosaburō, 1910, n.d.
115 Yamada 2009, p. 111.
116 Ibid. Yamada refers to: Tamura 1925, Ill. 40. Yamada 2009, p. 112.
117 Cf. Speidel 2007, p. 253.
118 Elsesser 2021, p. 183.
119 Tschumi 2004, p. 11.

It is not just in Japan that the story of its "rediscovery" is likely to be told differently: in August 1888, the historian and landscape architect Keijirō Ozawa 小澤圭次郎 (1842–1932), visited Katsura. As part of a series of articles on his research into the history of Japanese gardens published in the magazine *Kokka* between 1889 and 1905[120], he dedicated four articles to the Katsura gardens alone.[121] The very same Keijirō Ozawa went on to design the "Garden of Peace" in White City in London in 1910. Given the knowledge of his intensive study of these historical sites, the criticism that he was more concerned in London with the "appeal of the gardens to foreign eyes" than with "educating in Japanese culture"[122] must be strongly relativised. Ozawa identified Kobori Enshū as the author of the Katsura villa, referring to the *Katsura gobetsugyō meisairoku* (桂御別業明細錄, Detailed Account of the Katsura Villa) published in 1878 by Nishimura Kanebumi (西村兼文, 1832–1896).[123]

The architect and architectural theorist Benoît Jacquet, who has lived in Japan since 1999, adds a further perspective: "While often portrayed as the 'discoverer' of Villa Katsura, Bruno Taut is neither the first to mention the aesthetic qualities of this villa nor the first architect to comment on its modernity. A few years before Taut came to Japan, Japanese architects had already begun to take an interest in this villa."[124] The villa's reception in its native country was due mainly to Kishida Hideto (岸田日出刀, 1899–1966). According to Jacquet, it was the teaching of this architect and architectural historian, and not Taut's writings, that had first drawn attention in Japan to the modern qualities of Japanese architecture. Hideto highlighted the contemporary relevance of Japanese architecture in his book of photographs *Structures of the Past (Kako no kōsei)*, published in the late 1920s.

Parallel to the renaissance of Katsura arose a renewed fascination for the rock garden of Ryōan-ji. Katsuo Saitō, a student of Tsuyoshi Tamura (田村剛, 1890–1979), undertook a reappraisal of the garden, that despite being "rather a lone voice"[125] was all the more concise for it. At a time when the Kare-san-sui (the dry garden) of the Ryōan-ji was interpreted exclusively in the context of its name *oranoko-watashi-no-niwa* (京都 南禅寺 方丈庭園, Garden of Tiger Cubs Crossing the River), Saitō was the first scholar to interpret it, together with the Daisen-in, as the epitome of a Zen garden.[126] In 1923, he produced the first diagram to focus on the geometric configuration, breaking down the relationships between the stones and illuminating their "abstract, philosophical and aesthetic effects"[127].

The research on Katsura and Ryōan-ji became the prelude to a detailed, theoretical, and practical study of the garden design tradition, among others by Mirei Shigemori (重森三玲, 1896–1975) and Osamu Mori (森蘊, 1905–1988). More than anyone else before him, it was Shigemori who dealt with the physical legacy – some 250 gardens – as well as the written sources[128] and heralded the modernisation of Japanese garden art. From 1936 to 1939, he also compiled a monumental almanach on Japanese garden art, that was to become a canonical work, publishing an expanded edition in 1971.[129] Osamu Mori published numerous publications in Japanese, along with a work in English. He also co-authored Irmtraud Schaarschmidt-Richter's *Der japanische Garten – ein Kunstwerk*.[130]

While Shigemori referred to the *Sakuteiki* to lend historical legitimacy to his creations, he reserved the right to interpret the traditional conventions freely. He postulated the dry garden of the Zen temples as the epitome of Japanese garden art[131] and modernised it[132]. Mori's ideal was also the philosophy set out in the famous historical manual of garden making, and Katsura appeared to him the supreme expression of the embodiment of this world view.[133]

Zen as an agent

After Japan's discovery as a haven of paradisiacal plant varieties, after the exotic interpretation of its gardens through "authentic" imitation, and after the cross-pollination resulting from numerous complex interconnections and creative misunderstandings, a new form of cross-pollination arose, this time with zen as its agent. Analysing Saitō's study of the rock garden at Ryōan-ji, Loraine E. Kuck undertook an extensive examination of the harmoniously balanced and rhythmically moving composition of the stone settings and connected

120 See Ozawa 1889–1905, cited in Jacquet 2013, p. 99, note 3.
121 See Jacquet 2013, ibid.
122 Schöppler 2020, p. 268.
123 See Ozawa 1894, No. 60, p. 231; cited in Jacquet 2013, p. 126, note 86.
124 Jacquet 2013, p. 103.
125 Watanabe 2016, p. 22.
126 Cf. Saitō 1923, p. 7, quoted in Yamada 2009, p. 169.
127 Watanabe 2016, p. 22.
128 Cf. Tschumi 2004, pp. 8, 29, 30.
129 In 1936–1939, he produced the 26-volume *Nihon Teienshi Zukari* (Illustrated History of Japanese Gardens), which he expanded to 35 volumes 33 years later and published together with his son Kanto under the title *Nihon Teienshi Taikei* (Great Almanac of Japanese Garden History). See Shigemori 1936–1939; Shigemori 1971; cf. Tschumi 2004, pp. 29, 30.
130 See Mori 1945; Mori 1955; Mori 1962; Mori 1986; Mori 1988; Schaarschmidt-Richter/Mori 1979.
131 Cf. Watanabe 2016, p. 14.
132 Cf. Tschumi 2004, p. 7.
133 Cf. Watanabe 2016, p. 14.

← An intimate veranda in the private villa Murin-an built in 1894, captured on film by Richard Arioli.

→ Ernst Cramer and Richard Arioli, who took this photograph, were fascinated by the abstract representations of nature in the Ryōan-ji.

↓ An intimate courtyard in the garden of the Hōshun-in in Daitoku-ji.

→ Detail of the Katsura gardens as photographed by Walter Leder.

↓ In the garden of the private residence Murin-an, the hills of Higashiyama are represented as shakkei. In the gardens of the Swiss landscape architects, one sees similarly designed streams, as well as miniature islands, which here looks like an ishizuke (石付け, “attached to the stone”) – a style of bonsai design in which a tree has grown together with a stone.

134 Cf. Kuck 1940/2016, p. 153.
135 Cf. Kuck 1940/2016, p 152.
136 Taut 1937, pp. 298.
137 Cf. Natsumi 2008, p. 8.
138 Cf. Tunnard 2014 (1938), p. 91.
139 Tagsold 2017, p. 42.
140 Cf. Tunnard 2014 (1938), p. 164.
141 Taut 2004, p. 93.
142 Muthesius 1903, p. 306.
143 Cf. Kreiner 1993, pp. 18–26; Reichert 1993.
144 Cf. Knippschild 2020, p. 29.
145 Cf. Lehmann 1984, p. 762.

the story of the tiger cubs with Zen philosophy by interpreting the garden as a Zen parable.[134]

She also noted that a garden is best seen after it has rained, or when gently drizzling, and not in sunshine [→ pp. 130, 131]. The wetness brings out all the subtle tones and shadowy texture of the rock surface, lending the garden a quality of depth and mystery that corresponds to what the Japanese call yūgen.[135]

Bruno Taut, for his part, connected the Katsura residence with Zen Buddhism, making reference to Kobori Enshū who had mastered the task of "purg[ing] Japanese art from misunderstood Chinese influence and also to incorporate tea-culture created by Zen philosophy in the typical Japanese aesthetic sense".[136] Taut saw Katsura as the highest achievement of Japanese aesthetic taste: "A happy blend of the expansiveness of Heian pond gardens, the stark simplicity of Zen aesthetics, and the subtle and restrained elegance of the world of tea."[137]

From a garden design perspective, his views are echoed by the Canadian-born Christopher Tunnard (1910–1979), who emphasised the spiritual character of Japanese gardens. He interpreted them not so much as witnesses to an ancient, bygone era to be evoked nostalgically, but as archaic sources of modernity, which in his eyes should draw on the special qualities of the fundamentals of Japanese landscape art.[138] "The Japanese garden had become the space for an alternative modernity to the Western one."[139] While Tunnard revived the vision of a garden as a paradise[140] – with Katsura as its facsimile – for Taut the villa was "the little palace" and its garden the "Sanssouci of the Japanese emperors", even "the architectural wonder of the world of Japan"[141]. Hermann Muthesius went even further, including the entire island state in his ideal of a country "that comes closest to one's dream of paradise"[142].

Japan became – after the characterisation passed down in Chinese mythology, after Marco Polo's tales of his travels to the land of Zipangu[143], after its reputation as an Asian paradise in the medieval world view and cartography[144], and after the widespread 19th century portrayal of the country as a pre-industrial paradise, as the antithesis of Western, economic and militaristic society[145] – once again the paradisiacal "island of the blessed".

Walter Leder photographed the stepping stones in the Heian shrine and the largest of the stone settings in Ryōan-ji during a trip on the occasion of the 1964 IFLA Congress in Tōkyō.

Inspiration, imitation, integration

Japanese garden design as a catalyst

← p. 73: Reflections of Japanese tree and shrub species in Willi Neukom's circular water basin at the UBS Center for Education and Dialogue, Wolfsberg in Ermatingen.

pp. 74–75: In Ernst Cramer's Prina Garden in Dulliken, square paving slabs in front of the window and a line of six concrete blocks reflect the geometry of the house. As in Japanese ponds, which Cramer had taken slides of during his travels, plants have grown on the stones.

pp. 76–79: The Japanese miniatures in the inner courtyards of the Bernarda Seminary (now the cantonal school) in Menzingen by Ernst Cramer were reconstructed as faithfully as possible, as was the rock garden.

p. 80: In the courtyards of the Sisters of the Holy Cross Institute in Menzingen, Adolf Zürcher laid cobblestones in meandering lines like a brook between the plants, here with a predominantly green colour palette as was increasingly characteristic of the Japanese garden.

Let us return to the beautiful Katsura Garden! Ancient trees and a dense, tall bamboo forest shelter and envelope the garden so that no unbidden eye from outside can look into this peaceful, poetic paradise.[1] Cäcilie von Rodt

This endearing impression of "Katsura Garden" comes from the travelogue of Cäcilie von Rodt (1855–1929), who came from an old Bernese patrician family. It is a valuable and rarely cited documentary record, not least because it contains an unusually detailed account of the "Katsura Garden" – at a time when it was just being rediscovered in Japan at the turn of the 19th to the 20th century – along with an illustration of the "Little Tea House in Katsura Park":[2]

"A visit to the imperial garden Katsura, which immediately followed the Mikado visit, showed me the ideal of a Japanese park [...]. [...] In the "Katsura garden", there are steep hills and paths, artificial ponds with goldfish, stone arched bridges and wooden walkways, Matsu [pines] artfully cut in the shape of ships, monkeys and parrots, stone lanterns and numerous pavilions, which now stand dilapidated and deserted. The location of these little houses, where jovial tea parties were once celebrated, is chosen with artistic intent. From each of them one enjoys a beautiful view, a balanced, harmonious picture. At the edge of the pond is a large level area made of bamboo, christened Tsuki-mi-dai, which means moonlight platform. Princess Katsura and her court used to stand there and watch the moon rise behind the tall pine trees."[3]

Cäcilie von Rodt began her journey around the world on 30 May 1901. She first crossed the USA, travelled from there via Hawaii to Japan, where she stayed for two months, before travelling through China, Java, Siam, Burma, India, Ceylon, and passing through the Red Sea to reach Egypt. Of the almost 700 pages in which she recorded her impressions in words and pictures, she devoted 86 to the Land of the Rising Sun – almost as many as China with 89 pages. Except for announcements for some of her lectures[4], her wealth of experience was only acknowledged posthumously in 1955 in the form of a benevolent obituary.[5]

On 29 May 1901, just one day before she set off, an article on Japan – probably the first of its kind to be published in 20th century Switzerland – appeared on the front page of the *Neue Zürcher Zeitung*. Entitled "The Japanese Cherry Blossom Festival", its author gave not just an account of the country's most important national customs but also referred to the topos of paradise and praised the intimate connection between the landscape and private garden.

The diplomat and travel writer Ernst von Hesse-Wartegg wrote that when the world was divided up, the most beautiful part of the earth fell to the Japanese, and compared the "island paradise" to a "large stately park", a "kind of Prater or Bois de Boulogne or 'Tiergarten'": "[...] with what taste, what understanding of scenic beauty the Japanese have endowed their land; on the hills that rise here and there from the gloriously green plains stand picturesque groups of tall trees, in the shade of which stand the most graceful little temples, pavilions and torii (archway), usually painted red; individual boulders do not rise naked from the green of the rice fields or the mirrored surfaces of the lakes and coves, but are certainly crowned by picturesque spruces: the watercourses are rearranged to suit the landscape, revealing here a secluded waterfall, there a small lake, in which the beautiful lotus flowers are just as certainly resplendent. Dark, secluded coniferous forests cover the mountains, and here and there the snow-covered cones of the still active volcanoes rise above them from a distance. Every single Japanese seems to regard the whole country as his own private garden [...]."[6]

Switzerland was no stranger to the notion of being associated with a natural idyll, an Arcadian landscape, a place of longing or alternatively a "terra incognita",[7] as that is how Switzerland was already commonly perceived in the 18th century. "Grottoes, waterfalls and cliffs" found their way into garden designs by way of the alpine garden and were considered a Swiss phenomenon for a long time before they were identified as features of Chinese and Japanese garden design. Now they served as an interface between Switzerland and Japan[8]. "All reports agree that artificial rocks play a major role in China and Japan, that in some gardens they occupy the largest area together with water, and that some small

1 Rodt 1903, p. 166.
2 Ibid.
3 Rodt 1903, p. 165. Von Rodt visited the "Katsura Garden" after being received at the Shishin-den, the imperial palace in Kyōto ("Mikado visit"). Tsukimidai (月見台, platform for moon viewing)
4 Cf. anon. 1904a.
5 Cf. Lüthardt 1955.
6 Hesse-Wartegg 1901, p. 1.
7 Jost 2015, p. 111.
8 Cf. Nerfin 2016, p. 54.

gardens consist almost entirely of water, rocky banks and rocky islands. Caves, grottoes, and underground passages are very common in these rocks [...]."[9]

Like the spectacular, untouched alpine landscapes, lauded for their Rousseau-like scenery, people were equally fascinated by the plant world. The well-known English plant collector Reginald Farrer collected plants in Switzerland[10] – especially in the Bernese Oberland – and in Japan,[11] and in 1907 his two-volume *Pocket Atlas of Alpine Plants of Japan* was published.[12]

The topos of the island of paradise or the lost paradise was then taken up in the 19th century by the Japanese themselves, in Kume Kunitake's logbook of the visit of the Japanese special legation to Germany, Austria and Switzerland in 1873 as part of the Iwakura mission: "For all these reasons, Switzerland is praised in the other countries of Europe as the lost paradise",[13] wrote Kume, who was the personal secretary of Ambassador Iwakura, head of the mission.

Aimé Humbert was the Swiss honorary companion of the Japanese legation. The diplomat, who spent ten months in the "Land of the Rising Sun" as part of his mission to conclude a trade treaty between Switzerland and Japan, used his time there to travel around the country. The fruit of these excursions was the two-volume work *Le Japon illustré*, which Humbert published in Paris in 1870.[14] It was the first book on Japanese culture in Switzerland. The "476 engravings made from photographs and paintings"[15] captured a rich spectrum of traditional Japanese architecture – ritual buildings, teahouses, storehouses, factories, farms, and urban quarters, as well as details of interior decorations. In addition, they included views of landscapes and gardens, illustrating the connection between them as well as the relationship between house and garden, between inside and outside, between art (screens) and nature (garden). They illustrated stone settings, lanterns, ponds, streams, and bridges as well as Japanese plant species. "Riding on the wave of enthusiasm for Japonism that began in 1860",[16] the work was widely received after its publication in 1870[17] and was responsible for "a clear expansion of the Europeans' image of Japan"[18].

Evariste Mertens may have heard about the book. There is no evidence of this, but in 1870, he had just moved to Switzerland and founded the collective office of Neher & Mertens together with Arnold Neher.[19] With the exception of Mertens, when the first texts on Japanese garden design became available in the West, the oldest of the landscape architects discussed in this book had just been born: Gustav Ammann in 1885, Fritz Klauser in 1885, Walter Mertens in 1885, Oskar Mertens in 1887, Albert Baumann in 1891, Ernst Cramer in 1898, and Walter Leder in 1892.

Japanese garden art – a painting and a poem

At the beginning of the 20th century, journalistic interest in Japan rose significantly, fuelled not least by increasingly active economic relations, which focused primarily on the silk trade – conducted by companies such as Sieber Wolf u. Cie, [Siber, Wolff & Co.], Sulzer Rudolph & Cie, Nabholz & Cie, Strähler & Cie, Baiver [Bavier] & Cie, Otto Streuli – and the watch industry – represented by C. and J. Favre-Brandt, R. Schmidt & Cie, Schwob Frères & Cie.[20]

The first publication on the subject to be mentioned in a Swiss architectural journal was the book *Japanese Homes and their Surroundings*[21], by Edward Morse (1838–1925)[22] – though not until 1885, some 20 years after its publication, and then only as an aside in a review of Franz Baltzer's work *Das japanische Haus*.[23] The same applies to Josiah Conder (1855–1920). When the *Schweizerische Bauzeitung* paid tribute to Baltzer again four years later,[24] it once more failed to recognise the relevance of the American and British authors: while Morse treats the subject "from a layman's point of view", another "veritable expert, the English architect Josiah Conder, has written about it in the Yearbooks of the London Institute of British Architects", however both provide only a cursory overview without going into detail.[25] The fact that Conder had written a canonical work on the subject in 1893, *Landscape Gardening in Japan*, escaped the reviewer. This seems all the more surprising given that the author was obviously well aware of the importance of the connection between house and garden: "The design of the Japanese house still shows the intimate coexistence with nature, that is, first with the garden, which one also sees, for example, in the English house. In both cases, the house opens onto the garden and turns its back on the street [...]"[26]

These early reports still reflected the superficial knowledge and misunderstandings of the day. Japanese garden design was regarded as something of "a curiosity", its creators nevertheless as unique and unsurpassed in their art. While the relationship to painting was recognised, it was still misunderstood as in essence a "naturalistic painting – a highly deceptive imitation of nature of extraordinary perfection"[27].

In 1907, the works of the Greek-Irish writer Lafcadio Hearn (1850–1904) made their way into publication. His conclusions on the nature of the Japanese garden, however, seem contradictory. On the one hand, the author falls prey to the misconception that the Japanese garden is not an attempt to create "an impossible or purely ideal landscape", the intention being rather to "copy faithfully the attractions of a veritable landscape, and to convey the real impression that a real landscape communicates". On the other hand, he recognises in it "at once a painting and a poem; perhaps even more a poem than a painting.[28]

The Bernese journalist and German scholar Hans Bloesch succumbed to Hearn's enchanting descriptions. Although he acknowledges they are the impressions of "a poetic dreamer" and "enthusiast", seen through the

9 Hunter 1877, p. 194.
10 Cf. Farrer 1913, pp. 46, 70, 202, 205, 208, 214, 238, 249.
11 Cf. "Mr. Edward Arnold's List of New Books. My Rock Garden by Reginald Farrer", Appendix, in: Farrer 1913, p. 6.
12 See Miyoshi/Makino 1907.
13 Kume 2002, p. 358.
14 See Humbert 1870. The trade treaty concluded by Humbert on 6 February 1864 was the first trade treaty between Japan and a Western non-major power. Cf. Rüegg 2015, p. 52.
15 Rüegg 2015, p. 56.
16 Pekar 2000, p. 231.
17 Cf. Rüegg 2015, p. 53.
18 Baiemi 2003, p. 81.
19 See Ruoff 1993, p. 88.
20 Anon. 1904b. For details on the companies and their founders see http://www.meiji-portraits.de (retrieved on 4 Jan 2023).
21 Morse 1972 (1885).
22 Cf. anon. 1903.
23 See Baltzer 1903.
24 See anon. 1907c; Baltzer 1907.
25 Cf. anon. 1903.
26 Ibid.
27 Weh. 1908.
28 V. M. 1907.

Paul Schädlich, Perspective drawing of a view into the Japanese garden courtyard of the Wiederkehr-Schmid Park, 1920s.

rose-tinted glasses of someone who had found the answer to his lifelong yearnings, he also wrote that rarely have more beautiful things been written about Japan – "the Japan of those days, a Japan romanticised and intellectualised by a grateful admirer".[29]

Japan-British Exhibition 1910: Nature to Art, Art to Nature

Not long after, the *Neue Zürcher Nachrichten* reported on the "Japan-British Exhibition" in London in 1910 and was full of praise for the Japanese garden design on show there: the exhibition sought to give Europeans an idea of the beauty of the Japanese landscape, and in the creation of artificial landscapes, one sees "all the refined skill of the Japanese to adapt nature as far as possible to art and, conversely, to make the artificial appear as natural as possible".[30] Among those who visited the exhibition was Gustav Ammann (1885–1955), who reported his impressions in a letter addressed to his parents in mid-July 1910.[31] It is one of the two earliest records of the perception of the art of Japanese garden design by the garden designers whose estates and legacies are now archived in Switzerland. Ammann wrote his letter only months before Leberecht Migge (1881–1935),[32] one of the formative figures for Gustav Ammann, Walter Leder, and Albert Baumann, wrote in the journal *Die Gartenkunst*, "Shouldn't we actually also add quarries, railway embankments and chimneys to the landscape garden, at least in theory, if we are to be as faithful to the history of the earth as the fathers of that style on the meadow isle did in their day, or as the Japanese do today, who simply take an extract of their great landscape and liliputanise it spatially in its entirety?!"[33]

In the mid-1910s, Ammann commented publicly: "Then [after the triumph of the French garden in the manner of André Le Nôtre] came a tremendous turnaround. First at an intellectual level, which loosened the ground for new ideas. Rousseau's ideas whipped minds out of the stubborn slumber of tradition. At the same time, Albion's [ancient name for England] world sailors brought back the customs and traditions of foreign countries. China and Japan showed their culture for the first time in the West. Their artistic creations and their gardens became known, and as it was precisely the opposite of which people in Europe had slowly grown weary, it was proffered by enthusiastic prophets to the people of Europe as a new revelation. And with success."[34]

The natural scientist and art historian Hermann Röthlisberger, who was secretary of the *Swiss Werkbund* and editor-in-chief of the journal *Das Werk* from 1915–1922,[35] recognised around this time "the dawn of a turning point",[36] which he illustrated with a few select gardens by Ammann. He praised the gardens "as created by [the company] Froebels Erben, Zurich"[37] and postulated – quoting Rousseau – a shift away from the architectural gardens of the past. At almost the same time, Oskar Mertens struck a similar note, also citing Rousseau, diagnosing the "fortunate overcoming of both the imitation of natural 'landscapes' of the last century and the all too far-reaching reaction to it at the beginning of the 20th century"[38].

Towards a new architecture – towards a new garden

Much later, in 1949, Ammann himself pinpointed the moment at which this shift took place to the year 1927,[39] the same year in which the Weißenhofsiedlung opened in Stuttgart as part of the Werkbund exhibition "Die Wohnung". It "is the product and also the landmark of this epoch, which today we consider historically to be the first phase of Neues Bauen or modern architecture [...]".[40] It was also the place where many renowned architects and proponents of this spirit of "New Objectivity" rejected the idea of the architectural garden. Le Corbusier and Pierre Jeanneret's design for a semi-detached house in the Weissenhofsiedlung is simultaneously a statement on garden design: "The crystalline forms of concrete thinking do not extend beyond the architectural body, but here encounter opposing shapes that create suspense: the shapes of nature. Nature appears wild. In a garden you want to view it quietly, to walk around it."[41]

In 1927, under the title "Die gute Garten-Wohnung" (The Good Garden Dwelling), Leberecht Migge commented with unconcealed admiration in the journal *Die Gartenschönheit* on the "the Frenchman's [sic!] attack

29 Bloesch 1912/13, pp. 459–460. Bloesch also referred to: Hearn 1911.
30 Anon. 1910a.
31 Ammann 1910 gtaA.
32 Migge's estate was found in 2015 in the Archive for Swiss Landscape Architecture (ASLA) as a bundle in the estate of Walter Leder.
33 Migge 1910, p. 195. Migge was responding to Engelhardt 1910 in the article.
34 Ammann 1914/15, p. 580.
35 See Weese 1922, p. 76.
36 Röthlisberger 1917, p. 87.
37 Ammann was the head garden architect at Froebels Erben horticultural firm from autumn 1911. See Stoffler 2006a, p. 28.
38 Mertens O 1916, p. 80.
39 Cf. Ammann 1950, p. 2.
40 Joedicke 1987, p. 108.
41 Roth A 1927, p. 37.

on the organisation of dwellings", which struck "like a bombshell" at the heart of the "chaos in European housing". To illustrate his article, he showed an image of the "jardin suspendu" in the Pavillon de l'Esprit Nouveau of 1925 that clearly showed the connection between the house and the surrounding landscape, as well as an axonometric of a house in the Frugès experimental housing estate in Pessac, built in 1924–1927, which showed the roof garden as a miniaturised and sublimated image of the surrounding nature. What sets this "painting engineer-architect apart from many of his colleagues", wrote Migge, is the "self-evident way in which Le Corbusier draws air and light into the dwelling, how he groups windows, cuts out loggias and pushes forward terraces, how he lets stairs glide, how he draws the obligatory gardens into the upper floors and onto the roofs, in short, how he *anchors living space in nature in a homely way*".[42]

In the March edition of the same magazine, Migge even went as far as to paraphrase the headline of the 1926 German translation of Le Corbusier's *Vers une architecture, Kommende Baukunst*[43] (*Towards a New Architecture*) and titled it "Der kommende Garten".[44]

Le Corbusier and Migge met the following year in Frankfurt on the occasion of a conference on "The Flat Roof". According to Joseph Gantner, editor of the journal *Das neue Frankfurt*, "the architect [sic] Leberecht Migge from Worpswede followed up on Le Corbusier's ideas ['The Aesthetics of the Flat Roof'] in his [...] lecture on 'The Green Roof'".[45]

Light and air had been key concepts for Migge as early as the mid-1910s. The plan for a "Light and Air Bath and Tea House" for the Ballin-Trittau country estate built in 1906 by the Hamburg architect Fritz Höger for the German shipowner Albert Ballin (1857–1918) dates from February 1914.[46] Around 1920, Migge created a plan for a pergola with a "light and air bath" for an unspecified project.[47] Not only the theme of light and air but also Migge's concept of the growing house[48] found their echo in the designs by the proponents of Neues Bauen.[49]

Migge's article "The Good Garden Dwelling" was also notable for its opening image, a photographic reproduction of a garden scene described as a "Japanese country house in the evening". It showed a view of a pond, a partially shaded Japanese dwelling with a garden pavilion – in which a person only dimly discernible through the half-open sliding paper screen door appears to be in conversation with a woman dressed in a traditional kimono sitting on the narrow veranda. The light source inside the pavilion makes it look like a lantern, causing the scene to be reflected on the water of the pond in front of it.

For Migge this was not just an attractive opening image; he also began his piece by praising the Japanese garden and its function as an extraordinary role model. "*The Japanese*, who got it from the Chinese, can be called the people who, in centuries of practice, have brought the good garden dwelling to a degree of refinement that to this day has no true counterpart in European culture."[50]

"Conscious imperfection"

In Switzerland, it was Peter Meyer (1894–1984) who, as a contributor to the *Schweizerische Bauzeitung (SBZ)* from 1922–1929 and as editor of the architecture and art journal *Das Werk* from 1930–1942, promoted the reception of East Asian culture. In particular, it was a lecture by the ethnologist Ernst Grosse (1862–1927), given on 24 February 1926 at the Zurich Society of Engineers and Architects under the title "The Japanese House", that caught his attention. Meyer reported on it in the *Schweizerische Bauzeitung* on 15 May 1926[51] – "greatly delayed", as he noted, "by the considerable difficulties in procuring suitable illustration material"[52] and came to the following assessment: "Although the Japanese climate is not milder than the European, the Japanese house offers much less protection; the inhabitant is not enclosed by walls as if in a fortress. On the contrary, the movable walls of his house are wide open to the garden and a Japanese person's life takes place in much closer connection with nature. The garden side is the main side of the house [...]."[53]

It was shortly before Grosse's death in 1927 that the Swiss architect and art educator Alfred Altherr paid tribute to him,[54] and decades later the art historian Willy Rotzler (1917–1994) recalled Grosse's "inspiring lectures on Japanese domestic culture and utensils", which had exerted "the greatest influence on the creative people of Switzerland"[55]. He did not exclude garden design from this – on the contrary, garden design, which in the 1920s had begun to free itself from the shackles of convention and since then had striven for maximum possible uncontrived naturalness, "was (and still is) decisively influenced by the experience of Japanese garden art conveyed in numerous publications".[56]

For Heini Mathys (1917–2000), 1927 was also the milestone year in which a shift from architectural gardens to dwelling gardens became apparent. As part of his review of "Swiss Garden Architecture in the 20th Century", published in 1972, he noted: "The aim was to free garden design from what was perceived as a straitjacket of axis and symmetry.[57] He included Ammann's timeline, which the latter had sketched in 1949 and published in 1950.[58]

It was no coincidence that Mathys illustrated the article with two drawings by the Zurich garden architect Paul Schädlich (1885–1946), "which show excellent examples of gardens from the 1920s".[59] These were perspectives of a garden project for "Mr. H. Wiederkehr-Schmid, Zurich", one of which showed a "Japanese Garden courtyard" [→ p. 85]. The original drawings came into the possession of Albert Baumann and, with the donation of his estate by the Oeschberg Cantonal Horticultural School, to the Archive for Swiss Landscape Architecture (ASLA).

In 1928, Ammann also tried his hand at a Japanese garden, which he designed for the embroidery manufacturer Josef Eugster in Altstätten (SG) (see "Gustav Ammann – Japan as a module"). It was the year of the accession of the new Mikado Hirohito and the beginning

42 Migge 1927a, p. 10.
43 See Le Corbusier 1926.
44 See Migge 1927b.
45 Gantner 1928, p. 75f.
46 "Licht-Luftbad u. Teehaus", February 1914, 80 × 78 cm, ASLA NL MIG_4_22. Migge identifies himself in the plan label as an "architect for horticulture".
47 "Pergola Licht-Luftbad", no date, 41 × 30 cm, ASLA NL MIG_4_21. "Bad mit Wasserpflanzen Laubengang", 27 × 33 cm, ASLA NL MIG_3_26. Walter Leder, who is also credited with saving this plan, provided it with his own plan label and the professional title "Architekt und Lehrer" [Architect and teacher].
48 See Migge 1932.
49 See Wagner 1932; Lotz 1932. The conceptual pair of light or sun and air in connection with natural surroundings had been coined in 1909 by Georges Risler in a lecture, who saw in the garden city the potential for renewing man's connection with fresh air, sunlight and earth. Cf. Risler 1910, p. 404. As a trinity of "sky, trees, light", Le Corbusier took this up in his writings in the 1930s. See Morel 1987, p. 41. What was for Migge the "intimate interweaving of house and garden" in his "Sonnenhof" in Worpswede initiated in 1920 – Migge 1927a, p. 13 – was for Le Corbusier the development of the ground plan from the inside to the outside, so that the outside was the result of the inside, indeed "the outside was always also an inside". Cf. Le Corbusier 1923, pp. 143, 146.
50 Migge 1927a, p. 10.
51 See Meyer 1926.
52 To all appearances, the photographs were taken from: Baltzer 1903. Drawings were taken from the first edition of: Morse 1972 (1885).
53 Meyer 1926, p. 259.
54 See Altherr 1927, pp. XXVIII, XXXI.
55 Rotzler 1955, n. p. 1.
56 Ibid.
57 Mathys 1972, p. 8.
58 Sketch of the time diagram: "Development curve of garden design from 1900 to 1950 (chronology)", 4.12.1949, gtaA, NSL 2-T-3-9. Published in: Ammann 1950, p. 1.
59 Mathys 1972, p. 7.

Paul Schädlich, Japanese-inspired water terraces at the Züga, 1933.

of the Showa era. The correspondent of the *Neue Zürcher Zeitung* gave an intimate account of the locations of the celebrations, which also included the Katsura residence, and praised its distinctive art of composing "little bridges, ponds, the narrow gravel paths, pretty groups of trees, mossy ground, camellia hedges around delicate tea pavilions into a whole of singularly delicate charm". He also drew attention to "a single, final window in one of the teahouses", which – left unfinished – celebrated that sense of "conscious imperfection" peculiar to Zen philosophy.[60]

Züga 1933 – organic design with Japanese touches

In 1933, a fateful year in world politics, the Guild of Garden Designers, which had formed the Bund Schweizerischer Gartengestalter (Federation of Swiss Garden Designers, BSG) on 19 October 1925, was preparing for the Zürcher Gartenbauausstellung (Zurich Horticultural Exhibition, Züga). It, too, was not immune to National Socialist infiltration, and Johannes Stoffler has traced its impact and the conflicting debates.[61] But the reception of East Asian culture and its effect on creativity in Switzerland, which was promoted most prominently by Peter Meyer, was free of racial connotations.

The Zurich Horticultural Exhibition (Züga), which was held from 24 June to 17 September 1933, provided an opportunity to make up for lost time. Karl Egender and Wilhelm Müller were responsible for the construction of the Züga, Gustav Ammann for the garden design. Its design testifies to the change from Ammann's earlier architectural gardens produced while at Froebels to "a new 'organic' design".[62] But it was the 50,000 m^2 site[63] on the western (left) shore of Lake Zurich – "composed of the formal private park of the 'Schneeli estate', the seaward side of the Seestrasse between Muraltengut and Belvoirpark, and the area of the city's municipal nursery" – that was perceived as the "real discovery"[64]. Because the existing groups of trees and avenues loosely defined garden spaces, it was possible to create "living rooms of plants and lawns"[65].

Despite the general rule that large special gardens were to be avoided and smaller ones should respect the overall effect of the park as a whole,[66] the reporting was dominated by pictures of the park sections designed by Gustav Ammann, Albert Bodmer, Ernst Cramer, Eugen Fritz, Walter Leder, the Mertens brothers, Ernst Oettli, Walter Pohl, and Paul Schädlich.

Schädlich benefited from the situation in that he had the middle section of the old park at his disposal and could develop his design parallel to an existing avenue of chestnut trees. Thus, its "beguilingly rising series of terraced water lily ponds", bridged by two artificial stone arches, extended from north to south as a "much-admired water feature"[67]. The avenue determined the straightness of the layout, but despite its geometric design, it did not conceal its Japanese inspiration, which was far more pronounced than in his earlier design for Wiederkehr-Schmid.

The opposite counterpart, as it were, to the generously proportioned grounds was "a thoroughly original little garden by Otto Sproß [...], that through its consistent emphasis on miniature garden art was almost Japanese in appearance. The miniature river course, meandering happily through a landscape of the most delicate saxifrages, could easily have been a vignette from a genre painting by Karl Spitzweg."[68]

In addition to the gardens by Schädlich and Spross, the special gardens by Walter Leder and Ernst Cramer had a Japanese touch, which can be seen in the illustrations in the journal *Das Werk*, but not in the text.[69] Here it seems Peter Meyer chose to reserve his opinion, focusing instead mainly on the architecture.

Oskar Mertens, in the context of his coverage of the Züga in the *Neue Zürcher Zeitung*, observed "momentary glimpses of ideas brewing" that were to characterise "the third decade of the 20th century". He attributed these as much to the "experience of light, simplicity and openness now made possible by the relationship between house and garden" as to East Asian models: "Japan and China are sending their influence and even their assortment of plants."[70]

The Mertens brothers' original overall plan, drawn up in the context of the competition, also included a "tea pavilion".[71]

The editor of the *Neue Zürcher Zeitung* described the Mertens brothers' proposal, which they had designed

60 bw. 1928.
61 Stoffler 2006a, p. 184.
62 Mathys 1972, pp. 8, 10.
63 Cf. H. J. K. 1933, p. 124.
64 H. J. K. 1933, p. 115.
65 Meyer 1933, p. 194.
66 Cf. Ammann 1933a, p. 121.
67 Anon. 1933.
68 Anon. 1933, ibid.
69 Cf. Meyer 1933, pp. 206, 212.
70 Mertens O 1933.
71 ASLA NL GM_937_1-0002.

under the motto "Light and Shadow", as a "prime example of artistic landscape design". He praised the "sophisticated organisation of the overall picture" and the "deep artistic sensibilities" of its authors, and drew attention to the "progression from shadow to semi-darkness and finally into bright light". This choreography began from a watercourse "on which hanging bushes cast green reflections, via a charming swimming pool" to a "ribbon of lawn immersed in bright sunlight".[72] The design, which emphasised the sense of spatial depth had no hard dividing lines with light and shadow not conceived as opposites but as mutually dependent entities. Gustav Ammann added a reference to Otto Valentien's then newly published book *Zeitgemässe Wohngärten* to the review in the *Schweizerische Bauzeitung*. His standpoint, according to Ammann, echoed that of all those who "turn away from merely decorative, symmetrical work".[73] The book contained several photographs of gardens with Japanese features. Valentien depicted a section of a garden by Ammann or Froebels Erben, though he failed to notice its Japanese aspect, focusing instead on the "pond with interesting vegetation"[74]. Not so Ammann: he found the impression of the Japanese gardens not only "so strong that we feel involuntarily drawn to these creations", but also "in many respects akin to our present-day way of designing gardens".[75] The interpenetration of house and garden, the strongly landscape-oriented design, had for Ammann a direct connection with Japanese influences.

Nothing less than the "shift away from the architectural garden"[76] had been achieved largely through the use of Japanese design approaches – coupled with "wilderness garden art", as Karl Foerster (1874–1970) had called "the artful horticultural interpretation of aspects of the cultural and natural landscape"[77] in the modern age.

A key role in the reception in Switzerland was played by the exhibition *Japanese Architecture and Gardens*, which went on show at the Kunstgewerbemuseum (Museum for Decorative Arts) Zurich from 29 November to 23 December 1933 and was based on articles written by Hentrich and Heisinger in the journal *Die Form*[78], which existed from 1925 to 1934. It was the first exhibition in Switzerland to portray Japanese architecture and garden design on an equal footing and was accompanied by a booklet with a text by Walter Curt Behrendt.[79] Four of the eight illustrations showed sections of gardens with teahouses, including the Shimo-no-Onchaya tea pavilions in the garden of the Imperial Shugakuin Villa (修学院離宮, Shugakuin rikyū) (1655) in Kyōto and Shōkintei of the Imperial Katsura Residence (桂離宮, Katsura rikyū),[80] which Bruno Taut was to herald the following year in a much-quoted dictum as one of the "architectural wonders of the world".

In "the leading Swiss art magazine 'Das Werk'"[81], Marta Heisinger wrote a reflection on Japanese gardens,[82] that was deemed worthy of mention in the *Neue Zürcher Zeitung*. Peter Meyer endeavoured to learn first-hand of Japanese work by arranging for *Das Werk* to enter into an "exchange with the magazines 'Sinkentiku' and 'Kentiku Sekai'"[83].

Meyer also heralded Yoshida's publication on Japanese architecture as the book "we have been waiting for"[84] in a detailed and insightful review,[85] and was not afraid to take a side swipe at the blood-and-soil ideologues. Commenting on the caption to a photograph of a farmhouse in the prefecture of Gifu in the Chūbu region on the island of Honshū, a "post-and-beam building, in its overall impression astonishingly similar to European half-timbered buildings", saying, "If the house were in Europe, its builders would undoubtedly be suspected of a racial relationship with the Alemanni."[86]

At the same time, Meyer drew comparisons with Neues Bauen by associating Japanese principles with Le Corbusier's ideas: "Consummate and refined primitive reduction, the close interrelationship between interior and garden space, and of building form and natural form. The basic form, deriving from the archetype of the farmhouse, is broken down so that its walls dissolve into a backdrop for the garden – an idea that can also be seen translated into concrete in the work of Le Corbusier."[87]

1st Intermezzo – the 1937 World's Fair

That Junzo Sakakura was able to merge traditional design principles with modern technology in his design for the Japanese pavilion at the Paris *Exposition internationale des arts et des techniques appliqués à la vie moderne* (25 May to 25 November 1937) is no surprise. From 1930 to 1936, Sakakura had worked for Le Corbusier and had given him an album of photographs of traditional Japanese houses. In 1937, he returned to Paris to design the pavilion.[88] The anonymous reviewer of "The Pavilions of Foreign States" in the magazine *Das Werk* applauded the design: "Here the design has successfully managed to express the typical lightness of the national-Japanese house with the modern means of iron skeleton construction – an exhibition built entirely in this spirit has to be delightful – a foretaste of which we have already seen at the Zurich 'Züga'."[89]

Again, Meyer recognised the role that the Japanese model played "in the handling of the interpenetration of space, in the proportioning of abstract forms, and in the striking confrontation of natural forms with geometric forms".[90]

Johannes Schweizer (1901–1983) was involved as the creator of the grounds of the Swiss Pavilion designed by Breuning, Leu, Dürig,[91] where he designed, among other things, a rhododendron garden and an alpine garden.[92] At the congress of the International Federation of Landscape Architects (IFLA) held at the World Exhibition in Paris, Schweizer, Gustav Ammann, Walter Leder, and Walter Mertens met the Englishman Christopher Tunnard – the Bruno Taut of landscape architects – alongside their colleagues from Germany, Alwin Seifert, Hermann Mattern, and Otto Valentien.[93]

72 Anon. 1933.
73 Ammann 1933b, pp. 206–207.
74 Valentien 1932, p. 66. He made up for the omission in 1959: Valentien 1959.
75 Ammann 1934b gtaA.
76 Stoffler 2006a, p. 126.
77 Stoffler 2016a, p. 13.
78 See Hentrich 1933. Heisinger 1933a.
79 See Behrendt 1933.
80 See Behrendt 1933, pp. 5, 7. In the list of illustrations the two were confused, the photograph of the Shimo-no-Onchaya being shown as Shōkintei and vice versa. Behrendt 1933, p. 23.
81 h. er. 1933, sheet 4.
82 See Heisinger 1933b, p. 33.
83 Meyer 1936a, p. XXII.
84 Yoshida 1935b, p. 461.
85 See Meyer 1936b.
86 Meyer 1936b, p. 79.
87 Meyer 1936b, p. 81.
88 Cf. Allaer 2007, p. 113.
89 Anon. 1937, p. 329.
90 Meyer 1937a, p. 58.
91 See illustration in: str. 1937, p. XVI.
92 ASLA NL js.10662.11.2, js.10662.13.2.
93 Cf. Imbert 2009, p. 109.

Gustav Ammann, Water Courtyard in the “Twining” courtyard of the “Kleider machen Leute” Pavilion.

Eugen Fritz & Co., Japanese Garden at the Pavilion for Swiss Nationals Abroad at the 1939 National Exhibition.

Sketch by Ernst Cramer of his "Pavilion Garden" at the Züka, 1947.

94 Cf. Mathys 1972, p. 10.
95 Cf. Luchsinger 1997, p. 56.
96 Meyer 1939a, p. 193.
97 Meyer 1939b, pp. 208–209.
98 Meyer 1939c.
99 Meyer 1940, p. 251.
100 Schweizer 1939.
101 Ammann 1939, p. 216.
102 Mertens W 1939.
103 See Meyer 1939c, p. 211.
104 See summer 1977.
105 See Du Cane/Du Cane 1908, pp. 22, 28, 35–37, 50.
106 See Mappes 1936, p. 31. gtaA, NLS 2-S-34_M2/3.
107 Mertens W 1939.
108 Cf. Schmid 1947.

Landesausstellung 1939: "Playing with Japanese Stimuli"

At the Landesausstellung 1939 (National Exhibition), architecture and garden design were finally placed on an equal footing. Gustav Ammann was in charge of landscape design,[94] the chief architect was Hans Hofmann (1897–1957)[95], and the overall director was Armin Meili. The latter would embark on a trip to Japan, 20 years after planning began in 1937, that made a lasting impression on him.

The sense of unanimity was expressed by Peter Meyer, who emphasised the "friendship that connects architecture with the art of garden design" and credited the fact that "for the first time in decades, a major undertaking of modern architecture finds the undivided approval of both professional circles and all sections of the public" to the merit of the gardens.[96] He attributed its success to the fact that the landscape itself served as the raw material for the garden, which was "also the point of contact with the Japanese garden". The "so conspicuous 'Japonisms'", were, in his opinion, more than "mere fashionable play: Japanese gardens are also developed from the edges and from the topographical conditions, not according to axial symmetries".[97]

In the *Neue Zürcher Zeitung*, Meyer spoke even more resolutely in favour of Japan: "[W]hile in Europe the regimented palace garden type has been transferred to even the smallest gardens, where it is pointless, in Japan the non-monumental garden type has consistently been cultivated, refined and developed, so that today, when the best architects are also seeking the non-monumental form in Europe, we really have a lot to learn from East Asia in this respect."[98] A year later, he went even further, asserting that modern garden design was one step ahead of architecture in "satisfying a need for intimacy that cannot be explained away by any rationalistic rationale". He recommended "playing with Japanese influences not only in gardens but also in architecture".[99]

For Johannes Schweizer, Japanese plants had at least become established and already entered into dialogue with the alpine garden: "And for his own garden, whether already created or dreamed of, he [the reader] is planning to include the inevitable red-leaved maple and dwarf shrubs for his 'alpine garden' [...]."[100]

An actual Japanese garden was created for the Pavilion for Swiss Nationals Abroad by the company Eugen Fritz & Co.,[101] who endeavoured to create a faithful replica. An organically shaped pond was lined with a few stone arrangements and "foreign looking"[102] plants. At the narrowest point, it was bridged by two stone slabs offset to one another in the Japanese manner, that led to a path also made of irregularly laid broken slabs à la japonaise. Decoratively grown pine trees graced the surrounding lawn, and finally a stone lantern signalled unmistakably the origin of the gardens. Peter Meyer paid tribute to the garden with a full-page illustration.[103] Fritz would still be designing gardens in the Japanese manner in 1977.[104]

Gustav Ammann's three courtyards for Karl Egender's "Kleider machen Leute" fashion pavilion likewise drew clear inspiration from Asian models but were freer in their interpretation. The azalea courtyard in the "Ready-to-wear" section, the small atrium in the "Outfitting" section, and the water courtyard in the "Twining" section were striking miniatures, displaying extracts of Japanese motifs: rhododendrons, whose charm Ella Du Cane had captured so stunningly in her drawings,[105] green islands between broken granite slabs and circular stepping stones leading across the water to water lilies, irises, and papyrus. A possible model for the latter may have been the rounded stepping stones of a teahouse garden depicted by Michael Mappes in 1936 in *Die Gartenkunst*.[106]

Through the artfully orchestrated interplay of plants, sculptures, wall treatments, stepping slabs, and water, they were, as Walter Mertens noted, "elegant still lifes", "small oases", and "jewel[s] of modern garden design"[107].

He associated these images of Japanese gardens with Hugo Spross' "Garden Idyll on the Lake" at Wollishofen steamship yard.

2nd Intermezzo – Züka and AGA

A further notable example was created in 1947 by Cramer & Surbeck, collaborating with Willi Neukom, at the Zürcher Kantonale Gewerbe- u. Landwirtschafts-Ausstellung (Zurich Cantonal Trade and Agricultural Exhibition, Züka), which ran from 23 August to 19 October.[108]

↑ The ephemeral character of Ernst Cramer's installation for the Züka 1947 can be seen in this undated sketch by Willi Neukom.

→ Ernst Cramer's design for the Züka in 1947 maximises the interpenetration of house and garden.

← Cramer designed a counterpart to his installation at the Züka 1947 for the Aargau Horticultural Exhibition in the same year.

↓ At least one of the two circular water basins sketched for the Züka was realised.

Fritz Haggenmacher oversaw the garden design and Hans Fischli[109] the architecture.

Here they reduced the "house" literally to skin and bones: it took the form of a red canvas tent,[110] supported vertically by five poles anchored in the ground and horizontally by two round wooden poles. It was placed on a platform of rough, hand-hewn sandstone slabs, interspersed by strips of pink clinker mosaic[111] – a mix of materials resembling the "semi-formal" (gyō) style of Japanese path designs.

Another allusion to its Japanese provenance could be found in the pages of the *Neue Zürcher Zeitung*. The reporter Willy Schmid was enamoured with the oriental grace with which the red canopy stretched over the Garden Forum, "surrounded by the eternal fairytale whisper of huge bamboo bushes".[112] He was referring here, perhaps unconsciously, to a borrowed landscape in the form of Norway spruces that formed the backdrop, and that "keep alive the memory of the far-sighted park creators from the turn of the century".

Cramer & Surbeck leveraged the contrast between inside and outside by maximising the interplay between the house and the garden, as Le Corbusier had postulated in 1928: "The landscape, the lawn, the flowers, the trees should flow through the house [...]".[113] And this house corresponded to "the 'flying hut'", whose material lightness and "fleetingness in time"[114] Migge was quite taken with. It was not just the ephemeral nature of its "four walls", but the whole microcosm of house and garden, which had literally been merged into one. By raising it off the ground, the garden designer not only allowed the garden to flow through the house, but also the park area to as it were wash beneath the "garden stage"[115]. The idea of conceiving of the scene as an island was further heightened by the use of stone slabs laid like bridges. The installation was partly supported on stone blocks – echoing the placement of Max Bill's "Endless Loop" (1935–1937)[116], just a short distance away, on prominent boulders, so that the two appeared to hover in a seemingly unstable equilibrium.

A variation on their contribution to the Züka was the sun canopy at the Aargauischen Gartenbauaustellung in Aarau (Aargau Horticultural Exhibition, AGA) organised by the Aargau Master Gardeners' Association from 20 to 29 September 1947. Here too, Cramer & Surbeck reduced the pavilion to the absolute minimum and placed it on a platform of stone slabs, in front of which they arranged a small water basin, so that the area appeared like a shallow, dry ditch. They accentuated this by placing a stone slab over it like a bridge. The Japanese impression was further heightened by "weighting" it down with three stones. In its abstraction, it bore reminiscences of Ryōan-ji.

Ten years after the Züka, in 1957, *Der japanische Garten* by Tetsurō Yoshida was published posthumously.[117] Yoshida had concentrated "with all his strength and dedication [...] despite serious illness"[118] on completing it, as the designer Reni Trüdinger, who had met him in 1952, wrote in his obituary. Like his first work, it

Walter Leder, Special Garden at the Züka in 1947. The bronze cranes had a precursor. The Earl of Sandwich used similar sculptural birds to adorn the pond of his Japanese Garden at the Hinchingbrooke estate in Huntingdonshire, which was featured in *Country Life* magazine in 1907 (see Anon. 1907e, p. 635).

attracted the attention of both professional journals and public newspapers. The art historian and Japan connoisseur Albert Theile highlighted the close connection between house and garden on the one hand and between ink painting and garden art on the other: "Gardens are indeed ink painting realised in nature."[119] He carefully traced the shin, gyō, and sō styles defined by Yoshida and compiled exhaustive lists of the symbolically charged stone types – protective stones, child stones, turtle head stone, both hands stone, moon shadow stone, mist veil stone, mountain end stone, and stone of shadow.

While Theile found that the Japanese garden could help to make the domestic "dwelling-garden" a place of refuge and recuperation, for the Zurich writer Margot Schwarz, it went without saying that the form of the Western garden at that time would be unthinkable without the Japanese model – with the caveat that the West strives for perfection. "The Japanese, on the other hand, seek imperfection [...]" in order to "come closer to the principle of beauty"[120]. The architect Armin Meili was equally enthralled by the notion of "striving for the 'beauty of the imperfect'"[121], and classified the Japanese garden as "superior in artistic sensibility even to the intimate courts of the Spanish Arabs and the landscape design of the Italian Renaissance as well as the works of a Le Nôtre".

Yoshida gave new impetus to Japan's function as a role model. After his first book *Das japanische Wohnhaus*,

109 Cf. Haggenmacher 1947, p. 306.
110 Schmid 1947. Udo Weilacher assumed a white colour scheme. Cf. Weilacher 2001, p. 62.
111 Cf. Schmid 1947.
112 Schmid 1947.
113 Le Corbusier 1928, p. 156. Translated after: Vogt 1996, p. 75.
114 Migge 1927a, p. 10.
115 Schmid 1947.
116 See anon. 1938, p. 253.
117 See Yoshida 1957.
118 Trüdinger 1956.
119 Theile 1957, p. 199.
120 Schwarz 1957. Schwarz followed this up a short time later with: Schwarz 1961.
121 Meili 1957, sheet 4.

published in 1935, and his second *Japanische Architektur* in 1952,[122] which advanced to become a veritable "bible"[123] for the young architects of the time, for whom the Japanese influence was formative, his last book provided landscape architects with what was tantamount to a manual of garden design. Floor plans, elevations, sections and photographs, which documented sections of gardens as well as the details of gazebos, teahouses, ponds, bridges, stepping stones, and fences, were so compelling that they could almost be lifted from the pages and be "put straight into practice".[124]

Two key mentors in the field were Richard Arioli and Heini Mathys with their contributions to the journal *Anthos*. The self-taught Arioli wrote his alongside his work as the city gardener of Basel (1940–1970), and the former "manor gardener"[125] Mathys was the journal's editor from 1969 to 1995. Both were interested in the character of Japanese garden design and identified Japanese traits in the creations of their colleagues when these arose. Each of them bequeathed extensive photographic libraries as part of their estates that bear witness to their special area of interest. Arioli created a special portfolio with prints of photographs of Japanese gardens, Mathys one with reproductions of Chinese and Japanese pictorial works. He also devoted considerable study to the philosophical-spiritual background: "Ancient Chinese thinkers such as Lao-tzu, Chuangtse, Confucius, Mengtse, and Liä-Dsi were his models."[126] Numerous books, of which more than 40 ended up in the Archive for Swiss Landscape Architecture (ASLA), testify to the breadth and intensity of his field of study. The architecture of cult and secular buildings[127] and artistic forms of expression such as the colour woodcut[128] interested him just as much as horticultural subjects – from land cultivation[129] and garden art[130] to creations of bonsai (盆栽), bonkei (盆景), and bonseki (盆石)[131] as well as the arrangement of flowers.[132]

Articles in the popular media illustrate a similar fascination with the art of Japanese garden design – highlighting both the paradisiacal connotation and the tendency towards abstraction: "Whether the Japanese garden may be understood as an artificial dream landscape? As an evocation of a natural and yet precisely planned refuge, as a paradise *en miniature*? While this dream-like element may seem convincing, such gardens have a rich and very old tradition."[133] The author dressed this up in the formula "stone and water", alluding on the one hand to waterfalls and ponds, and on the other to dry landscapes. Their streambeds, he writes, are formed by naturally strewn stones to create the illusion of a fluid element. Finally, he cites the garden of the Ryōan-ji, an abstract world of sand and stone where no tree or shrub disturbs the sparseness of the impression, as the most famous example of this reduction.

The Swiss Pavilion at Expo 58, the World's Fair in Brussels from 17 April to 19 October, which is most widely remembered for the Philips Pavilion designed by Le Corbusier and Yannis Xenakis and the Atomium, was a precursor to the 1st Swiss Horticultural Exhibition in Zurich in 1959. It is also notable in that Werner Gantenbein, its architect, described his intervention not in terms of the buildings but instead saw the garden design by Hans Nussbaumer – a tree-covered natural space – as its starting point.[134] Its centrepiece was an artificially created expanse of water lined with 32 hexagonal pavilions, arranged staggered next to one another. The roughly rectangular surface of the water spread out in front of the delicately poised, seemingly floating buildings, flowing beneath them in some places and thus heightening the connection between inside and the outside. Practically devoid of any vegetation, the small lake was effectively an unbroken mirror on which the pavilions seemed to float, and the reflection of the sky in the water may have evoked the illusion of infinity. The outer edges of the water area were enlivened by perennials, and creatively arranged stones placed to strengthen the shoreline. Here and there, a group of three stones were gathered to form an island that rose out of the water, corresponding with the design principle of main stone, auxiliary stone, and supplementary stone outlined by Yoshida.[135]

3rd Intermezzo – SAFFA 1958

The tried and tested site of the 1939 Swiss National Exhibition was the venue for the 2nd Schweizerische Ausstellung für Frauenarbeit (Swiss Exhibition of Women's Work, SAFFA) from 17 July to 15 September 1958. Verena Dubach (1927–2002) had won the competition for the planning of the outdoor facilities and was responsible for the Wollishofen sector.[136] She had grown up in Münsingen (BE), attended the Hünibach School of Horticulture in 1942-44, worked for Walter Leder, and opened her own office in 1953. After her success at SAFFA, she became the first woman to join the BSG in 1959. In the mid-1960s, she left Switzerland for Denmark, where she worked in the gardens department of the city of Kolding, before retiring from working life to devote herself to painting and sculpture. Her oeuvre is comparatively small. In addition to the SAFFA and several private gardens, she designed the church grounds at Faulensee and Meggen, the cemeteries at Oberrieden and Kloten, and the outdoor areas of Wangen bei Olten schoolhouse.[137] She was also involved in the grounds of the Rapperswil Zoo[138]. Her estate, now managed by the Archive for Swiss Landscape Architecture, comprises just two boxes of plans and a folder. But it does contain some remarkable finds. In an unpublished manuscript entitled "Bamboo", which she originally wrote for the magazine *Anthos*, which she co-founded, there is a sliver of thought condensed into a haiku-like sentence: "Inhale – a stroke of the brush – exhale and the leaf is done – man and plant united."[139] A direct reference to East Asia can be read in the typescript of a lecture she gave in 1987 at an event organised by the Association of Swiss Architects (BSA). Using the Etzel as an example, she recommended incorporating this landscape as a background and using planting in

122 See Yoshida 1952.
123 "We grew up with the architecture of Japan," Felix Schwarz (1917–2013) stated in conversation with the author on 18 Oct. 2005. Hartmann Schweizer 2013, p. 119.
124 Theile 1957, p. 199.
125 Schubert B 2001.
126 Ibid. Present-day spellings: Laozi, Zhuāngzǐ, Kong Fuzi (Latinised: Confucius), Mengzi (Latinised: Mencius) and Lièzǐ.
127 See Yoshida 1935a [ed. 1954]; Blaser 1958, Ishimoto/Ishimoto 1963; Erdberg 1964; Franz 1978; Fujioka/Tsunenari 1982; Blaser 1985, Blaser 1988.
128 See Loewenstein 1922; Martin 1926; Benesch/Benesch 1938; Moriya 1953; Boller/Hokusai 1955; Hájek/Forman/Forman 1957; Brinker 1974.
129 See King 1984.
130 See Ishimoto 1958; Conder/Lancaster 1964; Kashikie 1964; Schaarschmidt-Richter 1977; Schaarschmidt-Richter 1979, Itoh 1985; Seike/Kudo 1986.
131 See Ishimoto 1956; Yoshimura/Halford 1957.
132 See Nakamura N 1958; Ishimoto 1959; Ishimoto 1960; Herrigel 1964.
133 thw. 1958.
134 See anon. 1958, p. 346; Gantenbein 1959, ill. p. 31–35.
135 Yoshida 1957, p. 11.
136 Cf. Lerch 2009, p. 55.
137 Cf. Schmocker-Willi 2002, p. 57.
138 Cf. Mathys 1991.
139 Dubach n. d. ASLA, p. 1.

the foreground to focus the view on this particular eye-catcher. Streams and ponds offered her a means of creating "the necessary balance"[140].

A Japanese undertone resonates in the note that Dubach left on her design for a "shaded perennials garden" designed for the SAFFA: "Dark, idiosyncratically towering trunks, creeping taxomyia and mountain pines, window-like views of the lake and the mountains, this was the setting for the shaded perennials garden. A resting place for reflection, closed off from the rest of the exhibition grounds by plants, that is only found by those who seek it".[141] She called it a "garden for the contemplative" – which may have served as inspiration for Willi Neukom's "Philosopher's Garden" a few months later at G59. Unfortunately, there is no photographic documentation of this garden, but the detailed plan suggests that it must have been a delightful miniature landscape à la japonaise. Similarly, it did not survive the test of time because the show, like the 1939 National Exhibition before it, was cleared afterwards and the entire site returned to its original form.

G59 – "A kinship with the art of Japanese garden design"

In contrast, the "primary design features of G|59",[142] which took place in Zurich from 25 April to 11 October 1959, were to form the basis for the ongoing expansion of the park landscape. Covering an area of around 143,000 m², the Swiss Horticultural Exhibition was divided into a section on the left bank and another on the right bank. On the left, it extended over 59,500 m² and included Belvoirpark and Schneeligut as well as a strip of land between Alfred-Escher-Strasse and Mythenquai as far as Hafen Enge. On the right, it expanded to 83,500 m² between the lake and Bellerivestrasse and between the Tiefenbrunnen Lido and Höschgasse.[143] The two areas were connected by shuttle boats and a cable car, which was dismantled in 1966. The overall direction of G59 was coordinated by Pierre Zbinden, garden inspector and head of the City of Zurich's Horticultural Office (today: Grün Stadt Zürich). The architectural design was led by the architects Werner Stücheli and Paul R. Kollbrunner, who devised a series of simple exhibition buildings based on a cubic system.[144] The grounds on the eastern, left bank of the lake were designed by Klaus and Walter Leder and Johannes Schweizer. Ernst Baumann and Willi Neukom were responsible for the western, right bank area on the Zürichhorn.

An impression of the extent of Japanese influences was given by Emil Steiner, who specifically examined "Japanese accents at the G|59".[145] The trained gardener and long-time editor-in-chief of the trade journal *Schweizerisches Gartenbau-Blatt* (from 1968 *Der Gartenbau*)[146] identified "a certain kinship" with Japanese garden design and attributed this directly to Tetsurō Yoshida's book. Some five decades later, Brigitt Sigel and Erik A. de Jong shed light on some of these Japanese traits.[147] One could see them in the overall design of the terrain and the layout of the pathways as well as in the individual thematic gardens – from the "Azalea Valley", to the "Medicinal Garden", the "Boulder Garden", the "Heath Garden", the "Philosopher's Garden", and the "Jardin d'amour", to the "Tea House Garden", the "Perennials Garden", and the "Water Courtyard."

On the right bank in particular, Lake Zurich formed the natural body of water, whose shores Neukom & Baumann lined with stone blocks in the manner shown in illustrations in Yoshida's book of the "Katsura Garden". They treated the lake as a borrowed landscape by causing it to bulge inland and creating the Nymph Pond near the Fischerstube as its artificial counterpart. An extensive "water garden with Mirabell fountain" and water lily pond of geometric design was the counterpart on the left bank. Both Neukom & Baumann and Leder & Schweizer treated the "paths and footbridges" as a separate theme, to which Heini Mathys devoted a separate article.[148] Mathys, who had obtained the recently published or reissued works by Tetsurō Yoshida, Werner Bischof, and Werner Blaser[149], was aware of such "Far Eastern-tinged"[150] material. And indeed, if one compares the photographs from the construction period against corresponding illustrations in Yoshida's book, there is no doubt about the source.

Along the lakeshore, Neukom & Baumann laid a path of sandstone slabs in a bed of riverbed rocks, which meandered, ascending, and descending across offset and angular slabs, interrupted now and then by stone blocks or stretches of the rock bed – an expression of the Japanese art of creating exciting paths. The concept served as the basic framework for the Seeuferweg lakeside path designed by Willi Neukom in 1963.

The repertoire also included paving of unjointed but carefully delineated bricks and accurately abutted cement slabs, interspersed with strips of small cobblestones and stones of various shapes pressed into the ground as if by chance. This graduation in the degree in which the materials were worked echoes the qualification of the tobi-ishi (飛石, stepping stones) in the "Katsura Garden" – shin (formal), gyō (semi-formal), sō (informal).

The patchwork-like arrangement of different-sized concrete slabs that appear to glide over a bed of rounded rocks to the Jardin d'amour, could have been inspired by a path in the garden of the Ura-senke family in Kyōto, also depicted by Yoshida.[151]

The cylindrical stepping stones in the Nymph Pond reminded Mathys once again of examples in Japanese temple and palace gardens.[152] In particular, they recalled the pond of the Heian shrine.[153] In that case, the stepping stones were formed from blocks taken from the plinth of a disused bridge pier, but here – as in Ammann's Twining courtyard at the National Exhibition – they were made of concrete tubes filled with gravel and capped with a cement top for the tread. In contrast to the Japanese model, Neukom & Baumann laid them in twin rows and in a straight line or at orthogonal angles. Individual

140 Dubach 1987 ASLA, p. 13.
141 Dubach [1958] ASLA.
142 Rohrer 2009, p. 12.
143 Cf. ghn. 1958.
144 Cf. Bucher 2009, p. 3.
145 See Steiner 1959, p. 789.
146 Cf. Schwerin 2014, p. 71.
147 See Sigel/Jong de 2010, pp. 41–48.
148 See Mathys 1959c.
149 See Bishop/Guillain 1958; Yoshida 1954.
150 Mathys 1959c, p. 201.
151 See Yoshida 1957, ill. p. 115.
152 Cf. Mathys 1959a, p. 109.
153 See Yoshida 1957, ill. p. 121.

→ Ernst Steiner's sketch of his Special Garden at the G59 shows a veranda cantilevered over a pond that is reminiscent of a Japanese moon-viewing deck.

↓ Perspective drawing of Neukom's Water Garden at the G59. The playground elements were not installed. A counterpart can be found in the green space of the Hiroshima Telecommunications Office designed by Fukaya Kōki in 1980, which Teiji Itoh shows on p. 166 of his book from 1999.

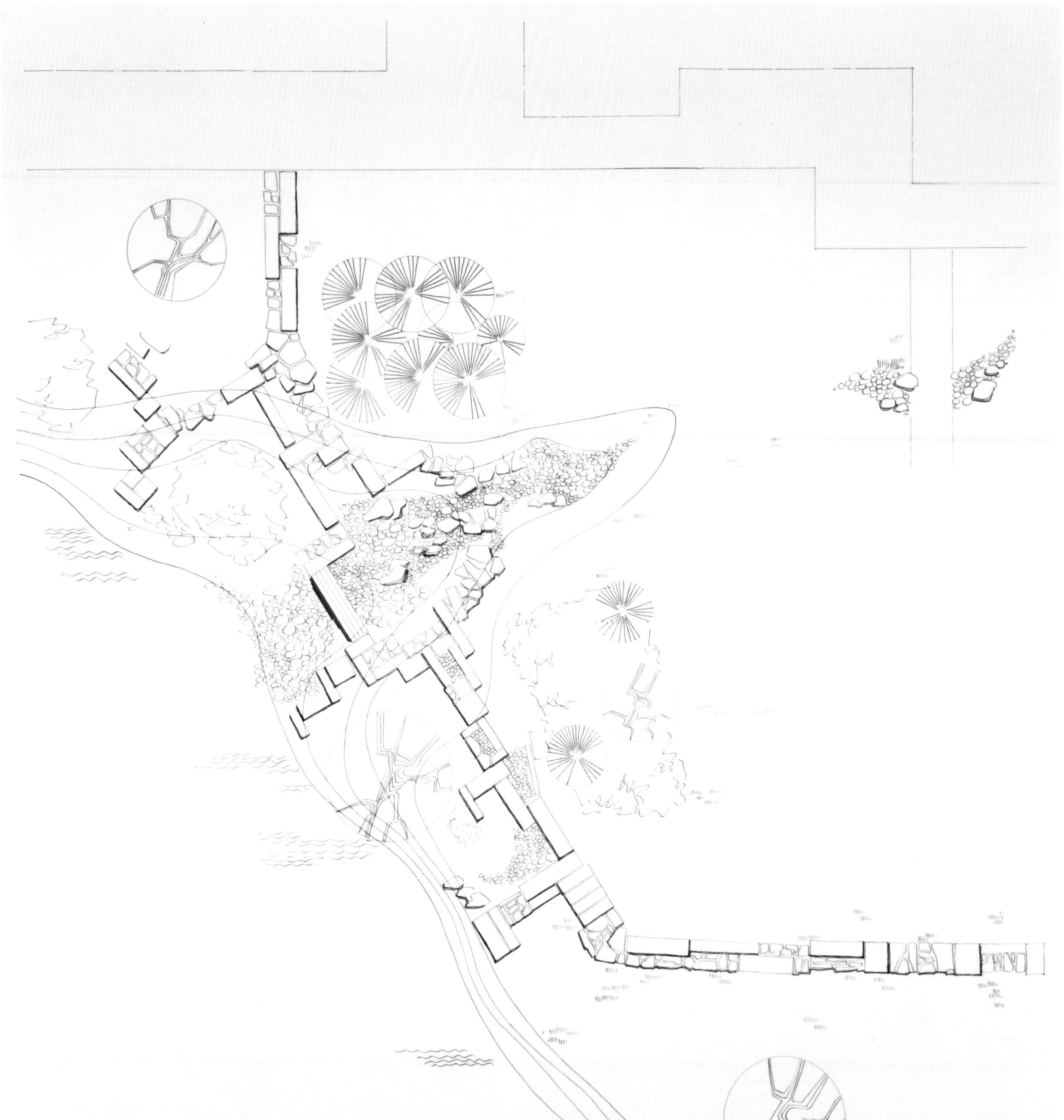

Ernst Baumann, Perennials Garden at the G59.

← This photograph of the waterside in a Japanese garden could have been taken on the shore of Lake Zurich and vice versa.

↓ Baumann's paths at the G59 clearly reference those in Japanese gardens.

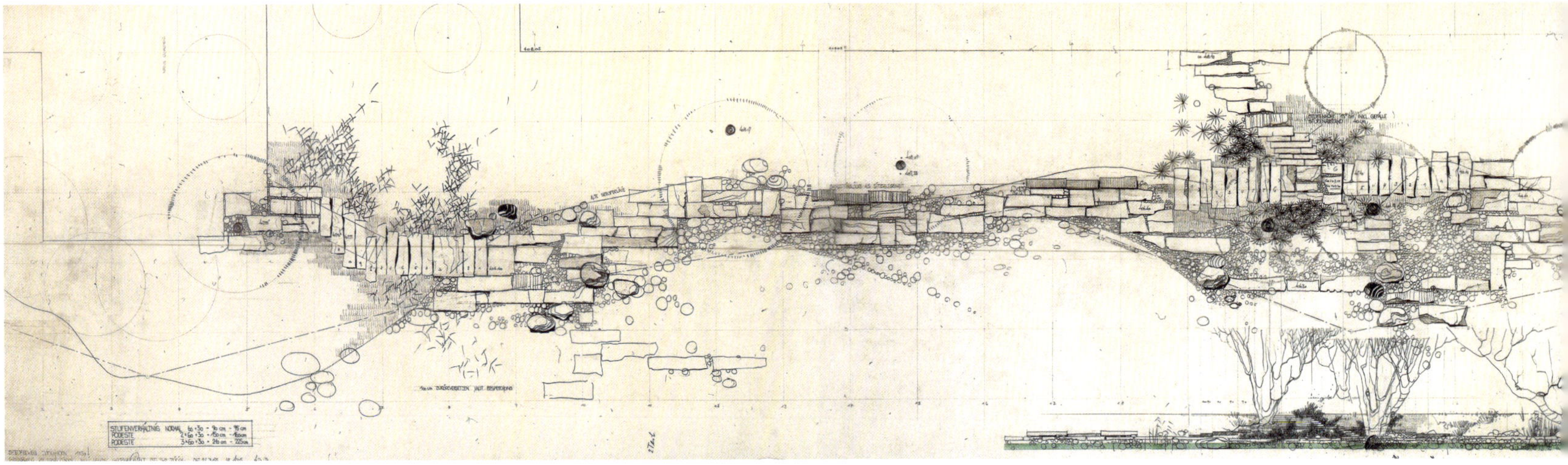

↑ Plan of the Seeuferweg lakeside path in Zurich by Willi Neukom, 1963.

← The design of the lakeside section of the G59 led to the Seeuferweg lakeside path created by Willi Neukom in 1963.

→ Willi Neukom's Philosopher's Garden at the G59.

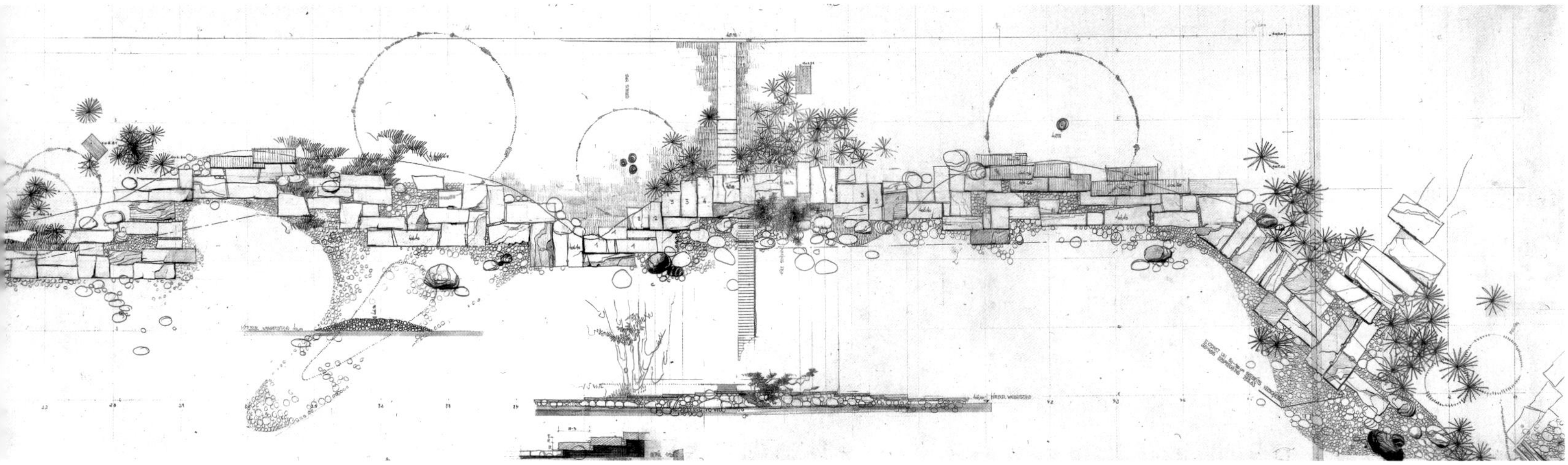

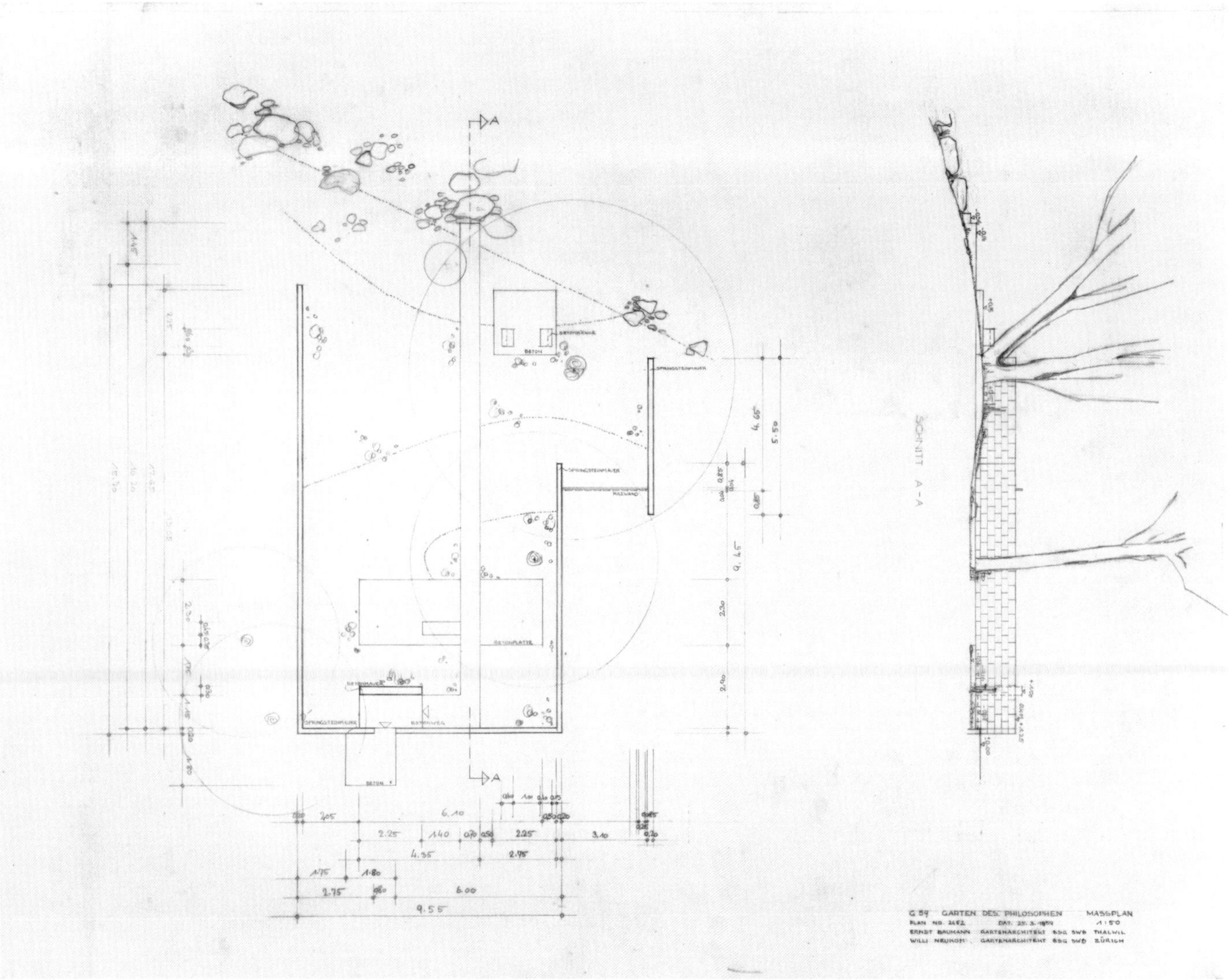

SCHNITT A-A
G 59 GARTEN DES PHILOSOPHEN MASSPLAN
PLAN NO. 2462 DAT. 25. 3. 1959 1:50
ERNST BAUMANN GARTENARCHITEKT BSG SWB THALWIL
WILLI NEUKOM GARTENARCHITEKT BSG SWB ZÜRICH

↑ Walter Leder's dwelling garden at the G59 with a Japanese-style water basin fed by a spring bubbling up between the stones.

→ The architectonic expression of the house facing Walter Leder's garden at the G59 bears the hallmarks of Neues Bauen.

Ernst Cramer's Poet's Garden with the cone-shaped hill that Arioli associated with a volcano.

← Willi Neukom's Nymph Pond was one of the most enchanting creations at the G59.

↓ A dry streambed of pebbles meandered through a valley of rhododendrons and was crossed by a gently arched bridge made of rough wooden planks that heighten its Japanese impression.

The glass “jewel” in the Jardin d’amour evokes
associations with the Buddhist wish-fulfilling pearl.

View across Neukom and Baumann's Crystal Garden towards Cramer's Poet's Garden at the G59.

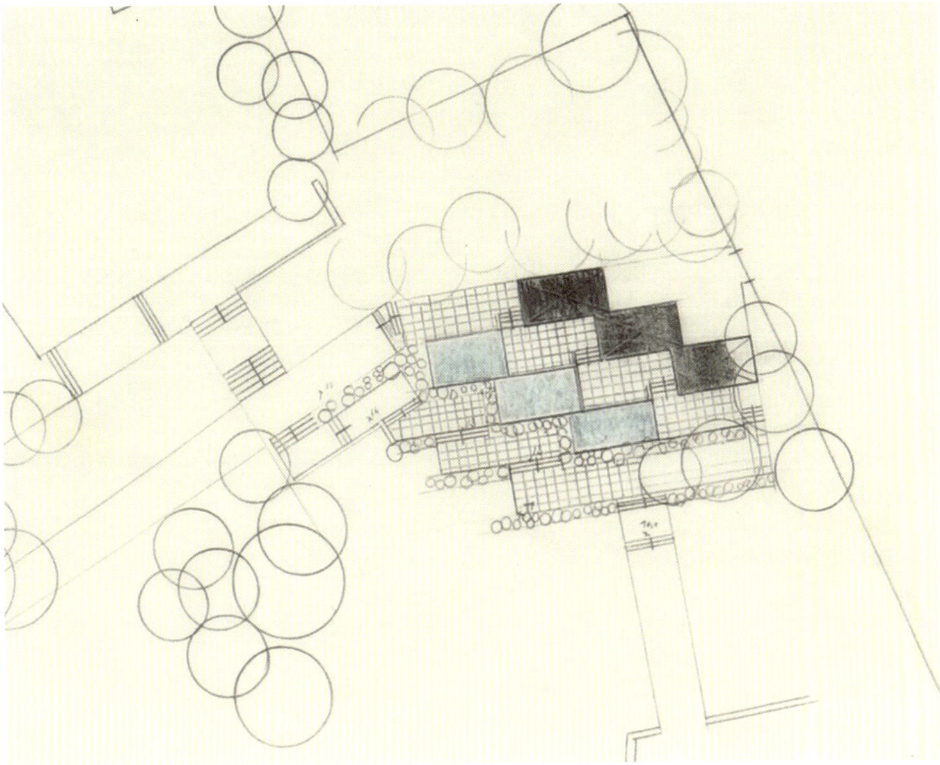

Walter Leder, Plan of the Teahouse Garden at the G59.

"outliers" – stepping stones placed off to the side – echoed the Japanese practice of placing stones so that they appear to "jump" slightly off the intended path, thus creating a sense of imperfection known as wabi-sabi (侘寂), an essential quality in the art of Japanese garden design.[154]

A *bijou* in the truest sense of the word was the Crystal Garden designed jointly by Ernst Baumann and Willi Neukom. Reduced to the Japanese formula "stone and water"[155], here it was not splendid floral compositions that vied for the visitors' attention but sparkling multi-coloured stones and colourful crystals lying in the shallow water.[156] Visitors could cross the water via cylindrical stepping stones so that they became immersed in light reflections, in effect creating a walk-in kaleidoscope. In a nod to their inspirational source, the designers encircled a square, which they then furnished with an imposing boulder and some bulrushes.

Ernst Cramer's "Poet's Garden" in turn figured as a shakkei, a borrowed landscape, for the "Crystal Garden", which framed a view of it. The elevations grouped around the rectangular water basin of this garden – four irregular pyramidal mounds and a staggered earth cone – were often cited and interpreted but not brought into connection with Japan. Although they focused in a reductionist manner on the components known in Japan as senzui (山水, modern pronunciation sansui), mountain and water, it was more commonly perceived as a standalone walk-in abstract work of art, and as a homage to Hermann von Pückler-Muskau's lake pyramid in Branitz Park created in 1855.[157] More recently, the landscape architect Udo Weilacher has offered a perceptive analysis that does indeed trace a Japanese trait. He associated the garden's abstracted representation of nature and landscape with the panorama in the background: "And so one gazed across his artificial earthworks to the moraine hills on the opposite shore of Lake Zurich, formed since the ice-age by the forces of nature".[158] Whether consciously or subconsciously, the artistic transformation of the natural phenomena of mountains and water testifies to a sensibility for shukkei translated into a modern formal language. In his 1928 edition of The *Gardens of Japan*, Jirō Harada depicted a sand cone in the garden of Ginkaku-ji.[159] His stylised silhouette of the Fuji-san could have inspired Cramer, as Harada's book was well-known in Europe.[160] For Heini Mathys, at any rate, the comparison with a "stylised volcanic cone"[161] seemed obvious. Another possible model could have been the miniature Fuji in Suizenji Park, which K. Takashima had depicted in 1937 in *Die Gartenschönheit* under the title "Japanese Garden Ideals II".[162] Alfred Kiener also alluded to its exemplary character, likewise publishing a picture of it in 1968.[163]

A dry garden à la japonaise was the area of rocks crossed here and there by sandstone blocks in Ernst Baumann's perennials garden on the right bank, reminiscent of the headland formed of pebbles in the "Katsura Garden".[164] Exposed to the tides, it already began to exhibit traces of the natural process of silting up as sand was washed up on the shore in the year of the exhibition. Unaffected by this was the perhaps more subtly conceived dry streambed of pebbles that meandered through the rhododendron valley, interspersed by a few rocks and timid grasses. Its Japanese flair was underlined by the slightly arched bridge made of rough wooden planks.

One of the most enchanting creations was without doubt the Jardin d'amour, which exuded an almost archetypal Far Eastern spirit. It was a three-dimensional concrete work of art, a riff on the theme of the "circle". This underlying geometry informed the path of the concrete wall that enclosed the island of love, the platform of this island, the stepping stones that led from the narrow passage across the water to the island, the concrete tubes in which the plant material was inserted, and finally the spherical glass "jewel" that sheltered the lovers within it[165]. The association with the Buddhist wish-fulfilling pearl was indeed obvious. Neukom originally planned to plant only water lilies, which are also associated with the Buddhist pearl, along with broad-leaved bulrush (*Typha latifolia*). Due to its cool and shady location, however, they did not grow quickly enough, and ultimately the basin was electrically heated and enriched with subtropical and tropical plants – sugar cane and papyrus perennials.[166]

Not mentioned in the review is Walter Leder's use of stepping stones with a circular cross-section for a sweeping passage across the water basin in front of the

154 An impressive example was captured in the photograph of the path from the covered veranda to the shore in the garden of the Katsura villa, which Yoshida depicted. See Yoshida 1957, ill. p. 112.
155 "[...] 'Stone and water', that would be a formula for the island kingdom of Japan, modified in the private domain of gardens with exquisite taste and ingenuity." thw 1958.
156 See Schubert M 1959, p. 153.
157 Cf. Weilacher 2001, pp. 114–115.
158 Weilacher 2001, ibid., p. 114.
159 Harada, Jirō, *The Gardens of Japan*, ed. by Geoffrey Holme, The Studio Limited, London, 1928, p. 135.
160 Tagsold, Christian, *Spaces in Translation. Japanese Gardens and the West*, University of Pennsylvania Press, Philadelphia 2017, p. 38.
161 Mathys 1959a, p. 117.
162 Takashima, K 1937, p. 109.
163 Kiener 1968b, p. 23.
164 Yoshida 1957, ill. p. 174.
165 Mathys 1959b, p. 197.
166 Cf. Mathys 1959b, p. 198.

"country house". Planted – shielded by a lightweight roof construction – with water lilies and irises, and probably also fir frond, arrowhead grass and feverfew, among other things, this pool probably resembled the Heian model even more closely.

A modern interpretation of the Japanese moment was the water garden, which Neukom depicted in a charming coloured perspective [→ p. 97]. Contrasting with the geometry of the elongated rectangular ground plan and the 1 m² square stepping stones were erratic boulders of varying dimensions. Grouped in arrangements comparable to Yoshida's description of the formula for the setting of stones,[167] they created a sense of overall balance while also establishing a tension between the two-dimensional and the sculptural elements. Neukom wanted to create the feeling of standing in water, and the water level was calculated "so that the surface of the slabs can just be walked on with dry feet, with water flowing in the spaces in-between [...]".[168]

Walter Leder also translated the teahouse garden in the Belvoir area on the opposite bank into a contemporary formal language, which Emil Steiner described admiringly as a "completely Japanese-informed watercourse"[169]. Picking up the stepped structure of the teahouse, Leder terraced the water surface and created a cascade of three vertically and horizontally offset rectangular water basins. He concealed the source of the pools in a bed of broken stones behind a screen of low bamboo.

Its charm and aura must have captivated visitors to Ernst Steiner's miniature. The terrace, which extended out over an oval pond, bore the hallmarks of a veritable moon-viewing platform. Bulrushes veiled an asymmetrically placed wooden footbridge; water lilies floated on the surface of the water from which a few stone islands emerged. An ornamental shrub in the background functioned as a borrowed natural element or alternatively could be interpreted in the manner of the solitary plant senzai (前栽, front plant), which in Japanese gardens denotes a favourite plant taken from the wild and transplanted close to the house.[170]

Iconic Ryōan-ji

The "Philosopher's Garden" was situated at a secluded spot on the shore between the Fischerstube and the theatre. The axial arrangement of the stone placements depicted in the design of 17 March 1959 is striking and recalls Katsuo Saitō's geometric, arithmetic, and proportional decryptions of the Ryōan-ji garden, which captivated garden historians, landscape architects, and artists alike.[171]

Neukom was also hypnotised by Ryōan-ji. Sifting through archive material for the 1964 Expo in Lausanne, which took place between 30 April and 25 October, unearthed a find par excellence: the breathtaking sketch of a courtyard garden modelled on the phenomenal Kare-san-sui. At first, it seemed to be a study of the original due to its deceptive similarity, but on closer inspection, it became clear that certain details differ. While the stone settings composed into five groups correspond amazingly well, deviations in the number of stones per group and their shape undoubtedly identify the drawing as composition by Neukom. Sadly, the trail of the sketch loses itself in the plans of the water landscape in which Neukom, working together with Walter Bischoff and Walter Brugger, embedded the show, which was under the supervision of the chief architect Alberto Camenzind.[172]

The terrain was predestined for Japanese-motivated green spaces, not least because André F. Desarzens, head of Lausanne's Municipal Garden Department, had created a Japanese garden the year before to mark Japan's participation as guest of honour at the Comptoir Suisse autumn fair. In the lines penned by its creator for the magazine *Anthos*, a self-critical note can be detected. Compared to the "centuries-old Japanese models", it was "too new". His garden lacked intimacy and failed to constantly create successive new impressions for the viewer.[173] Instead, the entire garden could be seen from the terrace of the teahouse. This shortcoming resulted from the need to channel the stream of visitors.

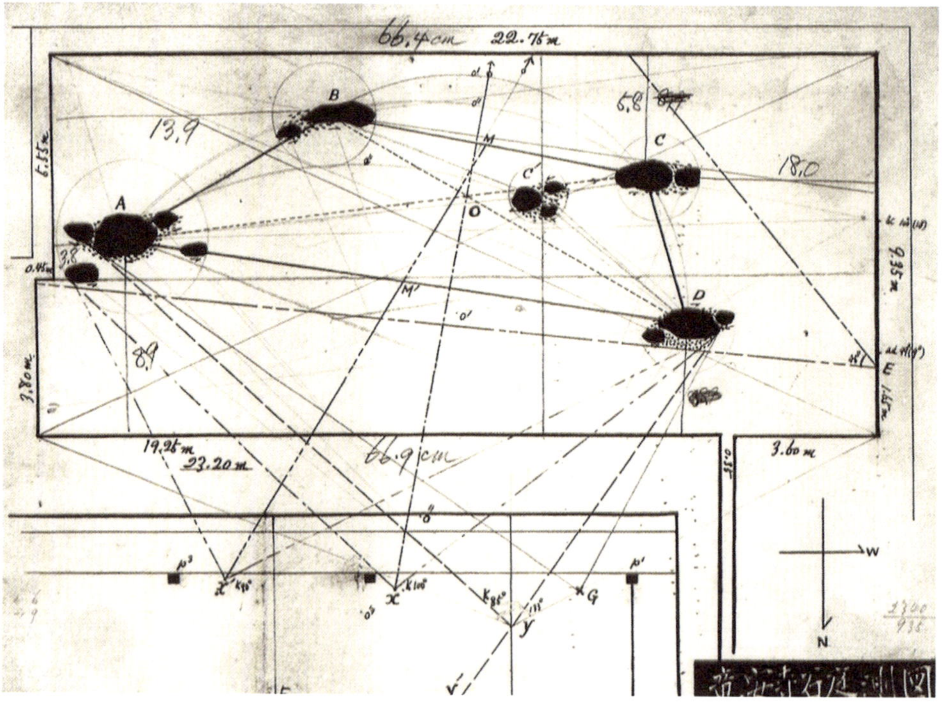

Diagram of the Ryōan-ji Garden by Noboru Kitawaki, 1939, ink on paper, 28 × 38 cm.

167 See Yoshida 1957, ill. pp. 10, 11.
168 Neukom 1959, p. 233.
169 Steiner E 1959, p. 789.
170 Cf. Takei/Keane 2004, p. 41.
171 See Yamada 2009: Saitō, Katsuo, 1923, p. 170; Okazaki, Ayaakira, 1931, p. 172; Shigemori, Mirei, 1933; p. 173, Eyama, Masami, 1935, p. 174, Kitawaki, Noboru, 1939, p. 176; Eyama, Masami, 1966, p. 175; Ōyama, Heishirō, 1970, p. 177, Ono, Masaaki, 1980, p. 180.
172 See anon. 1964b, p. 34.
173 Cf. Desarzens 1963, p. 32.

Photograph by Walter Leder of the Japanese Garden
designed by Akira Satō and Kuro Kaneko at the
International Horticultural Exhibition in Hamburg in 1963.

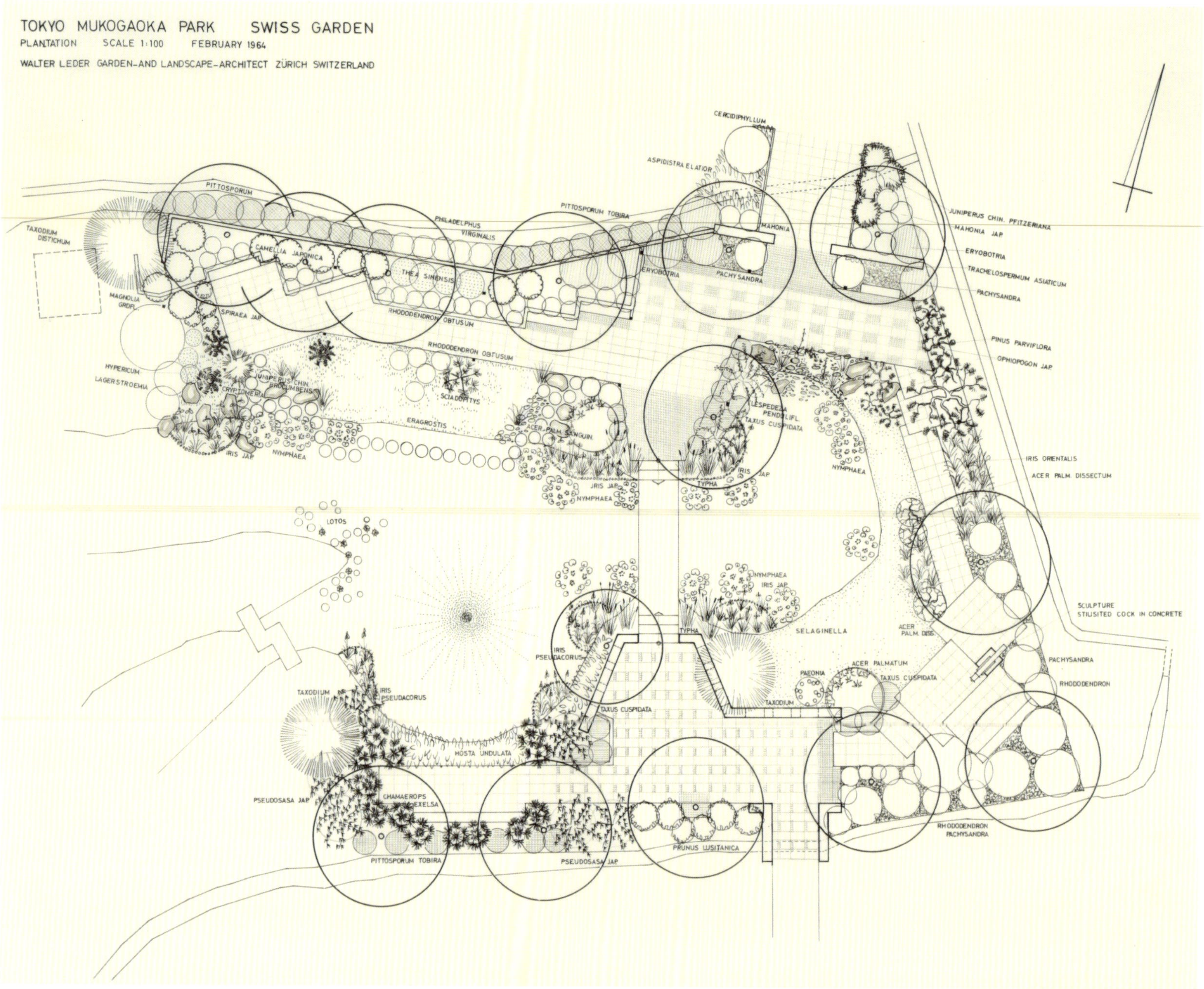

Walter Leder designed a Swiss National Garden for Mukogaoka Park that was quintessentially Japanese.

Neukom's sketch for the Expo 64, in which he divides the stones into five groups exactly as in the Ryōan-ji garden.

174 Zahner 1963, p. 20.
175 Cf. anon. 1964a, p. 32.
176 Leder 1964a ASLA, p. 5.
177 Cf. Arioli 1963b, p. 40. Arioli 1964a held out the prospect of a conference volume, which appeared in 1966. See Tandy 1966.
178 The Concert Hall built by Japanese architect Kunio Maekawa in 1961 in Ueno Park, which opened in 1876; the National Stadium built in 1958, which was the main venue for the 1964 Summer Olympics; the Parliament Building built in 1936; and the "Jindai Botanical Garden" built in 1961. See Arioli 1963b, p. 40.

IFLA 1964: "How to bring the garden into your room"

One of the most beautiful evocations of the experience awaiting the delegates at the Congress of the International Federation of Landscape Architects (IFLA) in Tōkyō was formulated by the artist Arnold Zahner (1919–2005), who from 1960 onwards undertook several trips to Japan: "And while you [...] look out into the fresh morning garden, the low sun's diagonal rays catching in a million pearly mirrors, you experience a moment of one the most beautiful things the land of the rising sun has to offer: magical, peaceful, dream-like sensations that are hard to put into words but feel all the more tangible. It smells of coloured moss, shadows of a few bamboo leaves play across the milky-white shōji, framed by the windows is a detail of the garden, a composition of gnarled branches, of dampness on shiny pebbles, of a shimmer of glowing light on even more mossy shadow; large slippery leaves and nervously trembling bamboo, their shafts rising abruptly from neatly even, firm sandy soil. This composition is especially beautiful when you sit upright on your mats. When you are standing, you are too tall and don't see the details in front of you. When seated, the garden comes to you into the room, the transition of the dark wood of the structure and the brown-golden shimmering straw mats does not demarcate a threshold, rather it forms an intimate connection."[174]

From the Swiss side, Walter Leder was the driving force behind the IFLA Congress, which was held from 13 to 23 May 1964.[175] The list of participants, preserved in the records of his estate, shows that 16 people from Switzerland registered for the trip: Richard Arioli, Ernst Baumann, Georges Boesch, Ernst Cramer, Fred Eicher, Eugen Fritz, Wolf Hunziker, Walter Leder, Hans Nussbaumer, Willi Neukom, Franz Vogel, Traugott Vogel, Emil Willimann – with Arioli, Baumann, and Hunziker accompanied by their wives.[176]

The conference theme "Landscape Architecture in Human Life" was covered from 13 to 16 May, with the final session scheduled for 23 May. Excursions were organised for the days in between along with the option of extending one's stay until 29 May.[177]In addition to viewing modern parks and gardens, the programme included traditional buildings and gardens created between the 7th and 20th centuries.[178]

In Tōkyō, the excursion included visits to the gardens of the Meiji shrine and Hamarikyū Park, on the main island of Honshū to Kōraku-en in Okayama and to Nikkō National Park with the Shintō shrine Nikkō Tōshō-gū, Shinkyo Bridge, Lake Chūzenji, and Kegon Waterfalls. In Nara, the programme included the Hōryū-ji, the Yakushi-ji, and the Tōdai-ji with its famous main hall Daibutsuden. The hosts had most of the stops ready in Kyōto: the Byōdō-in in Uji, the Shintō shrine Heian-jingū, the Nijō Castle, the Otowasan Kiyomizu-dera, the garden of the Jōju-in, the Sanjūsangen-dō, the Shugakuin Mansion, the Imperial Katsura Mansion, the Daitoku-ji, the Saihō-ji, and the Ryōan-ji. Finally, the

Japanese also showed their guests the beauty of the Seto Inland Sea, the inland sea between the islands of Honshū, Shikoku, and Kyūshū, which has always fascinated Japanese garden designers.[179] The extensive programme of excursions almost drowned out the opening of the national gardens of France, England, Italy, Austria, Germany, Portugal, Sweden, the USA, and Switzerland, which were located in Mukogaoka Park at the beginning of the congress.

Walter Leder was the only one among the group to keep up a rich correspondence that testifies to intensive contact with Japan over more than a decade. Initially, the correspondence revolved around Leder's planning of the "Swiss Garden" in Mukogaoka Park, which he was invited to design by Akira Satō,[180] who, thirty years earlier as a young researcher, had helped to set the reform of the reception of Japanese garden art in motion. Now he acted as president of the Japanese Institute of Landscape Architecture (JILA) and head of the organising committee. Satō chose the "most important members of the Association"[181] for the "Gardens of Nations" complex, which is divided into nine areas: Jean-Bernard Perrin (F), John Brooks (GB), Pietro Porcinai (I), Eduard M. Ihm (AU), Richard Schreiner (BRD), Francisco Caldeira Cabral (P), Ulla Bodorff (S), Sidney N. Shurcliff (USA), and Walter Leder (CH).[182] Contrary to Satō's original intention to preserve the gardens for the long term,[183] the park subsequently underwent a profound transformation.[184] The year before, Satō, together with Kuro Kaneko, who acted as Congress Secretary in Tōkyō,[185] had created the Japanese garden at the 1963 International Horticultural Exhibition in Hamburg, where Ernst Cramer had appeared with his theatre garden.[186]

Richard Arioli was highly critical of Leder's Japanese-style Swiss garden, though his target was less its actual design but rather "the absurd idea" of "making 'gardens of nations' and presenting them around the world at exhibitions like export goods".[187] The overall response to the Gardens of Nations was rather modest.[188] *Anthos* devoted most space to the historical gardens of Japan and their relationship to the environment. Like Gerda Gollwitzer in *Garten und Landschaft*[189], Arioli strove to give a genuine perception.[190] In contrast, passages in the commentaries by Geoffrey Alan Jellicoe and Garrett Eckbo still bore a rather "imaginary" appraisal of Japan. Jellicoe, President of IFLA until 1954, for example, describes the landscape design of England and Japan as being "so similar" to each other.[191] Eckbo for his part reduced the Japanese garden to "that typically 'Japanese arrangement' of lanterns, rocks, gravel, sand, bodies of water, carps, trimmed trees and pruned shrubbery" and even went as far as to claim that any layman would recognise a Japanese garden at first sight.[192]

While Leder's design in Mukogaoka Park received little attention in his homeland, another of his contemporaries was assured of attention for his work on local soil: Willi Neukom's lakeside design on the Zürichhorn, which was developed after the G59 and opened in 1964, had a palpable "vision of Japanese gardens": "Did it not happen recently that a world-famous Japanese garden architect walked critically along the new lakeside path on the Zürichhorn and declared in no uncertain terms that the layout of this path was a masterpiece and that he felt at home in these surroundings?"[193] These words were probably uttered by Akira Satō, who "had already visited Zurich several times",[194] and in 1954, "when visiting the Tiefenbrunnen Lido, took a great liking to its beautiful stone settings"[195]. That he was visiting the city on the Limmat at this time probably had to do with his work on the Hamburg Nation Garden[196]. The public baths, originally designed by the Mertens brothers and then built by Mertens & Nussbaumer, opened on 12 June 1965.

A return visit took place from 27 to 29 August 1972, ten months after the Japanese garden designed by Katsuo Saitō (1893–1989) on the theme of peace had been inaugurated in the courtyard of the headquarters of the World Health Organisation (WHO) in Geneva.[197] Its designer was the very same Katsuo Saitō who had once deciphered the Zen character of Ryōan-ji and Daisen-in. All the architectural elements – pagoda, lantern, zigzag bridge, and bamboo fence – came from Japan, the natural stones from the Geneva region. The pagoda was built on an artificially created hill that provided the slope for a waterfall. This flowed into a gently meandering stream, which in turn poured into a pond crossed by a zigzag bridge and planted with irises.[198]

The reason for the visit from Japan was not the Geneva project, but the IFLA Congress in Brussels, and the delegation of 30 Japanese landscape architects arranged to stop in Zurich on their way there. The meeting was again organised by Akira Satō, who, however, did not join them.[199] He did, however, surface once more in the broader environs of Switzerland in 1979, when he was designing the Japanese garden for the BUGA 79 Federal Garden Show in Bonn[200], which was built on a wasteland site that is today the Rheinauenpark.

Stones and woody plants imported from Japan – black pine (*Pinus nigra*), white enkianthus (*Enkianthus perulatus*), dwarf bamboo (*Bambusa nana*), Japanese azalea (*Rhododendron japonicum*), and bush clover (*Lespedeza*) – were draped over an artificially created hilly landscape replete with a pond and gravel beach. It was now the turn of the younger generation to be fascinated by its design. Hans Jakob Barth (1925–1984) kept a detailed review of Satō's creation, in which the stroll garden, the pond-spring appreciation garden or scroll garden, and dry garden were interwoven.[201] Adolf Zürcher (1934–2000) also visited the show and photographed the Japanese garden extensively.

179 Meiji shrine (明治神宮, Meiji Jingū, 1912–1920): Shintō shrine dedicated to the souls of Meiji-tennō and his wife Shōken-kōtaigo; Hamarikyū Park (浜離宮恩賜庭園, Hama-rikyū onshi teien): Imperial garden of the Hama residence; the Kōraku-en (後楽園) in Okayama: together with Kairaku-en (偕楽園) in Mito and Kenroku-en (兼六園) in Kanazawa, one of Japan's three famous gardens (日本三名園, Nihon sanmeien); Nikkō National Park (日光国立公園 Nikkō Kokuritsu Kōen) on the main island of Honshū: houses, among others. a. the Shintō shrine Nikkō Tōshō-gū (日光東照宮), Shinkyo Bridge (神橋, Shinkyō, Sacred Bridge), Lake Chūzenji (中禅寺湖, Chūzenji-ko) and Kegon Waterfalls (華厳滝, Kegon-no-taki); Hōryū-ji (法隆寺, Temple of the Buddha's Teaching) in Ikaruga: built in 607 to commemorate Emperor Yōmei Tennō (517–587); Yakushi-ji (薬師寺): originally built in 680 by Emperor Tenmu at Kashihara, moved to its present site in Nishinokyo when the capital was moved to Nara; Tōdai-ji (東大寺, Great Temple of the East): built in 752, the present structure dates from 1692, its main hall Daibutsuden (大仏殿) from 1708; Byōdō-in (平等院) in Uji (Kyōto Prefecture): built in 998 in the Heian period; Shintō shrine Heian-jingu (平安神宮, Heian-jingū): created in 1894 as a replica of the first Heian-period imperial palace; Nijō Castle (二条城, Nijō-jō): inaugurated in 1603 as the residence of Tokugawa Ieyasu, the first shogun of the Edo period (1603–1867); Otowasan Kiyomizu-dera (音羽山清水寺): origin in 798, present buildings date from 1633; Jōju-in teien (成就院庭園): garden of the abbot's residence of Kiyomizu-dera, built between a conflagration in 1629 and the first mention in 1735; Sanjūsangen-dō (三十三間堂): founded in 1164; Shugakuin Mansion (修学院離宮, Shugakuin rikyū): at the foot of Mount Hiei; Imperial Katsura Mansion (桂離宮, Katsura-rikyū): built between 1620 and 1645; Daitoku-ji (大徳寺): built between 1315 and 1319 Saihō-ji (西芳寺) with garden attributed to Musō Soseki: originally founded in 731; Ryōan-ji (龍安寺): rock garden dated around 1450.

180 Leder 1963a ASLA.

181 Satō 1991, p. 284. Leder was IFLA President 1954–1956, Shurcliff 1958–1962, Cabral 1962–1966 and Schreiner 1966–1968. https://www.iflaworld.com/ifla-past-presidents (retrieved 17 Sep 2012). Porcinai, like Leder, was one of the founding members of IFLA in 1948. Ibid., p. 241, note 59.

182 Satō originally had ten plants in mind. Leder 1963a ASLA.

183 Leder 1963a ASLA.

184 Cf. Satō 1991, p. 284.

185 Cf. Arioli 1964d.

186 Cf. Arioli 1963a, p. 39.

187 Cf. Arioli 1964b, p. 2.

188 The Italian contribution by Pietro Porcinai (1910–1986) found its way into the monograph on the Italian landscape architect, whose affinity with Japan manifested itself in some of his creations and about which he also made public statements. See Matteini/Porcinai 1991, p. 237; Porcinai/Mordini 1966, pp. 133–142.

189 See Gollwitzer 1964.

Adolf Zürcher took numerous photographs of Akira Satō's Japanese garden at the Federal Garden Show in Bonn in 1979.

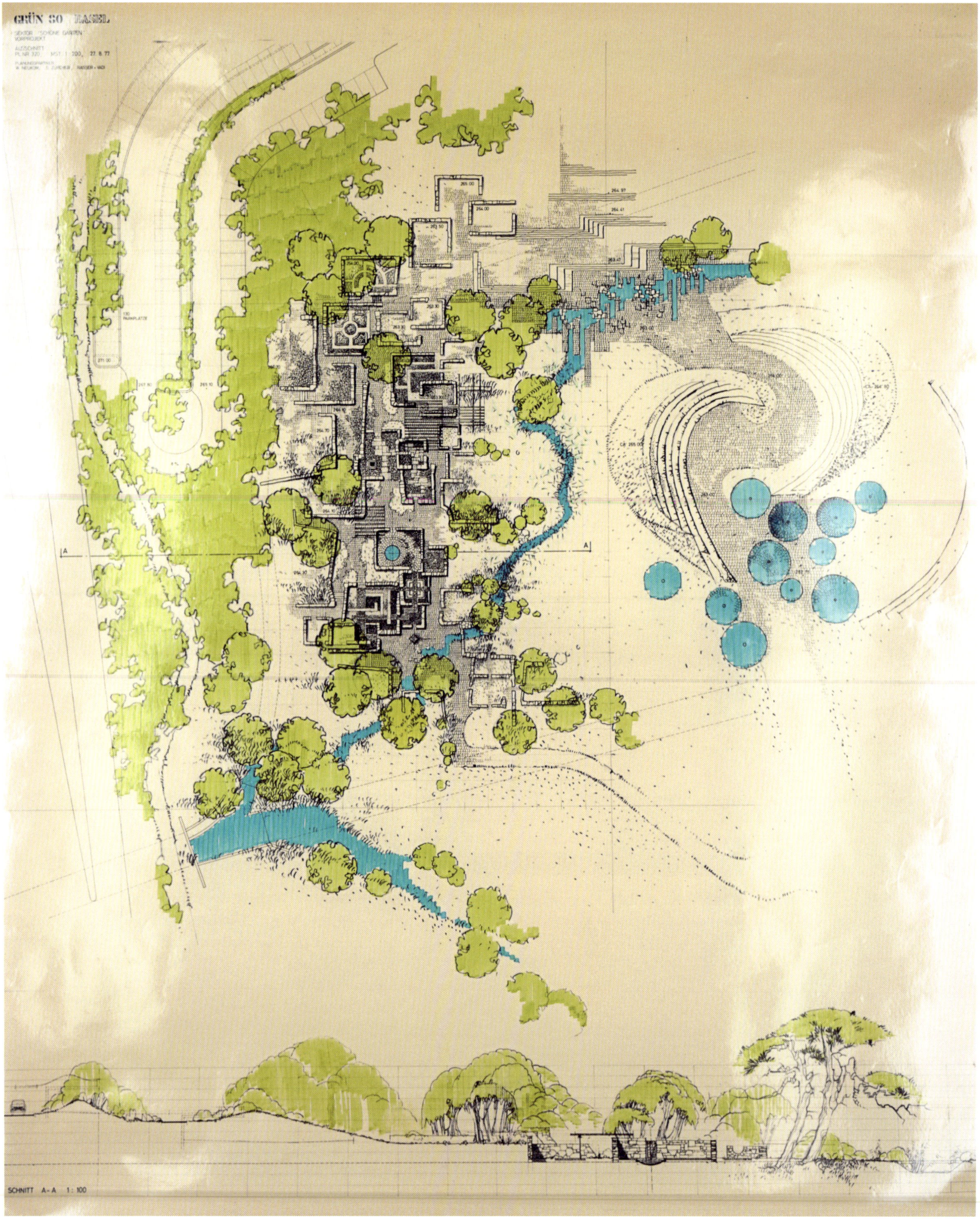

The geometrically delineated Spring Garden was designed to connect to an artificially created lake and St. Alban pond via a stream meandering through the garden courtyards.

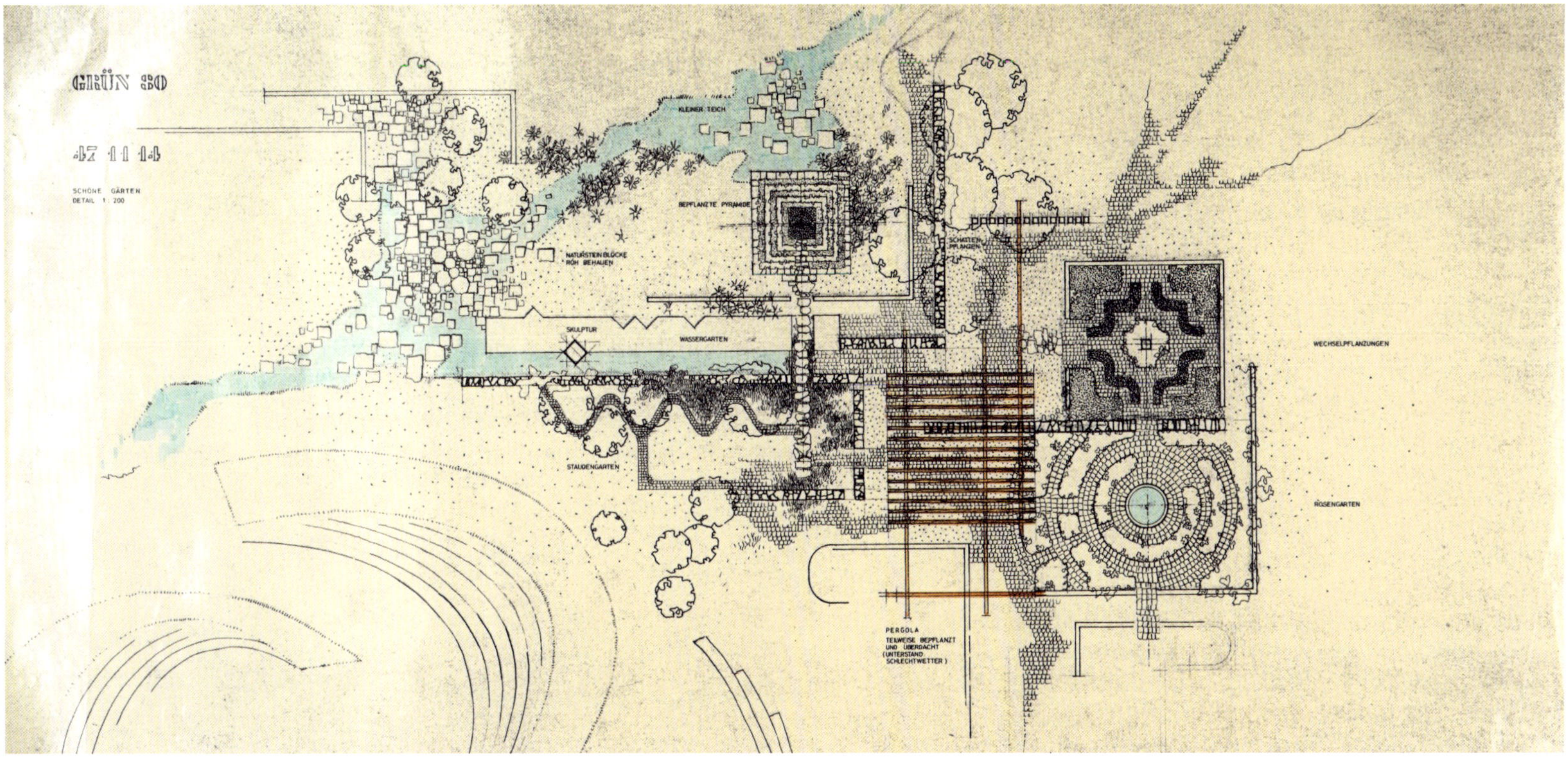

Section of an area of the themed gardens at the Grün 80 showing a naturalistic water landscape interspersed with sections of rocks and boulders.

Grün 80 – "Far Eastern atmosphere"

It was the run-up to the Grün 80, the 2nd Swiss Gardening and Landscaping Exhibition, which opened its doors on 12 April 1980 and closed again six months later. While the exhibition was organised by the Association of Swiss Master Gardeners (VSG) in collaboration with the Federation of Swiss Garden and Landscape Architects (BSG), the Christoph Merian Foundation also became involved as it owned the majority of the Brüglinger Ebene between St. Jakob in Basel and the municipality of Münchenstein on which the exhibition was sited.[202] For the Foundation, it presented an opportunity to press ahead with the wider conversion of the area from its former predominantly agricultural use into a near-natural area that could serve as a recreational park.[203]

Given the nature of the site, this area partially overlapped with the perimeter of the Grün 80. The designers Willi Neukom, Adolf Zürcher, as well as Max Rasser and Tibère Vadi therefore developed an idea that went beyond the six sectors outlined by the organising committee – "Sowing and Harvesting / Daily Bread", "Green University", "Market", "The Earth", "Beautiful Gardens", and "Land and Water" – and "were the only ones to also think in terms of an overall project".[204] Additionally, they submitted competition projects not only for "Beautiful Gardens"[205] and "Land and Water"[206], both of which they won, but also for "Daily Bread",[207] which they did not.[208]

The project "Beautiful Gardens" was roughly divided into three areas, one with a naturalistic, landscape-like character, one horticultural and urban, and one with an architectural focus. The landscape area encompassed the lake of St. Alban-See and its shore zones, the horticultural area was concentrated on the south-western part of the Brüglinger Ebene and the architectural area of the restaurant on its northern area adjacent to the "Merian Gardens".

The St. Alban-See, which was formed by damming the flow of water from the St. Alban ponds, can be read as an artificially created landscape of longing in the Japanese sense. While Lake Zurich had featured visually in the G59 show – as a borrowed Shakkei landscape – the newly created body of water in Basel figured as an imitation of an imagined natural scene à la Shukkei, to be explored by walking around it. The western shores bore a striking resemblance to those of the G59 and evoked a very Japanese atmosphere, especially with its small island of heaped round-washed rocks. Sand, rocks, and stone blocks interspersed with red Colchic maple (*Acer Cappadocicum Rubrum*), silver willow (*Salix alba 'Sericea'*), and bulrush (*Typha*) lined the water's edge, which was dotted with water lilies (*Nymphaea*). The team planned the opposite east side as a landscape of dunes, grasses, and heathland. The southern part was modelled into a dune landscape of sand and gravel – evoking the impression of a longed-for seashore in the Japanese style. The rich assortment of grasses, small shrubs, and woody plants – also originating from East Asia – is documented in the planting plan of 14 May 1979 (see plant catalogues).[209]

190 See Arioli 1964a; Arioli 1964b; Arioli 1964c; Arioli 1964d; Arioli 1964e.
191 Jellicoe 1964, p. 341.
192 Cf. Garrett 1964, p. 342.
193 t. 1964.
194 Leder 1972b ASLA.
195 Hofmann 1954.
196 See Arioli 1963a, p. 39.
197 See n.d. 1971. Saitō created around 400 Japanese gardens in Japan and overseas during his lifetime. He also wrote books: see Saitō 1964; Saitō 1969; Saitō 1966.
198 See The Japanese Garden at the Headquarters of the World Health Organization (WHO), http://www.nodaigarden.jp/garden/europe/switzerland/geneva/japanese-garden-at-world-health-organisation-headquarter/, 2022 (retrieved 15 Jan 2023).
199 Leder 1972a ASLA.
200 For a general overview see Hansjakob/Hansjakob 1979.
201 See St. 1979; Satō 1979.
202 Cf. Hunziker/Voss/Meier 2000, p. 14–15.
203 See Zürcher 1980b, p. 48.
204 Zürcher 1980a, p. 19.
205 See anon. 1978.
206 See anon. 1977c, p. 32. The second-placed Atelier Stern was entrusted with the realisation of "Land and Water". See Badcja 1980.
207 See Neukom/Zürcher/Rasser and Vadi won 3rd place. anon. 1977b, p. 28f.
208 At the very least, Neukom also outlined ideas for the "Green University", ASLA NL wne.144.1.2-II. Wettbewerb-12- wne.144.1.2-II. Wettbewerb-18.

The heathland area covered with Scots pines was also less a garden than an imitated landscape. It formed a transitional area to the northern zone, which was enriched with bog soil, and again featured an array of plants including some Far Eastern species. The same applies to the wetland, which was incorporated here as a miniature landscape.

Wooden platforms and walkways extended out into the lake, sometimes widening to form broad platforms, sometimes narrowing to form footbridges, whose staggered arrangement was reminiscent of the zigzagging bridges of Japanese models. Their layout also resembled the system of verandas of a Japanese residence. Neukom, Zürcher, Rasser & Vadi extended these terraces, which act as a link between the interior and exterior spaces, far into the landscape.

In keeping with the motto "Beautiful Gardens", they classified St. Alban-See as belonging to the man-made garden. Zürcher also conceived its eastern shore in the border area of the two sectors "Beautiful Gardens" and "Land and Water" as a "transition from the artificially created natural landscape to the horticulturally enriched living areas".[210] The charm of the pond landscape also derives from the way in which it borrows from nature. This "borrowing" is in essence much more than visual incorporation or artificial imitation. It is physical incorporation and has more in common with what Wybe Kuitert and also Xiao, Jing, and Charlie Q. L. Xue identified in their study of the Jiejing than with the traditional concepts of shakkei (借景) and shukkei (縮景).

"Jing", as Ji Cheng (計成) discusses in the tenth and last chapter "Jiejing" (借景) of his work Yuanye (園冶), is not just the passive optical capture of a static tableau, but a processual interaction between the landscape – which is constantly changing with the seasons, the weather, and human intervention – and the viewer.[211] The process of borrowing the scenery is therefore a ritual of interaction, through which the borrowed scenery and the original landscape influence one another.[212] Neukom & Co. managed to achieve this by capturing the two natures of the Birs: the way it might have behaved before its straightening – as an unruly stream meandering through the landscape – and the way its artificial levelling into a tight bed forced it to behave.

Where the footbridges dock on land, they transitioned into stone paths that lead away into the laid out horticultural areas. These were arranged fanned out according to historical, geographical, and thematic aspects. The historical areas included traditional gardens laid out in the manner of peasant and baroque gardens. The geographical gardens had a regional focus – the "Jardin romand", "Ticino Garden", and "Jura Garden" – while the thematic areas were divided into a "Water Courtyard", the "Garden of Encounters", "Residential Garden", "Flowering Perennials Garden", and "Sandstone Garden" as well as the "Bamboo Courtyard".

Compared to the original ideas that can be seen in the numerous preliminary studies, some of the Japanese influences were lost in the built result. This becomes particularly obvious when comparing the plans. The waterfall with stream was a case in point. In the original plans, it even featured as the centrepiece of "plan drawing 5", which roughly corresponds to the competition project.[213] In this drawing, St. Alban pond served not just as a source for the lake, but was also tapped for a stream that meandered through the entire area of the garden courtyards – in a bed whose path was markedly more organic than the Birs, which was canalised in the 19th century, and which was to embody the theme of water's originally unruly course. The canal that was diverted from it in the Middle Ages inspired the designers to also highlight this layer of time by sublimating it in the "Quellengarten", which is divided into small channels. Neukom, Zürcher, Rasser & Vadi's intention was to highlight different "aggregate states" of the found water landscape: its original, free-flowing natural state, its artificiality channelled by human hand, and its artistic refinement. The succession of spaces not only revealed current, directly experienceable natural beauty but also condensed the wealth of experience to be found in the memories of the history of the place, a characteristic of Japanese gardens that Teiji Itoh had described can be experienced by walking through them.[214]

In the Bamboo Courtyard, by contrast, the "Far Eastern atmosphere"[215] was expressly intended. It is evident in the relationship between inside and outside, in the sense of perspective focus resulting from the framed view, as well as in its original planting of green powder

209 ASLA NL wne.144.1.2-Planting Plan-26.
210 Zürcher 1980b, p. 42.
211 Cf. Kuitert 2015, p. 32.
212 Cf. Xiao/Xue 2015, p. 224.
213 See anon. 1978, p. 22.
214 Cf. Itoh 1973, p. 108.
215 Zürcher 1980b.

← The original foliage of the Bamboo Courtyard formed its green walls. Today, one sits among a grove of stems and is covered by a forest of leaves [→ p. 121].

→ Neukom and Zürcher's original plan for the Grün 80, showing a Japanese-style waterfall above the geometrically arranged garden courtyards. The Japanese impression is heightened by an arrangement of platforms that resembles the mesh of verandas in a Japanese residence, and a dry landscape with striking rock formations on the far bank.

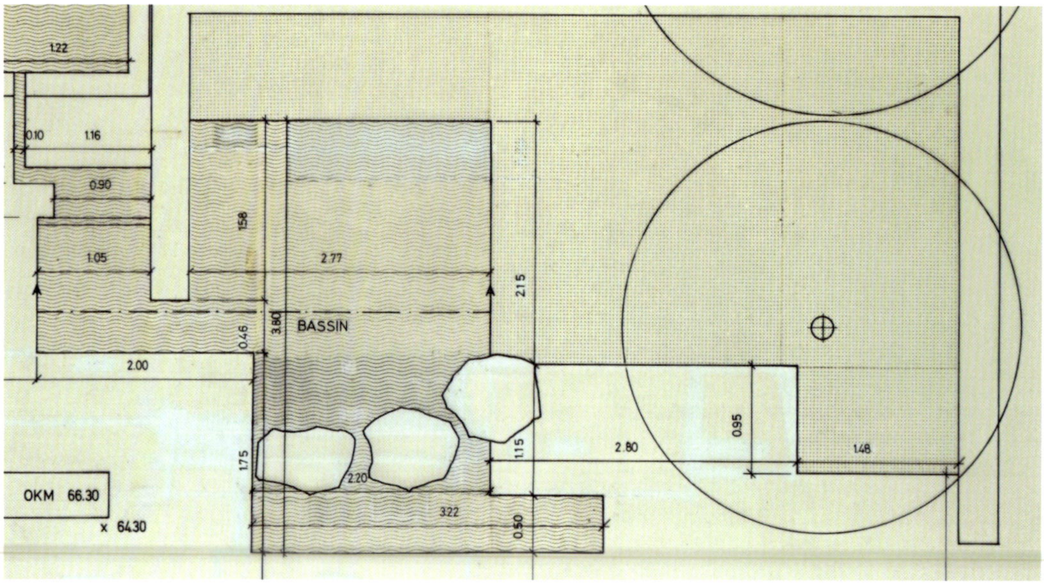

↑ In a later variant, Neukom and Zürcher moved the Spring Garden, crossed by small bridges à la japonais, to the shore, which they furnished with boulders and rocks in the manner of the Katsura garden.

→ Detailed plan of the shallow basin that still exists today [→ p. 125], with irregularly cut stepping slabs in the family garden of the Suisse romande section.

Perspective sketch of a pond situation designed for the Grün 80 with pebbles lining the shore and bridge. The scene has a strongly Japanese impression.

bamboo, Japanese arrow bamboo, umbrella bamboo, and dwarf bamboo.[216] Bounded by whitewashed cement stone walls, the courtyard can be entered only via wooden walkways that lead through two narrow openings into the courtyard and on to a square platform of diagonally laid boards. The elegance with which the relationship between inside and outside was arranged can be seen in a drawing dated 5 March 1979. The similarity of the conception to Japanese motifs is striking. Not only is the bamboo courtyard placed directly on the shore, as it was in the executed version, but also it is washed by water on two sides, making it appear to partially float on the water's surface – almost like the moon-viewing platform (Tsukimidai (月見台) in the Katsura residence, which is located "at an extreme end of the villa, where the ground plan projects into the garden like the prow of a ship".[217]

A photograph in Neukom's estate – similar to the one which Arioli made [→p. 65] – of the famous Eboshi-iwa (烏帽子岩, literally equivalent to head-covering stone, eboshi = priest's head-covering) in the Jōju-in teien (成就院庭園), which he had taken on the excursion as part of the IFLA Congress, leaves little doubt that he had more than just passing familiarity with Japanese lunar symbolism. Certainly, the significance of the stone had been made clear to the guests on that trip, not least because only a few years earlier, towards the end of the 1950s, it had been transferred back to its original location as documented between 1735 and around 1890, after having spent the first half of the 20th century relegated to the northern periphery of the garden.[218] Since being relocated as an eye-catcher in the centre of Turtle Island in 1964, it acts as a "moon catcher", a projection screen for the light of the moon, which cannot be observed directly because it cannot be seen from the temple. In other words, it is only thanks to the stone that the earth's satellite – or rather its reflection – is visible.[219]

Without doubt, the Swiss landscape architects found, what their German colleague Gisela Fleig-Harbauer called an appropriate "mode of translation" by which to evoke images of Japanese garden design while integrating them into a regional and contemporary modernist manner.[220] Rainer Zulauf, who, towards the end of the 20th century more by chance than design created a garden with Japanese aspirations, came up with an expression that would not be out of place in a Japanese poem to describe this process of transcription: "Instead of tackling the impossibility of creating a Japanese garden, an attempt was made to allow the space for the dream of one to emerge."[221]

216 *Phyllostachys viridiglaucescens*, *Pseudosasa japonica*, *Sinarundinaria murielae* or *Fargesia murielae*, *Sasa Pumila* or *Pleioblastus pumilus*.
217 Grütter 1976, p. 466.
218 Cf. Kuitert 2010, p. 2.
219 Cf. Kuitert 2010, p. 6.
220 Cf. Fleig-Harbauer 1981.
221 Zulauf 1996, p. 12.

Neukom and Zürcher's Bamboo Courtyard for the Grün 80 in Münchenstein near Basel.

From Ammann

to Zürcher

← pp. 121, 125, 126–127: Bamboo room, pool in the Family Garden and Japanese-style pond in the grounds of Grün 80 (now Park im Grünen) in Münchenstein by Willi Neukom and Adolf Zürcher.

pp. 122, 123: Gustav Ammann's pond in the garden of the Oerlikon, Bührle & Co. Wohlfahrtshaus (now Gustav Ammann Park) in Oerlikon can be crossed by circular stepping stones and has an island whose shape resembles the outline of a turtle.

p. 128: The pond in the Guyer Garden in Zumikon designed by Willi Neukom still exists in its original form, albeit now somewhat overgrown.

pp. 129–131: In the former Hegner Garden (now Mirani) in Meilen, designed by Ernst Cramer, the Japanese maple, bamboo and stone lantern have survived. The miniature pond and waterfall were modelled on the original.

p. 132: Evariste Mertens designed a replica in miniature – of the Klöntalersee and its tributary – in the Schuler-Ganzoni Park.

p. 133: A group of three stones similar to the Buddhist model in the pond of Ernst Cramer's Prina Garden in Dulliken.

pp. 134–136: Japanese influences abound in Willi Neukom's design of the pools, planting, stone arrangements and especially the rippling waterfall in the grounds of the IBM International Business Machines Corporation headquarters (now IBM Research – Zurich) in Rüschlikon.

In 1945, Ammann created an enchanting Japanese miniature in the "Deller Garden" in Winterthur-Wülflingen. From a shallow depression in the rock, as if washed out by water, a small waterfall flows into the pond planted with water lilies, Japanese swamp irises, Japanese primroses, and bulrushes. Japanese ornamental quince and threadleaf maple accentuate the Japanese colouring.

Gustav Ammann 1885–1955

← Rectangular water pools, offset from one another and delineated at their intersection with stepping stones, characterise the Japanese influence in the "Schleiniger Garden", created in 1934–37.

↓ "Garden of the Wohlfahrtshaus" at Oerlikon, Bührle & Co. in Oerlikon: Ammann interlocks the organically shaped pool with two islands with the landscape using the irregular contour of the edging slabs.

Japan as a module

The art of Japanese garden design, which astounds us today with an exquisite collection from its gardens, is born of a spiritual connection to nature that we as Europeans can only begin to comprehend. Yet the impression these garden creations have on us is so strong that we feel involuntarily drawn to these creations, which are in many respects akin to our present-day way of designing gardens.[1]

Gustav Ammann set off on his journeyman years immediately after completing his gardener's apprenticeship with Otto Froebel in Zurich between 1903 and 1905 and moved frequently – quite literally because nowhere held his attention for very long. In 1907, he joined Reinhold Hoemann (1870–1961) in Düsseldorf, where he worked with Paul Schädlich (1885–1946) and met Robert Froebel, who ran the firm Froebels Erben in the third generation. In 1908–1909, he enrolled at the Kunstgewerbe- und Handwerkerschule (School of Arts and Crafts) in Magdeburg. After a brief spell in England from July to October 1910, where he worked for Milner & White in London, he found employment with Ludwig Lesser in Berlin. In February of the following year, he switched once more to Jakob Ochs in Hamburg, where he met Leberecht Migge, then the firm's head of garden design. In Migge he found not just a like-minded colleague but also a friend, but nevertheless he only stayed for eight months before returning to his original company in autumn 1911, which he ran as head garden architect until the dissolution of Otto Froebels Erben in 1933.[2] He founded his own office immediately thereafter in 1934.[3] Over the years, some of the most renowned Swiss garden designers passed through his schooling: Richard Arioli, Ernst Cramer, Hans Epprecht, Ernst Graf, Johannes Müller, Hans Nussbaumer, and Pierre Zbinden, as well as his son Peter Ammann.

Arioli later became the City Gardener of Basel, Cramer one of the most innovative garden designers of the time, Epprecht partnered with Willi Neukom on the design of the grounds of the ETH Zurich,[4] Graf together with Fred Eicher 1959–1961 produced an exquisite design interspersed with Japanese elements at St. Michael teacher training college in Zug,[5] Müller became City Gardener of Winterthur, Nussbaumer the successor of the Mertens dynasty, and Zbinden Garden Inspector in Zurich. Finally, Ammann junior, who began working in his father's business in 1942, followed in his footsteps in 1955.

Of significance for Swiss landscape architecture is the period that Richard Neutra spent working for Gustav Ammann in Zurich from 1918 (see "Ernst Cramer – Japanese tempered"). "He learned there how to use plants according to size, growth, colour, and flowering time so that they would fit into a country garden or a castle garden. He stayed in Zurich for three years. In those three years, he learned an infinite amount about nature and how to embed architecture in nature."[6]

Decades later, Neutra would honour his teacher by writing the foreword to Ammann's book *Blühende Gärten (Flowering Gardens)*.[7] Ammann, for his part, included photographs of Japanese-inspired garden designs for Neutra's Rourke, Brod, and Perkins Houses in his book, which was published in the year of his death and became, as it were, his written legacy.[8] In it, he emphasises the importance of Japanese sources both in writing and using photographs,[9] including some from the archives of Werner Bischof, who travelled through Japan in 1951.

Gustav Ammann first encountered Japanese gardens in 1907 at the "Jubiläums-Ausstellung – Internationale Kunst- und grosse Gartenbau-Ausstellung" in Mannheim (Jubilee Exhibition in Mannheim), which he visited together with Paul Schädlich. As far as the Japanese elements of the exhibition were concerned, Ammann's contemporaries had quite contradictory opinions. The *Schweizerische Bauzeitung (SBZ)* describes the special garden by Henkel as exhibiting a certain "restlessness" in its "central motif", an extensive "water plant pond surrounded by a layout strongly influenced by Japanese garden design", which "brings together a rich range of planting materials in diverse groupings within a relatively constrained space, with the result that none comes into its own."[10]. By contrast, the editor of the *Neue Zürcher Zeitung (NZZ)* was not entirely disapproving of the charisma of the grounds: "A neat, amusingly laid-out garden by Heinr. Henkel from Darmstadt which, with little bridges and rocky sections, picturesque garden houses and a bamboo pergola, turns its stylistic face towards Japan".[11] While the *NZZ* showed no pictures at all, the visual material in the *SBZ* conveyed a rather meagre impression. *Die Gartenwelt* did, however, provide a better indication of two sections "from the Japanese garden with pond and bridge" and "with pergola in the background"[12]. Ammann certainly had access to this magazine, as well as to *Die Gartenkunst*, in which Henkel's garden was reviewed and illustrated with five photographs,[13] not least because Hoemann was the editor in charge at the time. Ammann had probably also become exposed to Japan through Hoemann, as the latter had adapted the "Fukiage Garden" at the Internationale Kunst-, kunsthistorische und grosse Gartenbau-Ausstellung Düsseldorf (International Art, art historical and Great Horticultural Exhibition in Düsseldorf) three years earlier, based on an illustration in Conder's *Landscape Gardening in Japan* (see "From Zukai teien zōhō to Sakuteiki").

Ammann had also visited the Japan-British Exhibition in White City, London, in 1910 and was smitten by the "attraction of the foreign culture"[14]: "I have been to the exhibition that Japan and England have set up nearby three times already. How far this nation has come in the most modern enterprises, in shipbuilding and weaponry! And in architecture and art, too, it is far superior to the English. Such attention to material, such rich, simple and large forms in the decorations and surfaces of the house, here and there bizarre, colourful forms, without

1 Ammann 1934b gtaA, p. 1.
2 See Stoffler 2006a, pp. 8 (Hoemann), 13 (Magdeburg), 22 (1910 London, Milner), 24 (1913 England again), 25 (Lesser), 26 (Ochs), 28 (back to Froebel 1911).
3 H. B. 1945a.
4 Epprecht posthumously gave a detailed eulogy of his former teacher's work. See Epprecht 1956.
5 Gadient 2009, pp. 162–169.
6 Klemmer 1993, p. 74.
7 Ammann 1955, n. pag. [p. 7].
8 The properties not specified by Ammann are the houses for George and Elizabeth Rourke, 1949, for Earl Brod, 1951, and for Constance Perkins, 1955. See Ammann 1955, pp. 126–127; pp. 124–125; pp. 122–123.
9 Ammann 1955, p. 10, ill. pp. 13, 15, 17, 19, 27.
10 Anon. 1907d, p. 118.
11 Anon. 1907b, p. 2.
12 Hesdörfer 1907, p. 413.
13 Schulze 1907, ill. pp. 215, 217, 218.
14 Stoffler 2006a, p. 149, note 425.

a grand plan or style, just as much and conspicuous as possible."[15]

In addition to the "Garden of Peace" by Keijirō Ozawa and the "Garden of the Floating Isle" by Kinkichirō Honda, he almost certainly saw the two miniatures of a teahouse garden and an exact reproduction of the Kinkaku-ji in Kyōto, which were arranged on two terrains measuring 2.15 × 3.65 metres each and were copied faithfully down to the last detail.[16]

Japan – the start of a dissolution of form

"The initial impetus to break away from the formal rigidity of the day came through a commission to design a Japanese Garden," explained Ammann in a lecture to the Basel Horticultural Society at the Bernoullianum in 1945.[17] He designed it for the embroidery manufacturer Josef Eugster in Altstätten in the Rhine Valley [→ pp. 41–42, 124], and the project would provide him with an answer to the question of "Should we dispense with form altogether?"[18] In his article in the journal *Die Gartenkunst*, he pitted the natural garden against the architectural garden. "When we first started working on architectural gardens, it felt counter-intuitive" to apply the rational-constructive principles of architecture to plant life, he wrote. Now, "paths are broken down into pieces of slabs, [...] walls are broken open, water basins have no distinct form or edge and merge into their surrounding plantings of beautiful Japanese or American species".[19] He underpins his line of thinking with an allusion to William Robinson's *The Wild Garden*[20] and predicts a development towards a "wild garden" in which nature is free and entirely unconstrained.[21]

In 1931, Ammann posits the gardens for the Keller-Meyer family in Castagnola on Monte Bré and for Josef Eugster in Altstätten as built examples of these two poles: "There, terraces staggered one on top of the other, water contained in strictly delineated basins, a panel of lawn in the axis of the house; here in the valley there is also a certain connection between the garden and the house [...], but otherwise plants and water are arranged entirely freely. A natural garden? Maybe – and today that is no longer a derogatory term!"[22]

At the heart of the "Eugster Garden" is an organically shaped water basin. Bordered by regularly laid slabs, the edges merge in places with the waterside planting. A stone lantern denotes its Japanese origins, as does a small island planted with arrowhead and papyrus. Another Japanese motif is the stepping stones that cross the pond in sweeping lines at its narrowest point. Their almost square shape is disguised through their dispersed arrangement. A swamp rose mallow (*Hibiscus moscheutos*) adds a distinctive splash of colour.

The pond, dotted with "colourful water lilies", is lined with catalpa (*Catalpa bignonioides*), black pine (*Pinus austriaca*), willowleaf sunflower (*Helianthus salicifolius*), knotweed (*polygonum*), marram grass (*Ammophila*

15 Ammann 1910 gtaA.
16 Brown E 1910, pp. 224–225.
17 p. 1945.
18 Ammann 1926, p. 81.
19 Ibid, p. 83.
20 See Robinson 1870.
21 Ammann 1934, p. 116.
22 Ammann 1931, p. 82.

← Father and son Ammann sketched an adaptation of the "Garden of the Wohlfahrtshaus" in Oerlikon for Winterthur Cantonal Hospital in 1949–1953 and 1955–1958.

↓ The pond, stream, and boulders are elements whose Japanese influence can still be felt in the Waid Hospital in Zurich.

arenaria subsp. Arenaria), and reeds (*Arundo donax*). He also intersperses plants originally from East Asia: Japanese angelica (*Aralia elata*, also *mandshurica* or *chinensis*), Japanese umbrella pine (*Sciadopitys verticillata*), hydrangea (*Hydrangea serrata*), and bridalwreath spirea (*Spiraea prunifolia*).[23] There were probably many more, and some were added afterwards, as "the garden later underwent an extension that was designed and planted in a similar way to the original garden"[24]. A second water basin was created, this time laid out at right angles to create a contrast. Two Japanese maples added a distinctive Japanese accent.

When the joint heirs of Josef Stadler, who had taken over the property from Eugster, sold the property in 2020[25], they handed over a dossier of photographs to Karl Zünd Park AG that showed the existing woody plants and perennials, many of which were originally native to the East Asian region.[26]

Even in 1950, when Ammann looked back on half a century of garden design, he cited the "Eugster Garden" as a built example of the changed conception of garden design.[27] He also used it as an advertising motif. In 1938, he placed an advertisement in the magazine *Das ideale Heim* showing a photograph of the "Eugster Garden" with its water basin, island, stone lantern, and stepping stones.[28]

Although he treated Altstätten and Monte Brè as contrasting models – a mountain and a valley garden – it was not so far-fetched to choose a Ticino and a Japanese garden for his publication. Other garden designers had also used combinations of "typical" Ticino motifs with Japanese set pieces. An exceptional synthesis of the two is the garden created for the textile art producer Carl Eschke in Zurich in 1936. It combines a water lily pond with a stone lantern in front of the living room, an arbour of woven bamboo adorned with a sculpture of Buddha, with a dry stone wall, a pergola supported on granite columns and an alpine garden.[29]

The exhibition "Japanese Architecture and Gardens" at the Kunstgewerbemuseum (Museum of Decorative Arts) in Zurich in 1933 was for Ammann a revelation. In the accompanying publication, Walter Curt Behrendt (1884–1945) emphasised the role of the garden for the relationship between man and nature and the interpenetration of house and garden as an expression of this. "The most striking feature of the Japanese house" in his opinion was "its intimate connection with the surrounding nature [...] such that it ultimately appears to be a piece of grown nature itself"[30].

Ammann's fascination with the show, for which he also organised a guided tour,[31] is documented in the typescript of an article he wrote for the journal *Das Werk* in 1934 in response to Marta Heisinger's article *Japanische Gärten* (*Japanese Gardens*), which its editor Peter Meyer had published a year earlier[32]: "The art of Japanese garden design, which astounds us today with an exquisite collection from its gardens, is born of a spiritual connection to nature that we as Europeans can only begin to comprehend. Yet the impression these garden creations have on us is so strong that we feel involuntarily drawn to these creations, which are in many respects akin to our present-day way of designing gardens. The Japanese building is also closely connected to the nature that surrounds it, almost as if it has grown organically like the trees that surround it. We sense the magic without immediately knowing – we think we need to have a reason – why we find it so appealing. House and garden are an inimitable creation of an image, of a God-blessed landscape design."[33]

Ammann recognises in the Japanese model the possibility of freeing himself from the "all too rigid fetters of the architectural garden" and at the same time avoiding the pitfalls of pure "imitation of nature". "The Japanese know how to imitate nature without violating the laws of the organic, and yet also to stylise it. This stylisation, however, is more subtle, closer to nature, but heightened or softened and then blended imperceptibly into real nature." He wishes to adapt this "delicate sense of touch" so that he may place less emphasis on "outward formality" and instead strive for "a greater inner connection to the self and devotion to nature, imbuing nature and dead material with spirit and idealising life".[34]

Ammann's reference is to *Japanese Gardens* by Josiah Conder, *The Book of Tea* by Okakura Kakuzō (1862–1913), and *Things Japanese* by Basil Hall Chamberlain (1850–1935), published in six editions between 1890 and 1936, but above all to *The Gardens of Japan* by Jiro

The unhewn wooden support of the arbour in the "Müller-Keyser Garden" looks like the tokobashira (床柱) of the tokonoma (床の間). The traditional portrait niche impressed Ammann both due to the emptiness it expresses and due to its "appreciation of elegance and simple refinement".

23 Ibid, pp. 74–78.
24 "Garten in Altstätten", documentation by the heirs to Stadler Lina Marie, today owned by Karl Zünd Park AG, n. pag.
25 Heirs to Stadler Lina Marie ["Lilly" Stadler-Lüchinger, 1918–2018], to Karl Zünd Park AG, with registered office in Altstätten: property no. 4356, Bahnhofstrasse 3, 7,384 m² site area, garage vers. no. 3539, residential building vers. no. 423. "Veröffentlichung der Eigentumserwerbe an Grundstücken in der Stadt Altstätten März 2020", in: *Handänderungen in der Stadt Altstätten*, March 2020, https://publikationen.sg.ch/fileadmin/ekab/files/2021/06/00.047.314/attachments/Handaenderungen_Mai_2021.pdf (retrieved 14 Dec. 2022).
26 Asian flowering dogwood (*Cornus kousa*), slender deutzia (*Deutzia gracilis*), sweet olive (*Osmanthus burkwoodii*), gingko (*Gingko biloba*), royal paulownia (*Paulownia tomentosa*), several Japanese maples (*Acer palmatum, Acer palmatum 'Atropurpureum', Acer palmatum 'Dissectum', Acer palmatum 'Dissectum Atropurpureum'*), Japanese snowball (*Viburnum plicatum 'Mariesii'*), Japanese azaleas (*Rhododendron japonicum*), Japanese flowering cherry (*Prunus serrulata*), star magnolia (*Magnolia stellata*), Thunberg's barberry (*Berberis thunbergii*), Japanese crabapple (*Malus floribunda*), and winged spindle shrub (*Euonymus alatus*).
27 Ammann 1950, p. 7.
28 Ammann 1938.
29 See Streubel 1936. Eschke ran the company "Schweizer Textile Kunst Carl Eschke", for which female artists such as Ursula Schneider created designs. Schneider was married to Hermann Schneider, the architect of Eschke's villa. https://www.emuseum.ch/en/people/120603/ursula-schneider (retrieved 22 July 2023).
30 Behrendt 1933, p. 3.
31 "Unfortunately, I was unable to take part in your guided tour of the Japanese Gardens exhibition." Ammann 1934b gtaA.
32 Heisinger 1933b.
33 Ammann 1934b gtaA, p. 1.
34 Ibid.

↑ The view of the “Jordan Garden” from 1947–1948 recalls a photograph of the interior of the Katsura residence shown by Peter Meyer from Yoshida’s book. Japanese maple and wisteria heighten its Japanese ambience.

↓ The wisteria, a popular plant in Japan, defines the view back from the garden to the Jordan House.

↑ To complement Max Frisch's architecture of the Letzigraben outdoor pool in Zurich, Ammann designed a Japanese-inspired pond.

↓ In his own garden, Ammann chose the standard organic geometric form of a fish and marshy plant pond with water lilies, lotuses, and Japanese irises.

Harada.[35] From this he quotes excerpts of a key scene, the full passage of which reads[36]: "In another instant, Rikyu, in his garden at Sakai, obstructed the open view of the sea, by planting a grove of trees in such a way that only when the guest stooped at the tsukubai [stone water basin] to wash his hands and rinse his mouth preparatory to entering the cha-seki, he caught an unexpected glimpse of the shimmering sea through the trees – a glimpse of Infinity – thus suddenly revealing the relation of the dipperful of water lifted from the basin to the vast expanse of sea, and of himself to the universe."[37]

Ammann goes into lyrical raptures: "The Japanese listen to the growing of moss, the dripping of the rain, they rapturously contemplate the moon, the setting sun, and know with keenest sense how to heighten such contemplation with suitable plantings".[38] Ammann expresses an intention to publish a catalogue of these plants, which will become part of his repertoire (see plant catalogues).[39] However, it remains unpublished, as does the planned article for *Das Werk*.

He pays particular attention to the "typification and stylisation" that he describes as the division into "Hill Garden" and "Garden of the Plain",[40] and its design rules, which have been taught "unchanged for a thousand years in garden books" and have ensured that "its artistic qualities are preserved".[41] He highlights the guidelines for the design of streams and waterfalls and the creation of hills, islands, bridges, stones, stepping stones, and bodies of water.[42]

Ammann pursues his own interpretation of typification. He develops standardised two-part basin forms, the division of which he emphasises with stepping stones. Some basins are geometrically arranged, but most of them consist of an organically formed and a rectangular section, which he uses like a module in many of his garden projects. He often arranges planting around the organic section – preferably water lilies – while the geometric one serves as a pond for fish and/or for bathing. This layout recalls that described by Georg Lasius for the Chinese garden: "The most essential thing in any garden, however small, is two ponds, one to grow fish in and the other for nelumbium, a kind of water lily."[43]

Respecting the laws of the organic but also stylising them, as he has admired in the Japanese,[44] Ammann harmonises nature and art by inscribing a golden spiral – a special form of the logarithmic spiral common in nature (sunflower, nautilus) – in the pond, as seen in the detailed plan of the pond for the "Spinner-Debord Garden".[45]

In Sennhauser's garden, which he designed together with his son Peter in the final years of his life in 1952–1953, a spiral also serves as a means of freeing the obligatory rose bed from traditional geometry and yet giving it formal definition.[46] As a Japanese "derivative", he marks the garden with his "standard" water lily pool and the planting of Japanese flowering cherry (*Prunus serrulata*), Japanese kerria (*Kerria japonica*), Japanese spirea (*Spiraea japonica*), winged spindle tree (*Euonymus alatus*), summer lilac (*Buddleja davidii*), fire maple (*Acer ginnala*), Japanese maple (*Acer palmatum*), threadleaf Japanese maple (*Acer palmatum* 'Dissectum') as well as corkscrew hazel (*Corylus avellana* 'Contorta') and pine (*Pinus*).

His review of Tetsurō Yoshida's 1935 book *Das japanische Wohnhaus* in the journal *Das Werk* occupied Ammann for a long while. He sketched some of the photographs of the imperial Katsura Palace in Kyōto contained within,[47] among them "the oldest section with an open bamboo veranda" and "detail of a house entrance from the garden and veranda".[48] He was impressed by the "neat but exceptionally refined primitiveness, the close interlocking of indoor and garden spaces, of building form and natural form".[49]

Between 1941 and 1943, Ammann's design for a garden of the Wohlfahrtshaus (workers' facilities) for the machine tool factory Bührle & Co. in Oerlikon was implemented. Ammann published it several times[50] and, like the "Eugster Garden", cited it in 1950 as an example of the radical change that began at the end of the 1920s.[51] A particularly eye-catching Japanese detail are the cylindrical stepping stones that cross the curved pond [→ p. 122]. They are modelled on those in the garden of the Heian shrine in Kyōto, which Ammann had already emulated at the 1939 National Exhibition in the water courtyard of the "Twining" section of the "Kleider machen Leute" fashion pavilion [→ p. 89].

Two islands – one planted with water lily (*Nymphaea*), club-rush (*Scirpus lacustris*), and marsh trefoil (*Menyanthes trifoliata*), the other with water lily, bulrush (*Typha*), and yellow iris (*Iris pseudacorus*) – complete the Japanese repertoire. Plants with a striking appearance – umbrella plant (*Saxifraga peltata*, today: *Darmera peltata*), willowleaf sunflower (*Helianthus salicifolius*), *Eulalia* or *Miscanthus* – line the banks.[52]

Ammann's interest in Japan stayed with him throughout his life. He kept a newspaper article from 1954 sent to him shortly before his death by Leder and his wife.[53] It concerns coverage of Akira Satō's appearance at the Bund Schweizerischer Gartengestalter (Federation of Swiss Garden Designers, BSG), in which he presented "Colour Photographs of Japanese Gardens" (see "Walter Leder – Japan at heart").[54]

35 As Christian Tagsold noted for Europe, Harada also reached a wide readership in Switzerland. See Tagsold 2017, p. 38. Ammann penned a handwritten overview of the historical development of the Japanese garden after Harada; Ammann n.d.b gtaA.
36 Ammann 1934b gtaA, p. 2.
37 Harada 1928, pp. 38, 39.
38 Ammann 1934b gtaA, p. 3.
39 Ammann n.d.a gtaA.
40 Ammann 1934b gtaA, p. 1.
41 Ammann 1934, p. 117.
42 Ammann 1934b gtaA, p. 1.
43 Lasius 1881, p. 358. *Nelumbo* is the scientific name for lotus. In contrast, water lilies are *Nymphaea*.
44 Ammann 1934b gtaA, p. 1.
45 Garden Spinner-Debord, 1935 / 1940–1941: "Bassin-Detail 1:20. Bassin u. Pergola zum Garten Dr. Spinner. Zollikon." gtaA NSL 2-0247-5.
46 Ill. see Ammann 1953b, p. 409.
47 Ammann, n.d.c gtaA.
48 Meyer 1936b, p. 81.
49 Ibid.
50 See Ammann 1943; Winkler 1943.
51 Ammann 1950, p. 10.
52 See Ammann 1952, p. 305.
53 "Von Frau Hofmann Journalistin / Grüsse W. u. E. [Emmy] Leder" ("From Mrs Hofmann Journalist / Greetings W. & E. [Emmy] Leder"). gtaA, NSL 2-S-34 M1/3.
54 Cf. Hofmann 1954.

Albert Baumann 1891–1976

→ The Japanese garden in Oeschberg in a rare colour photograph.

↓ In the drawing of a Japanese garden, the fountain appears to be a precise copy of the water basin depicted by Kinkichirō Honda on plate 24, while the lantern resembles the one on plate 17.

Baumann also used Honda's plate 17 as a model for the light source he drew in the foreground on the second sheet and plate 35 of the catalogue for the smaller lantern behind.

Japan as a motif

Surrounded by shrubbery walls of snowberries, cut-leaved elder and conifers, the Japanese garden is self-contained and lies next to the alpine garden. Models from early Japanese garden designers served as the basis for the design.[1]

Born in Arbon, Thurgau, the son of a ribbon weaver, Albert Baumann initially trained at St. Gallen embroidery drawing school.[2] His interest in ornamental plant motifs led him to embark on an apprenticeship as a gardener in 1909 at the École cantonale d'horticulture de Châtelaine (Châtelaine School of Horticulture) near Geneva, where he met Walter Leder, with whom a friendship developed. Realising that his interests lay less in becoming a gardener than a designer, he entered the intercantonal horticultural school in Wädenswil in 1913.[3]

After graduating, Gottlieb Friedrich Rothpletz (1864–1932), director of the City of Zurich's Gardens Department, employed him in the department's drawing office, where Baumann remained until 1919[4] before going freelance and travelling to France and Germany. In autumn 1920, the state council appointed him head teacher at the newly founded Oeschberg Cantonal Horticultural School in Koppigen, Bern,[5] where he worked until 1957.[6]

It was Albert Baumann's estate that gave rise in 1981 to thc Stiftung Archiv für Schweizer Gartenarchitektur und Landschaftsplanung (Archive for Swiss Garden Architecture and Landscape Planning), the precursor to today's Archiv für Schweizer Landschaftsarchitektur ASLA (Archive for Swiss Landscape Architecture) at the University of Applied Sciences Rapperswil (today OST Ostschweizer Fachhochschule).[7]

For a long time, Baumann was known primarily only as a teacher but, as the garden historian and landscape architect Steffen Osoegawa has related, he played an important role as one of the "leading protagonists of the Swiss reform garden movement"[8]: "For him, the architectural garden was not just an art form, but an expression of his world view informed by the Lebensreform movement, which helped him liberate himself from many conventions he disliked."[9]

The reform character of the dwelling garden style, however, was not, in his view, sufficient to constitute "an independent development of garden culture".[10] Rather he saw their character as having themes, which is why he coined the term "motif garden" for them.[11]

Osoegawa pinpoints the time at which Baumann began to devote himself exclusively to the motif garden and the young discipline of landscape design to the mid-1930s, noting that by then, the peasant garden and the stately baroque garden had almost completely disappeared from his design repertoire.[12]

The "motif",[13] for example, of a meandering stream created using stones to simulate its natural course, or of a bathing pool filled with water lilies, as well as "a new

1 Baumann A 1931/1932, p. 32.
2 See Osoegawa 2020b, p. 22.
3 See Osoegawa 2020b, p. 23.
4 Ibid.
5 See Osoegawa 2020b, p. 24.
6 See Wettstein 1977, p. 37.
7 See Osoegawa 2020b, p. 24
8 See Osoegawa 2020c, p. 22.
9 Osoegawa 2020c, p. 24.
10 Ibid.
11 Ibid.
12 Cf. Osoegawa 2020a, p. 17.
13 Osoegawa 2020c, p. 24

Plates 27 to 30 in Honda's work show sections with stepping stones that Baumann, who owned the book, adapted for the garden.

quality of landscape",[14] in which the surroundings are incorporated into the garden, provided an opportunity to loosen the corset of symmetry and rigid divisions.[15] Osocgawa attributes this change, among other things, to Baumann's experience of a show garden at the Solothurn Horticultural Exhibition in 1933.[16]

A discovery in Baumann's estate offers a further interpretation. The shift in his design approach coincides with the time in which he became interested in Japanese garden design. In Baumann's estate was a list of books compiled by Maruzen Company of Book Traders in Tōkyō at around 1930, the most recent of which was *Juraku. A Collection of Photogravure of the Elegant Buildings and Artistic Gardens in Japan, Tokyo, 1929–1930.*[17]

The list includes 29 publications on Japanese garden design. Along with *Juraku* was *The Gardens of Japan* by Jirō Harada[18], another Japanese publication that was widely read in Europe.[19] Together with works by Issotei Nishikawa, Matsunosuke Tatsui, and Tsuyoshi Tamura[20], the 1956 edition of this book can be found in the library that Baumann bequeathed to the archive.[21]

Among the Western authors also included in his library are works by Josiah Conder, Reginald Farrer, Florence Du Cane, Lafcadio Hearn, and Charles Geoffrey Holme, among others.[22]

What makes this list even more invaluable is where it was found, namely in a book whose discovery may be considered a minor sensation. It is a 1926 Japanese-language edition of the work *Zukai teien zōho* (図解庭造法, *Illustrated Horticultural Methods*), first published in 1890 by Kinkichirō Honda (本多錦吉郎, 1850–1921).[23] None of the estates of other Swiss landscape architects includes an original Japanese-language edition. Baumann evidently studied it closely, even translating individual terms into German, while conversely adding the Japanese characters for "Juraku" to Maruzen's list.

Kinkichirō Honda was none other than the creator of one of the two gardens – the "Garden of the Floating Isle" – at the Japan-British Exhibition of 1910 (see "From Zukai teien zōhō to Sakuteiki"). Honda's affinity with European gardens was probably helpful to Baumann. Some of the illustrations showed geometrically designed gardens that deviated from tradition. Plate 40, for example, documents a focally placed hexagonal bed featuring rock and tree arrangements, plate 41 a baroque-style fountain in a circular pond enlivened with stone islands, and plate 12 a garden whose view is obstructed by a meandering box hedge of tall rectangular blocks.

The temporal coincidence with Baumann's change in direction is striking. The suggestion that this new direction is connected with his interest in East Asian sources of inspiration seems all the more plausible given that in 1931, he concluded the ten-year work on the park of the horticultural school[24], which began in 1921, with a Japanese garden, the design of which drew on "model[s] from early Japanese garden designers".[25] Fittingly, he placed it next to the alpine garden but created an introverted space by surrounding it with "shrubbery walls of snowberries, cut-leaved elder and conifers". The feathery foliage of *Sambucus nigra* "Laciniata" was undoubtedly a reference to the fan maple of Japanese origin. The garden's 2-metre-high stone lantern, the bird fountain, and the stepping stones, although "cast from reinforced concrete", only heightened its Japanese character. The latter ran in a "channel-like depression" and "moving line" over "slightly sloping" terrain "from the alpine garden along the stream" and, after a branch at "the massive lantern [...] through the miniature valley and on to the bird fountain", before they led into the alpine garden path. *Cerastium tomentosum* (snow-in-summer), *Globularia willkommii* (common globe-flower), *Sedum spurium* (Caucasian stonecrop) as well as low and high irises were interspersed with "naturalistically arranged firethorn bushes, suitably cut box forms, peonies and ornamental grasses", which Baumann used for their "sculptural" effect for preserving the spatial image.[26]

The importance of the garden can be gauged not least from the fact that in 1932, the newspaper *Der Bund*, citing its annual report for 1931/1932, placed it top of the year's new noteworthy gardens.[27] In addition, Baumann refrained from making alternations to the Japanese garden, unlike all the other garden sections, which he redesigned with his students almost as soon as they

14 Ibid.
15 Ibid.
16 Ibid.
17 Baumann, A n.d. [1930–1932] ASLA. At that time the publisher had "I Series: in 12 parts" with 12 plates each and "II Series: in 12 parts" with 12 plates each; "III Series" was "in preparation". By 1933, all four volumes had been published. The compendium seems to have enjoyed a certain degree of popularity in Switzerland. In any case, the journal *Das Werk* published illustrations from it in 1936: "All four illustrations after: 'Juraku', Pictorial of Japanese Architecture and Gardens. Four volumes, 144 plates. The Zauko Press. Tokyo 1933 (in the Zurich Museum of Decorative Arts)"; Anon. 1936, p. 39.
18 Harada 1928. An excerpt from it had already appeared in 1922; Harada 1922.
19 Tagsold 2017, p. 38.
20 Nishikawa 1936. Tatsui 1936. Tamura 1937.
21 Harada 1956.
22 Conder 1893/1994. Conder 1886, pp. 119–175. Farrer 1904, pp. 378–384. Du Cane / Du Cane 1908. Hearn 1892. Holme 1893.
23 Honda 1890–1926. The book was found on a shelf with stacks and boxes of as yet unseen documents – rolls of plans, photographs, drawings – by various garden designers. Since Albert Baumann's plan label is stamped on the inside cover, one can be sure that it was once in his possession.
24 See Osoegawa 2020c, pp. 22–23.
25 Baumann, A 1931/1932, p. 32.
26 Ibid.

One of Baumann's famous "tabular classifications" gathers several examples of bridges of Japanese (A, C, D, H, K, L) or pseudo-Japanese (B, E, F, G, I, K, N) origin.

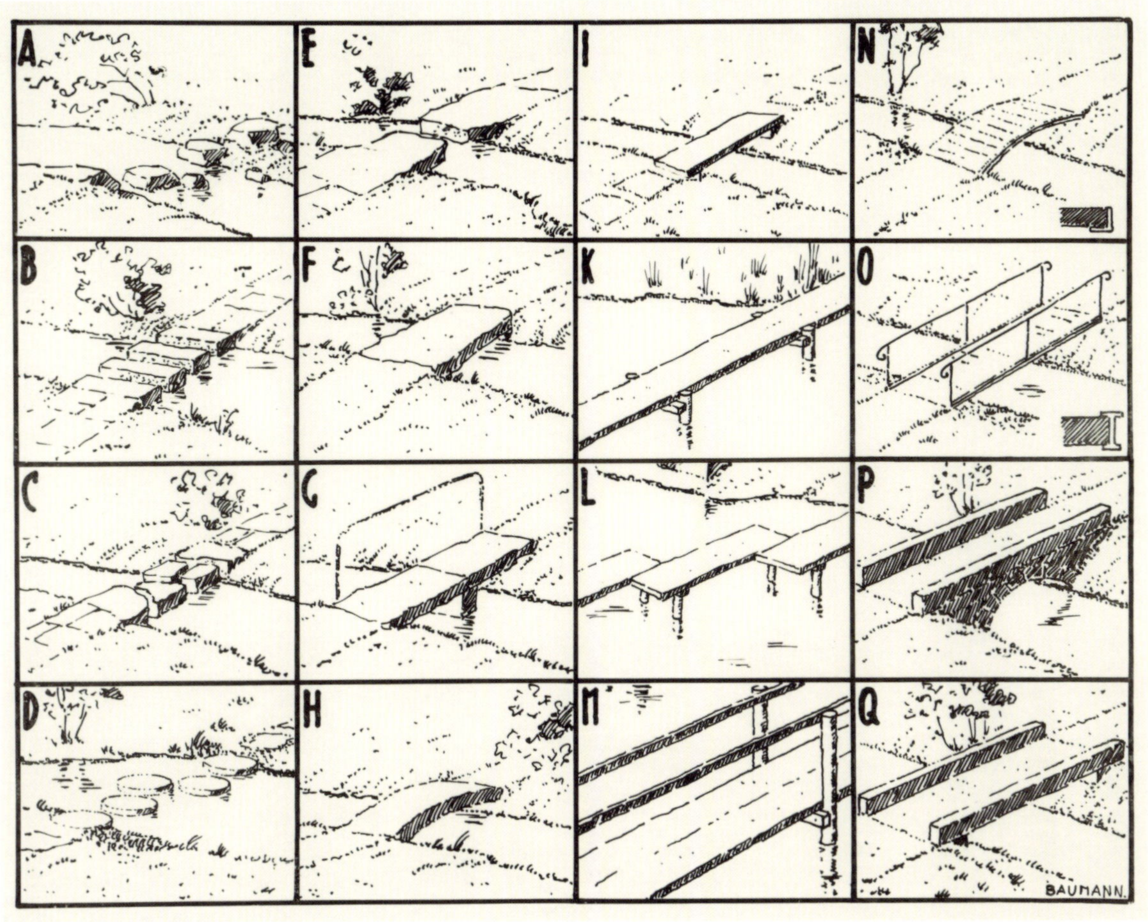

were completed.[28] By the 1990s, however, all that was still readily identifiable of it was a stone lantern, and the garden has since been integrated into an overall maintenance concept.[29]

Two perspective drawings by Baumann show that he was directly inspired by Honda's illustrations [→ pp. 148, 149]. In addition to lanterns, fountains, and irregularly laid stepping stones, he added elements seen as being typical of Japanese gardens: swamp irises, azaleas, and presumably a Japanese maiden pine (*Pinus parviflora*).

Baumann's study of Japanese influences can also be seen in his "tabular classifications", display boards of between 6 and 16 ink drawings, each detailing thematically related, horticultural subjects, which he used in the classroom, alongside photographic reproductions of numerous gardens at home and abroad.

He contrasted channelled water with meandering streams, orthogonally directed fountains with springs bubbling up from between stones, and circled garden squares with moving green spaces. He even claimed the yin-yang symbol as a design variant for circularly shaped planting areas.

The references also appeared in some of his garden projects. Bridge motif A, for example, can be seen in an undated design for Hans Nyffenegger in Kirchberg, a rivulet flitting over small rapids is sketched in a design for a garden for A. Gerber in Lyss, and in 1938, a widening and narrowing stream for the Bichsel bakery in Lützelflüh-Goldbach was proposed.

The fact that the Japanese garden at the school in Oeschberg was still considered a motif garden years later is documented in an article in the magazine *Schweizerisches Gartenbau-Blatt* from 1945. The photograph of a garden plot by Paul Fisch, who worked in Muttenz in the Basellandschaft, was captioned: "The Japanese garden, like the one today, is a motif garden".[30]

Baumann also treated individual areas of the G59 as motifs, which he captured photographically. Certainly, he was not unaware of the Japanese connotation that his former student Emil Steiner (1922–2018) had highlighted (see "Inspiration, imitation, integration")[31] and which, as the new editor-in-chief of the *Swiss Horticultural Journal*[32], encouraged Baumann to resume his publishing activities in 1952 after two decades of abstinence.

Baumann's interest in Japan apparently did not wane. In any case, he is said to have travelled to the IFLA Congress in Tōkyō,[33] although his name does not feature on the list of participants kept by Walter Leder, only that of his namesake Ernst Baumann.[34]

27 See a. v. 1932.
28 See Osoegawa 2011, p. 76.
29 Information provided by the landscape architect Christof Wenger on 7 March 2023, who developed the maintenance plan for the school's parks in 2000–2005 for the company Xeros, which he runs together with Pascale Akkermann.
30 t. 1945, p. 253.
31 See Steiner, E 1959, p. 789.
32 See Osoegawa 2020a, p. 17.
33 Osoegawa dates the congress to 1960, but as the IFLA event was first held in Japan in 1964, this must have been the occasion. See Osoegawa 2011, p. 76.
34 Leder 1964a ASLA, p. 5.

Ernst Baumann 1907–1992

Watercolour by Baumann of the "view of the garden from the living room", a photograph of which Tetsurō Yoshida shows in his book on page 7.

> We must give him credit for a sense of form and structure that guided the Japanese Zen garden masters. One can see an expression of something similar to what the seven characteristics of Zen art calls 'stillness in movement'.[1]

Ernst Baumann was one of Albert Baumann's many "pupils". After completing an apprenticeship as a gardener and working for Fritz Haggenmacher (1896–1924) in Winterthur and Ernst Klingelfuss (1878–1938) in Zurich, he attended the one-year course at Oeschberg Cantonal Horticultural School in Koppigen in 1926. Just two years and several study trips through Germany, Holland, France, Italy, and Spain later, he founded his own horticultural business in Thalwil in the canton of Zurich in 1928.[2]

Heini Mathys almost involuntarily identifies in Ernst Baumann's work the same catalytic function that Japan was having on Swiss garden design at that time, while writing in the same article associating him with a Zen master: "At first glance, nothing seems to have been formed in these gardens, and yet nature did not put them there. Everything about them, [...] the whole garden space has been worked through to the last detail and has become an organic whole. This sense of a whole reveals how much earlier garden literature and garden exhibitions fragmented the garden into motifs."[3]

The piece, written in 1982 and illustrated with 46 photos and plans, focussed on Baumann's fascination with stone, which Mathys praised as a veritable "language of stone".[4] As early as 1968, *Anthos* editor Alfred Kiener had "quoted from the book *IZUMO. Glimpses of Unknown Japan*" to express Ernst Baumann's sensibility for the material: "Until you can feel, and keenly feel, that stones have character, that stones have tones and values, the whole artistic meaning of a Japanese garden cannot be revealed to you."[5]

Baumann's most prominent work was undertaken together with Willi Neukom at the G59 (see "Inspiration, imitation, integration"). Alongside the famous "Perennials Garden", he created an "Ideal Garden", although this was only later noticed by the reviewer. Particularly striking was its "contrapuntal composition" resulting from the juxtaposition of lush plant life and "almost ascetically simply built rooms".[6] Baumann divided the architecture into modularly structured horizontal and vertical surfaces – floors of clinker bricks and walls of sand-lime brickwork – and used no roofs or reduced them to individual wooden beams.

A wooden frame marked the boundary between inside and outside, between the surrounding park and garden, and recalled the symbolic entrance gate, torii (鳥居), of a shrine or temple. In the Japanese context, this signals the transition from the profane to the spiritual sphere; in Baumann's "Ideal Garden", it indicated the change from "untouched" to artistically designed nature, from the "wilderness" to a paradisiacal realm.

This kind of homage is typical of Baumann's rich stock of Japanese references. In some cases, the similarity is so direct that one is inclined to confuse the original with a "replica" – as in the detail of an unspecified residential garden [→ p. 154]. The opening in the wall that offers a view of the outer garden space is remarkably similar to the entrance to the "Katsura Garden", which Tsayoshi Tamura depicted in 1964[7] and which the Swiss delegation was able to see for themselves during their visit to the IFLA Congress in the same year.[8]

His repertoire also included garden pavilions with wooden supports made of untreated logs anchored in unhewn stones, water troughs made of hollowed-out boulders modelled on Japanese hand-washing basins, and fountain covers made of bamboo canes.

Photographs of his trellises and latticework set into walls can be compared against images of the wisteria trelliswork in the Imperial Sentō Palace and of wickerwork window openings documented by Tesurō Yoshida.[9]

Baumann's affinity for working with stone, which Mathys had already commented on in the context of the G59,[10] is also apparent in his designs for private gardens. The photograph of a section of path [→ p. 158, top] would hardly be out of place in a book on Japanese gardens. Baumann heightens the effect by interspersing the irregularly laid pattern of broken slate slabs with individual off-white sections. His treatment of the gaps is also noteworthy: what is usually regarded as a necessary or unavoidable evil in our latitudes is elevated to a source of beauty by filling them with strips of brick fragments.

A converse reading is also possible: Baumann creates the impression of a path damaged by earthquake tremors that has caused the slabs to crack. His intervention is a means of repair. From this perspective, the light-coloured slabs can be seen as later insertions bridging the sections between the broken edges of the dark grey slate. Rather than concealing the flaws, they highlight them: the "new" slabs take on an aesthetic role inviting the viewer to decipher their purpose and turning the path into a stimulating mental game. The strips of brick likewise serve as a later refinement. Unlike the slabs, they are man-made and deliberately broken to size to repair the fractures.

Baumann's "repair" echoes the principle of wabi-sabi in the way it draws attention to the points of "damage" and takes pleasure in the imperfect. His treatment of the gaps evokes the technique of "golden joinery" in which fractures in Japanese ceramics were repaired using urushi (lacquer) mixed with gold dust. Cracks repaired in this way celebrate the fracture, or rather the joint and void between two broken pieces.[11]

While Baumann was familiar with the respective publications on Japanese gardens and the numerous illustrations of walkways they contain, as his designs for the G59 confirm, there is no evidence in his legacy at ASLA of any involvement with Japanese ceramics. As a participant at the 1964 IFLA congress in Tōkyō, however, he would have taken part in tea ceremonies[12], and therefore would also have drunk from raku bowls – drinking

1 Mathys 1982, p. 24.
2 See Mathys 1992.
3 Mathys 1982, p. 24.
4 See ibid.
5 Kiener 1968a, p. 3. The original quotation is from: Hearn 1894–1895, vol. 2, p. 346.
6 Cf. o. A. 1962d, [p. 14].
7 Tamura 1964, p. 338.
8 Franz Vogel showed three photographs of the "Katsura Garden", including the aforementioned wall opening. See Vogel 1968b, p. 1334.
9 Yoshida 1957, p. 57.
10 See Mathys 1959c.
11 Oskar Münsterberg had described the technique long ago, without, however, calling it by the term it is known by today: kintsugi (金継ぎ). He justified it as a virtue born out of necessity because as shard and glaze do not shrink at the same rate during firing, cracks can occur. "The Japanese have often very skilfully covered such flaked spots and veins with gold lacquer, thus giving the defective pieces a peculiar charm." Münsterberg 1904–1907, vol. 3, p. 30. Otto Kümmel emphasised the legendary background: "Special mention should still be made of [...] Fujishige Tōgan [Togen], who is said to have repaired the meibutsu chaire in the collection of Hideyoshi found under the ruins of the destroyed Osaka castle with gold lacquer and thus invented this most tasteful way of repairing clay vessels." Kümmel 1911, p. 14. The term "kintsugi" is not found in his work either, nor almost a century later in Charly Iten's detailed treatise. See Iten 2004.
12 See 1964b, p. 30; Vogel 1968a.

vessels that date back to Sen no Rikyū and are an established expression of Zen culture.

The ceramic artist Sophie Lechner-Laur offers an enlightening description of the appeal of imperfection in an article published in 1967 in the magazine *Werk*[13], which includes a vivid passage on the subject of cracks: "One specifically uses a clay that is susceptible to cracking; a crack in a tea bowl increases its individuality. [...] It is a sign of quality when traces of the tongs used to remove the raku bowls from the red-hot kiln are still visible [...]. A crack in an otherwise perfectly harmonious bowl is cherished, especially when artfully applied. Sometimes, therefore, the crack [...] is patched with gold lacquer to give it special emphasis. In Japan, everyone is familiar with the tale of a potter who, on retrieving a perfectly formed ceramic from the oven was inconsolable. He went to the crack mender and had a crack painted on it, and thereafter the ceramic became famous."[14]

Interpenetration of house and garden

Alongside the artistic accentuation of primitive qualities, Baumann was also a passionate proponent of connecting the house and garden. In a floor plan he drew for the architect Adelbert Stähli in Lachen in 1964, the garden literally flows through the house. Reflecting the client's division of the architectural space into two structures, Baumann divided the garden space into courtyard-like compartments. But these are not arranged to serve the architecture. Rather, the house takes up residence in the garden, which has primacy. It is almost as if Stähli had carefully lowered the buildings into the green space as if on a cushion of air.

If one were to spirit away the house, the garden would still be a continuous space. The architecture instead serves as a backdrop for the scenery of a dry pond that he calls a "rock garden". On one side, the pond breaks through the wall, where a few stones suggest a dry waterfall. On the narrow side, it washes out the corridor connecting the two buildings, pours into another atrium, and spreads out diffusely into the entrance area, where it again flows seemingly unhindered through the architectural lines.

If Baumann had made a coloured perspective of the view from the large window fronts, it would evoke the imagery of Tosa Mitsunobu (see "From the plenty of paradisiacal islands to the beauty of emptiness"). The "passage" of the pond is also reminiscent of the body of water in the Daisen-in [→ p. 53] that springs from a dry waterfall and also frames the architecture in an L-shape and flows beneath a crossing point. In 1961, Baumann designed a garden for the Leutert House in Zurich, built by Ernst Gisel, that diffuses into the interior in a similar way.[15] Complementing the architectural composition of rooms flowing into one another, the garden appears to flow through the house [→ p. 161, top].

From time to time, Baumann rendered perspectives that evoke illustrations of Japanese origin [→ pp. 164, 166]. And while his colleagues took photographs during the excursions accompanying the 1964 IFLA Congress in Japan, he produced watercolours of specific emblematic motifs [→ p. 152].

In addition to his talents at merging the house and garden, Baumann was similarly artful at mastering the incorporation of the surrounding landscape. It always pervaded his gardens. In some gardens, it is incorporated conceptually, as in the "Speiser Garden" in Oberrieden in 1963, where the stream and pond enter into a dialogue with his stone settings. In others it is literally borrowed to feed a pool. He sketched this with characteristic strokes in a drawing for the "Dieterle Garden" in Zollikerberg in 1966 [→ p. 160]. The same drawing shows Baumann's self-assured handing of miniature landscapes captured in atria. The same themes can be found in his plans of the gardens "Braendlin-Blumer" in Jona, 1958; "Staub" in Thalwil, 1969; "Fuchs und Moos" in Illnau, 1969; as well as the schools in Oberrieden, 1963; Thayngen, 1967; Zurich, 1968; Schlieren, 1969; Greifensee, 1970; Matzingen, 1971; the vocational training centre in Weinfelden, 1977; and the SUVA in Lucerne, 1982.

13 Sophie Lechner-Laur trained in Japanese pottery during a stay in Japan. She is the wife of the architect Max Lechner, who, together with Rolf Keller, Walter Moser, and the couple Esther and Rudolf Guyer, designed the Seldwyla housing estate (see "Willi Neukom – Japan as a basso continuo"). From 1961–1963 Lechner was involved together with the office of Guhl + Lechner + Philipp and Yoshitaka Akui in the construction of the cathedral in Tōkyō designed by Kenzō Tange (see Lechner 1963). The Lechners hosted Akui during his stay in Zurich when the Japanese delegation visited in 1972. Detailed in a handwritten note by Leder, Walter, no signature.
14 Lechner-Laur 1967, pp. 347–348.
15 See anon. 1962e; anon. 1965a; Inventory Supplement 2013, pp. 12–13.

← The wall opening bears remarkable similarity to that of the entrance to the garden of the Katsura residence.

→ Baumann emulated Japanese examples of trellises and climbing frames in the “small dwelling garden”.

↓ The barely treated wooden post resting on a similarly unhewn stone in the “small dwelling garden” is reminiscent of the tokobashira (床柱, pillar) of a traditional Japanese tokonoma (床の間, picture alcove).

← The inclusion of the imposing backdrop of the "Rossel Garden" and its rock garden are clear indications of its inspiration.

↑ The boundaries between inside and outside in the Rossel House in Baar are fluid. The lack of windows gives the space the character of an arbour, while the curtains denote it as a living space.

→ The framed landscape is incorporated into the "Rossel Garden".

→ The irregular paving slabs with brick infill in the gaps echoes the principle of artfully repairing – and refining – the shards of a cracked Japanese tea bowl.

↓ A water basin carved out of a boulder reminiscent of a Japanese model.

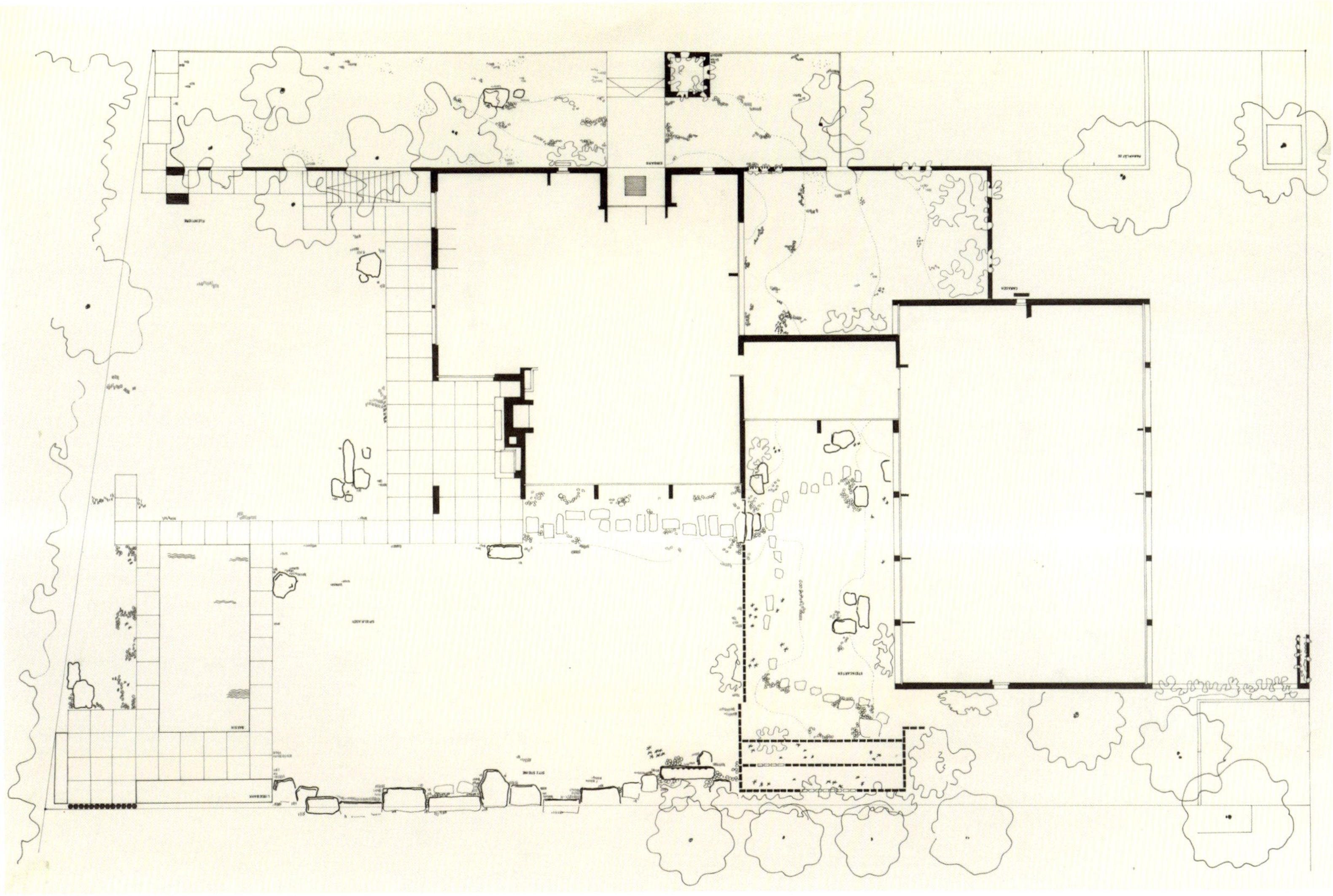

Bridged by stepping stones and accentuated with stone settings, the dry pond in the long rectangular courtyard of the "Stähli Garden" in Lachen flows through the entire house.

1966 sketch of the courtyard of the "Dieterle Garden" showing distinctly Japanese features. Baumann pulls in the landscape outside the boundary using a small river and pond to feed a stream and a basin. Stone settings and a distinctive solitary shrub emphasise its East Asian aura.

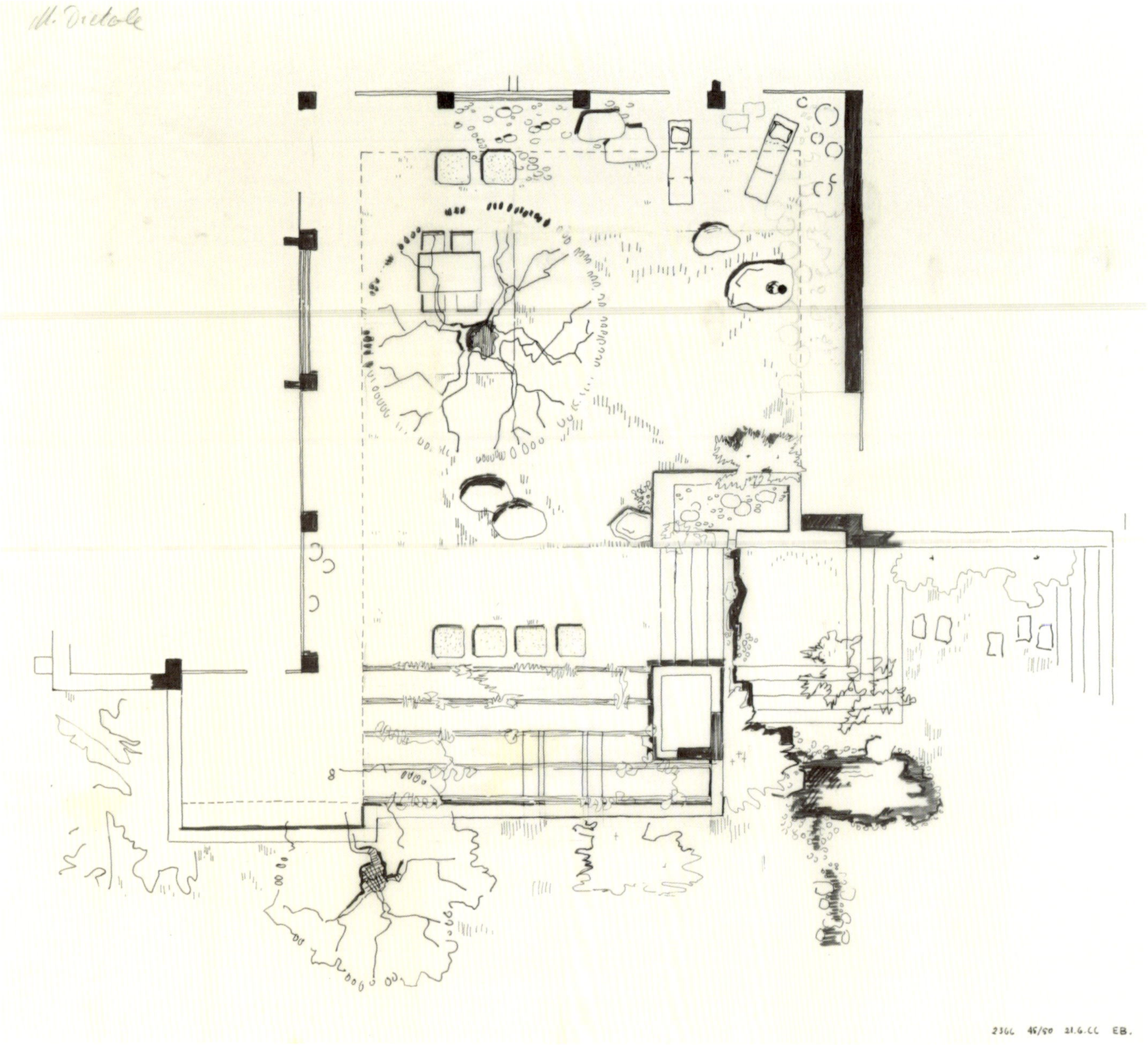

↑ The permeability of the Leutert House is matched by that of the garden, which passes through precisely placed openings. Sections of Japanese-influenced rock garden are the connecting link. While Baumann used dense plantings of hornbeam, field maple, hazel, hawthorn, catkin willow, and yew at the boundaries of the property, he placed ornamental cherry trees and lilacs as solitary trees near the house in the Japanese style.

→ The garden detail is just as compelling as a Japanese one – such as the one shown on page 128 of Tetsurō Yoshida's book from 1957.

↑ The watercolour of a dwelling garden shows a wall opening similar to the one that marks the threshold to the Gepparō (月波楼) "Moonwave Teahouse" in the Katsura residence – likewise accentuated by stone steps.

→ For the Sihlweid school building in Zurich-Leimbach, Baumann sketched a water landscape with a stream, pond, and rough stone blocks.

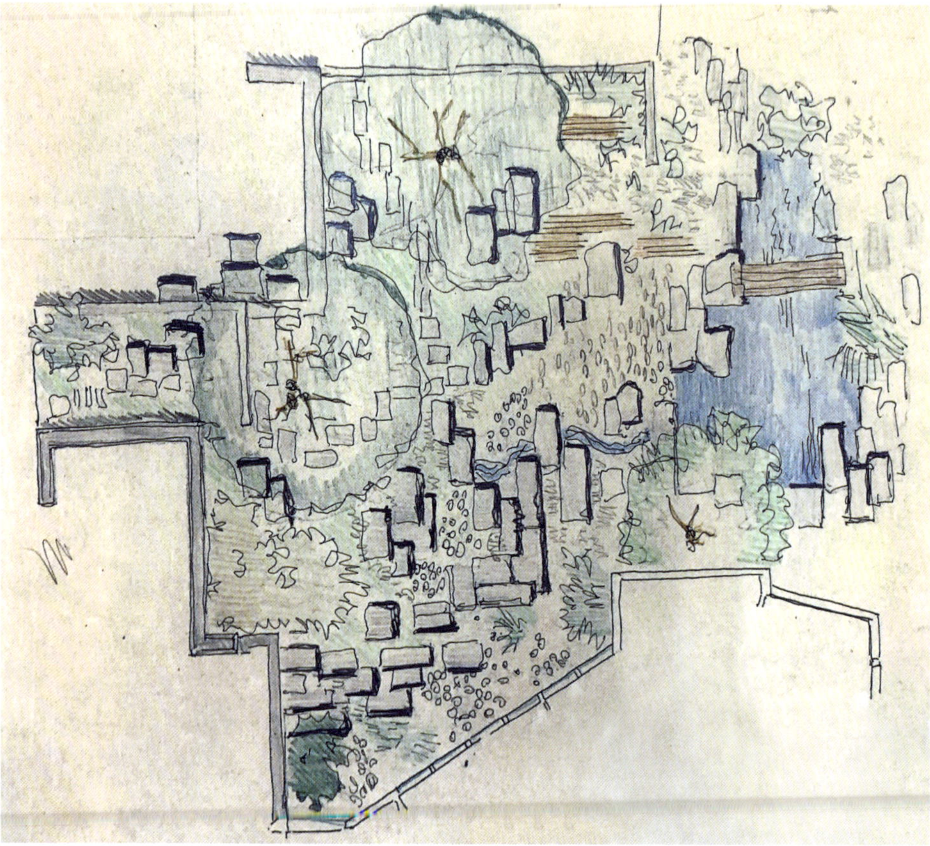

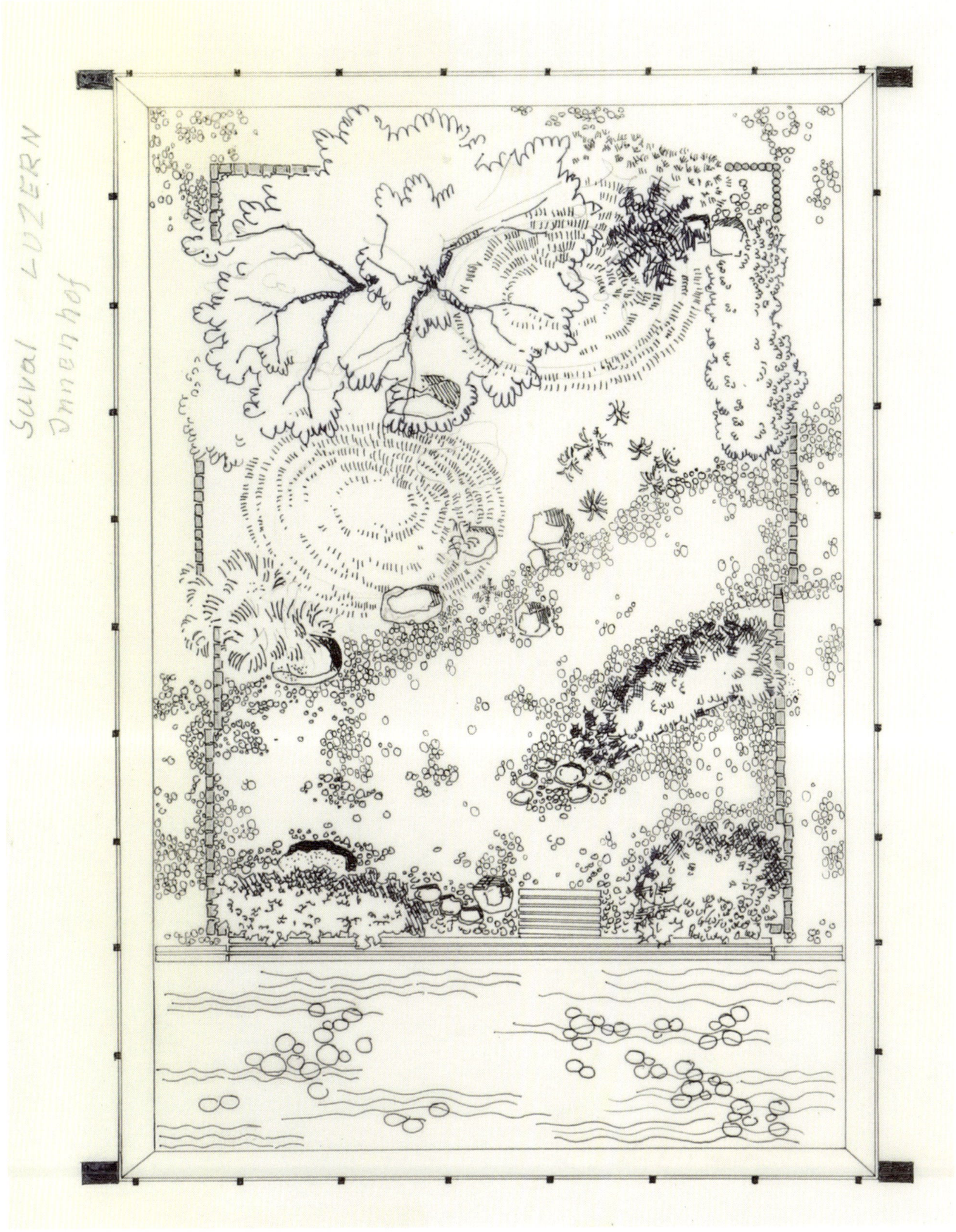

For the insurance company Suva in Lucerne, Baumann designed a distinctive miniature landscape in which a “stream of rocks” flows from the park area to the west through the building into the inner courtyard and from there towards the east.

Ernst Cramer 1898–1980

The inner courtyard in the Bernarda Seminary in Menzingen (now a cantonal school), designed by Cramer between 1955 and 1958. The tree in Cramer's drawing appears almost to dance in its naturalistic twisting form.

From the raised terrace with bamboo decking, the Japanese devote themselves to attentively contemplating the wonders of nature. In the background, a high concrete wall shows that this garden lies within the sphere of influence of Western technology.[1]

Gustav Ammann may have told Ernst Cramer about his visits to the "Jubilee Exhibition – International Art and Great Horticultural Exhibition" in Mannheim in 1907 and the Japan-British Exhibition in London in 1910 during his gardener training at Otto Froebels Erben in 1914–1917. During his wanderings in 1918–1922 through Switzerland, Germany, and France, he worked, among others, with Henry Correvon (1854–1939)[2], the "godfather of the alpine garden"[3], which became a "stepping stone" for Japanese-inspired implants of stone settings and waterfalls.

At the time when Cramer attended Oeschberg Cantonal Horticultural School in 1922–1923, it is doubtful that Albert Baumann's interest in Japan had flourished to the point that it had featured in the curriculum. But during the time of his employment in horticultural businesses in Winterthur and the municipal gardening department of Schaffhausen 1924–1929, Ammann's design for the Japanese-inspired "Eugster Garden" in Altstätten 1928 can hardly have escaped Cramer's attention. Twenty years later, after opening a nursery in Suhr in 1938, his garden design office in Zurich in 1942, and joining forces with Ernst Surbeck, he himself realised a garden whose Japanese aura glowed even more intensely.

He created it for Wolfgang E. Hegner, co-owner of the import-export company founded in 1865 in Yokohama as Siber & Brennwald, trading from 1900 as Siber & Wolff, and renamed Siber, Hegner & Co. in 1910, which traded both in watches and silk fabrics. Hegner had a retreat built in Feldmeilen on the shores of Lake Zurich by the architect André E. Bosshard, which was modelled on traditional Japanese buildings. It comprised a main building that was used as a summer house, and an annex reminiscent of a teahouse, which was intended for use in the winter months. The client must have commissioned the office of Cramer & Surbeck to create a Japanese-infused garden, which they did in collaboration with Willi Neukom.[4] Hegner found in them a partner with the requisite sensibilities both in theory – through reading the relevant publications – and in practice – through the garden at the Züka in 1947 (see "Inspiration, imitation, integration"). Cramer's acquaintance with Max Bill may have contributed to this. Bill's sculpture "Endless Loop" must have fascinated Cramer, as he kept numerous photographs of it – more than of any other object, including his own gardens. Bill, in turn, had an affinity for Japan: he visited the *Exposition internationale des arts décoratifs et industriels modernes* in Paris in 1925[5], which included a Japanese pavilion, and designed the poster for the 1933 exhibition "Japanese Architecture and Gardens" at the Kunstgewerbemuseum (Museum of the Decorative Arts) in Zurich[6], as well as an entirely Japanese-inspired "Residence and Dental Practice in the Countryside"[7] in 1943. In 1955, he also reviewed the second edition of Tetsurō Yoshida's book *Das japanische Wohnhaus*, published the year before.[8]

In contrast to the comparatively ascetic conception at the Züka, the garden in Feldmeilen had a romantic touch. It had an organically shaped, two-part water basin, crossed at its narrowest point by broken stone slabs and surrounded "by summer ragwort (*Ligularia clivorum japonica*), pampas grass (*Cortaderia selloana*), and bulrush (*Typha latifolia*)" with stone settings along the shore zone.[9] A stone lantern near the water's edge was flanked by creeping juniper (*Juniperus horizontalis*) and red Japanese maple (*Acer palmatum* 'Dissectum Atropurpureum'). Other East Asian plants included Chinese silver grass (*Eulalia japonica*) and Chinese boxthorn (*Lycium chinense*).[10]

Cramer, who was always motivated to creating something "different to what he had already made" and was exceptionally "driven by a lifelong restlessness and self-destructive self-criticism"[11], was in this case still convinced of his creation twelve years later, even after having caused a sensation with his geometrically abstract "Poet's Garden" for the G59. In 1960, he had the garden portrayed in the magazine *Das ideale Heim* under the title "Influences of Japanese Garden Culture", and its publication was met with superlatives. "Through the delicate balancing of the elements of stone, plants and terrain", Cramer had "demonstrated his artistic finesse", along with a "deep sensitivity to foreign garden forms" [12], so much so that even "the Japanese" would be impressed by the design of the garden.[13]

Although the property has since been altered, the original impression of the garden remains. After Hegner sold the property, the new owner divided it into two plots, keeping the former winter quarters and selling the former summer house. He drained the pond on his plot, but its outline remained, as did parts of the planting. His neighbour, by contrast, cherished the Japanese atmosphere, which stems from the pines, bamboo, blue vine, and several maples – especially a red cut-leaved Japanese maple. He again created a wetland area as a paradisiacal landscape *en miniature*. It features a two-part biotope divided by a small waterfall. A slightly curved stone slab functions as a miniature arched bridge. Water lilies accentuate the Japanese flair. A path of irregularly broken slabs leads to the garden gate, which dates back to Hegner's time, who brought it with him from Japan. In the entrance area, bamboo stands tall, with a stone lantern ducking at its base.

Gardens as images

That Cramer's designs bore an affinity with Japanese models was implied in an article in the magazine *Das ideale Heim*. Preceding the article and photographs of the Hegner Garden were pictures from Japanese gardens, accompanied by an introductory text describing

1 M-r 1960, p. 373.
2 See Weilacher 2001, p. 16f.
3 Nerfin 2016, p. 54.
4 In the years 1942–1950, when Surbeck worked with Ernst Cramer, the two traded as Cramer & Surbeck. Willi Neukom, who worked for Cramer from 1939–1951, was listed as an employee. See Weilacher 2001, p. 39; Rotzler 1950, pp. 69, 70. Rotzler, who shows some of Cramer & Surbeck's gardens, prefaced his article with a photograph of a Japanese garden epitomising the connection between house and green space.
5 Cf. Koella 2008, p. 29; Thomas 2008, p. 89. The Japanese exhibition building at the *Exposition internationale des arts décoratifs et industriels modernes* was located on the Cours la Reine near the Pont Alexandre III. It was designed by the architects Shichigoro Yamada and Iwakichi Miyamoto. All materials for the house, which was made of bamboo and equipped with sliding partitions and screens, were of Japanese origin and imported especially for the show. See Fecht 1995.
6 Berehndt 1933.
7 Cf. Hartmann Schweizer, p. 121.
8 Bill 1955.
9 M-r 1960, p. 373.
10 M-r 1960, p. 373.
11 Rotzler/Barth 1987b, p. 8.
12 M-r 1960, p. 374.
13 M-r 1960, p. 376.

the Japanese garden as a "work of art" and a "scenic image".[14] Speaking to Stefan Rotzler in 1984, Hans Jakob Barth, who had worked with Cramer from 1946–1951, testified to Cramer's singular skill in creating images.[15]

Barth also recalls his former teacher's particular liking for woody plants with an eccentric character. He was more inspired by their form than "the colour of the blossoms, the flowering time or leaves".[16] "If a tree moved in a certain way, rose out of the ground crooked and then twisted, then it would interest him."[17] While Cramer's ground plans showed curved basins and winding pathways, which Barth called "perpendiculars"[18], the woody plants would rotate in dancing formations. One has an impression of his partiality for "interesting tree forms" in the swinging and interwoven treetops depicted in a garden plan drawn in 1952 for an eight-family house in Aarau.[19]

Flowing layouts characterised Cramer's designs until the early 1950s, and where they had Japanese facets, he interpreted them organically. His pond outlines were circular, sickle-shaped, bubble-shaped, kidney-shaped, and amoeba-like, his stepping stones strung together in curved lines, his boulders placed as if at random to make their constellations appear true to nature. The plans of the houses for Bechtler in Freienbach, 1947; Rubli Mattenberger in Aarau, 1947; Vock in Wohlen, 1947; Fehlmann in Aarau, 1948; Härdy in Oberentfelden, 1950; Ingold in Herzogenbuchsee; Frey in Olten, 1953, as well as the Rüti School, 1948; and the undated projects for Ammann in Zurich, Angst in Herrliberg, Baumann in Lenzburg, Brändli in Zurich, Fehlmann in Schöftland, Käser in Aarau, Klameth in Küsnacht, Regis in Zurich, Scherrer in Schönenwerd, and the Parkweg Building in Buchs, as well as a project in an unspecified location for an anonymous client, all follow this pattern.[20]

Crystalline clarity and natural splendour

In the mid-1950s, Cramer differentiated between structural and plant components and flirted with "material appropriateness" – an emblematic and also problematic term for both Neues Bauen and the reception of Japan in the West. While Alfred Roth defined this as being characteristic of contemporary architecture and Max Bill as an aspect of "beauty out of function and as function", Gaudenz Domenig criticised the "Japanese treatment of materials" as being "by no means always 'materially appropriate' in the modern-Western sense".[21]

Cramer called for an appropriate material treatment of man-made elements. "They should reveal themselves with crystalline clarity and not, through false romanticism, be made to fade into nature". Plant material, on the other hand, "should be allowed to reveal the splendour of their natural growth form".[22]

His drawings for the "Meister Garden" in Zollikon, designed in 1966–1967, visualise this transition in an almost transitory manner. On 8 August 1966, Cramer sketched a planted water basin in front of the conservatory with a sinuous outline and rectangular, irregularly staggered stepping stones; two days later, he constrained it in a rectangular form. The alpine garden, which emerges slightly raised beyond the lawn appears, on the other hand, like a reminiscence of his early period and a nod to the once popular association with Japanese design.

The naturalistic aspect, which – as Barth recalled in 1984 – characterised the inherent beauty of his groups of trees, fascinated the restless designer, as did stones "in their natural, untouched form" as seen "most expressively" in the Japanese garden. There, "in the sand garden of Kyōto", their "contemplative symbolism" can be felt most impressively.[23]

These were the words Cramer used in a lecture in 1972 – three years after giving the layout of Hugo Mann's Park in Baden-Baden a distinctly Japanese character. At its core was a "teahouse", a magnolia, and a "Japanese garden" in the atrium of the house, featuring "exquisite plant forms" [→ pp. 182, 183].

After participating in the excursions accompanying the 1964 congress of the International Foundation of Landscape Architects (IFLA) in Tōkyō, he gained first-hand experience of Japanese gardens in their country of origin. He documented the trip in detail, and his estate includes numerous slides showing temples reflected in lakes and residences such as the Kinkaku-ji and the Katsura Mansion. They range from studies of the twisted trunks of pine trees and stone islands from which spindly plants grew, to details of elements such as water basins and stone lanterns. He was especially fascinated by the stones in the gravel bed of the Ryōan-ji and miniature gardens in atriums.

Cramer had previously designed a miniature garden in this ilk for the "Schmidlin Garden" in 1961. The client and architect Josef Schmidlin had already entrusted him with a garden design in 1954 for the Amsler-Merz House in Aarau. There, Cramer sketched a stream whose zigzag course was a product of the stones set in its path. It flowed into a pond with an angular form that responded to the geometry of the house, which also turned an angle.

He took a similar approach for Schmidlin's cubic single-family house. In contrast to the earlier project, where the garden was positioned a distance away from the house so that today it lies on the neighbouring plot, here the house and garden are interlinked. The approach echoed Schmidlin's own preferences, who was influenced by Neues Bauen and especially by Richard Neutra (1892–1970).[24] Cramer's client, however, tended to define the garden, like the house, geometrically[25], and in this respect deviated from the model.

Neutra, for his part, tended towards biomorphic garden layouts – influenced by Japan, whose culture, architecture, and garden concept he had been fascinated by since a trip to the country in 1930.[26] Three decades later, he reflected: "Everything was so unbelievably different from my own background, and yet so close to my feelings of treating space and nature. [...] I had been

14 M-r 1960, as note 1, p. 372.
15 Rotzler/Barth 1987a, p. 8.
16 Rotzler/Barth 1987a, p. 6.
17 Rotzler/Barth 1987a, p. 8.
18 Ibid.
19 ASLA, CRE_6870_1-0002.
20 The style of the drawing and/or the joint authorship of Cramer & Surbeck allows us to date its creation to the beginning of the 1950s at the latest.
21 See Roth A 1938; Bill 1949; Domenig 1965, p. 5.
22 Cramer 1987 [1955].
23 Cramer 1972, p. 106.
24 Weilacher 2001, p. 137.
25 Weilacher 2001, ibid. (p. 137)
26 Neutra had been invited to Japan by the architect Kameki Tsuchiura. See Helfrich/Raymond/Whitaker 2006, p. 269.

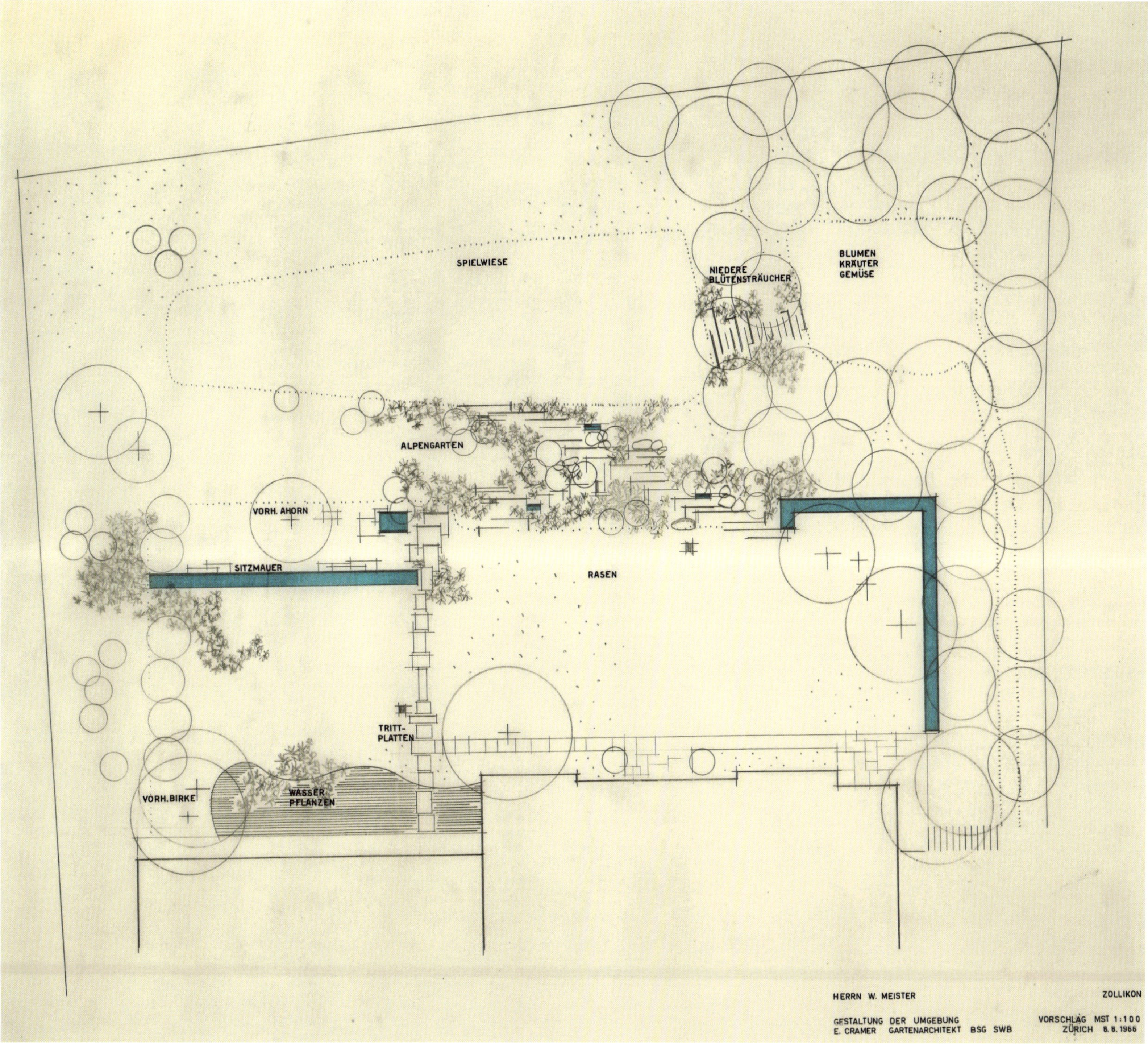

"Meister Garden", Zollikon: The pool that in this drawing still has Japanese connotations gave way to a swimming pool in the final plan from 1967.

↑ Conservatory for Haus Frey in Wangen, 1953: The water basin with an hourglass-shaped floor plan straddles the edge of the conservatory, blurring the relationship between inside and outside in a similar manner to Richard Neutra's Nesbitt House.

→ An early example of Japanese inspiration in the garden for Rudolf Siegrist in Aarau, 1937.

27 Neutra 1962, pp. 259–260.
28 See Lods 1947, pp. [315, 316]. A more comprehensive overview can be seen in the book *Richard Neutra – Buildings and Projects*, published in 1951; see Boesiger 1951.
29 Cramer 1946, p. 88.
30 "In 1920 Ernst Freud [...] invited him to Berlin, where he had now founded an architectural office. Ernst Freud designed the buildings and Richard Neutra planned the gardens." Klemmer 1993, p. 76.
31 The Shinji-ike design had been described previously in a guidebook from the 19th century. See Murray/Chamberlain/Mason 1891, p. 89. Jirō Harada also noted it. See Harada 1928, p. 27. While Cramer worked on the project for Spirig, Heinrich Dumoulin's seminal work was published, which also cited this phenomenon. Dumoulin was a religious scholar specialising in Zen Buddhism, and his work is still referred to today. See Dumoulin 1959, p. 190.
32 Adam 2019, p. 107.
33 Anon. Chronological review of the activities of the Roman Catholic Religious Association Dulliken-Starrkirch; https://niederamtsued.ch/assets/manager/dulliken/Gruppierungen/Kultusverein/chrono.%20R%C3%BCckblick%20Kultusverein%20Dulliken.pdf (retrieved 15 April 2023).
34 Braun 1954. Two colour and one black-and-white photograph of the Moore Villa – albeit from different perspectives – are reproduced in: Lamprecht 2000, pp. 252–253.
35 Lamprecht 2000, p. 38.
36 Mathys 1978, p. 32.
37 Weilacher argues that Cramer did not "deal intensively with the theoretical foundations of modern art" and describes his approach to Mondrian as a "purely formal interest". 2001, p. 123. The fact that Cramer only chose the colours red and blue for individual glass infills, but not yellow, could indicate that he was particularly inspired by Mondrian's "Composition en blanc, rouge, bleu", which had been pictured in the magazine *Werk* in 1944. See Meyer-Benteli 1944, n. pag. [p. 133].
38 For further details on the varied relationships between Mondrian and Japan, see Johnson/Hart 2019, p. 31; Hartmann Schweizer 2013, pp. 76–82.
39 Cramer 1964, p. 369.

striving for all that [...]."[27] An article by the French architect Marcel Lods in the magazine *Werk* in 1947 offered a compelling explanation of how nature permeated Neutra's buildings. It showed, among other things, the 1943 Nesbitt house in Brentwood, California, using photographs of two circular pools in front of the entrance, adorned with a Japanese white pine, and a curved water lily pool that passed through the threshold to the living space, effectively dissolving the boundary between inside and outside.[28] Cramer also sought to "[merge] house and garden in a harmonious, uncontrived whole of the kind appropriate to modern-day sensibilities". It should "be structured and reveal itself with sensitivity and openness" and take into account the "inherent qualities and innate diversity" of nature.[29] The fact that Neutra had learned the use of plants from Gustav Ammann (see "Gustav Ammann – Japan as a module") and later also designed gardens himself cemented his credibility as a reference for landscape architects.[30] Accordingly, some of Cramer's projects seem to reflect Neutra's approach [→ p. 168, top].

One such project was a park for the pharmaceutical entrepreneur Hugo Spirig in Will near Olten, which partially overlapped with Cramer's commission for Schmidlin from 1958–1964. The villa, which occupied barely a tenth of the plot, was based on a three-wing layout – a variation on Neutra's windmill wing arrangement. Its principal element was a 960 m² pond that partially surrounded the villa [→ p. 176]. It had a distinctive heart-shaped contour that involuntarily recalls the once popular adaptation of pond shapes to the Chinese character for "heart" in the Japanese tradition.[31]

Another design in this vein that still exists today in Dulliken, Solothurn, was designed by Cramer for the architect Aldo Prina. Like Schmidlin, Prina was also an important local figure. Belonging to the "Solothurn School" of architects associated with Alfons Barth, Fritz Haller, Max Schlup, and Hans Zaugg,[32] he gained recognition together with Nino Gervasoni for the design of the Church of St. Wendelin in Dulliken-Starrkirch 1969–1972.[33]

In contrast to the Aarau example, where the geometric disposition of the building defines the garden, the architect Aldo Prina envisioned a fusion of house and garden for his own home in 1965 that largely echoed Neutra's conception [→ pp. 74, 75, 177]. For the house, a conglomeration of three cubes, Cramer designed a landscape in the Japanese-style dominated by an extensive pond with an amoeba-like outline, which he furnished with stone islands and water and marsh plants. To this day, the water extends right up to the house as it did in Neutra's 1950–1952 villa built for James D. and Orline Moore in Ojai, California. In an article on Richard Neutra in 1954, the *Tages-Anzeiger für Stadt und Kanton Zürich* showed a photograph that bears a striking resemblance to later photographs of Dulliken.[34]

Neutra correlated his Japanese-infused creative sensibilities with Piet Mondrian's compositional method.[35] Heini Mathys, writing about Cramer, ties together like strands using a similar analogy: "Compared, for example, with the painter Piet Mondrian", he had "his romantic beginnings [...] in which he created 'painterly' gardens", "and like Mondrian in turn, he gradually shed all that was merely naturalistic and advanced to a strong and clear form".[36]

Cramer not only occasionally employed the primary colours red, blue, and yellow, for example, in his 1964–1965 design for the Kantonsschule Wetzikon – but also borrowed compositional tricks from Mondrian, as, for example, in 1963 at the Frawa AG house in Zollikon [→ p. 184]. He flirted most intensively with Mondrian's compositional approach in a 1958–1962 design for Blözen cemetery in Pratteln, the entrance gate of which he subdivided in the manner of a Mondrian.[37] Not long after, during his trip to Japan, he became fascinated by the artful divisions of Japanese windows, and photographed through the filter of windows, effectively segmenting the view of the garden into a set of framed vignettes.[38]

An article on Blözen in the magazine *Garten und Landschaft* shows a photograph of Buddhist saints in a Japanese cemetery – could Cramer have had a hand in this choice of motif?[39]

↑ The water basin in the Hegner Garden is bridged at its narrowest point by two broken stone slabs and lined with leopard plant, pampas grass, and bulrushes.

↑ The gate Hegner imported from Japan still exists today.

→ Hegner also gave the buildings a Japanese appearance with infill panels reminiscent of paper shōji screens.

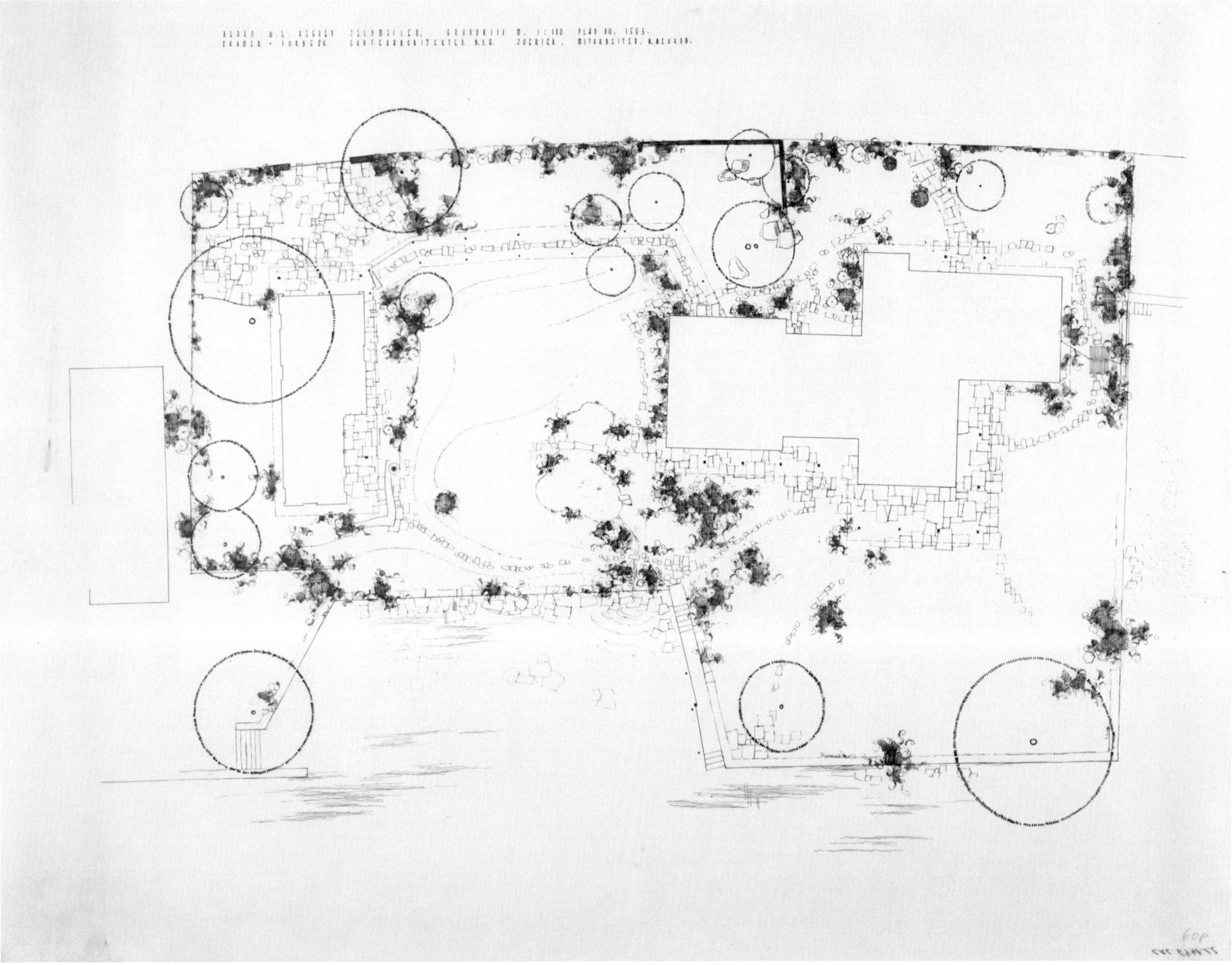

Plan of the garden for Wolfgang E. Hegner in Feldmeilen.

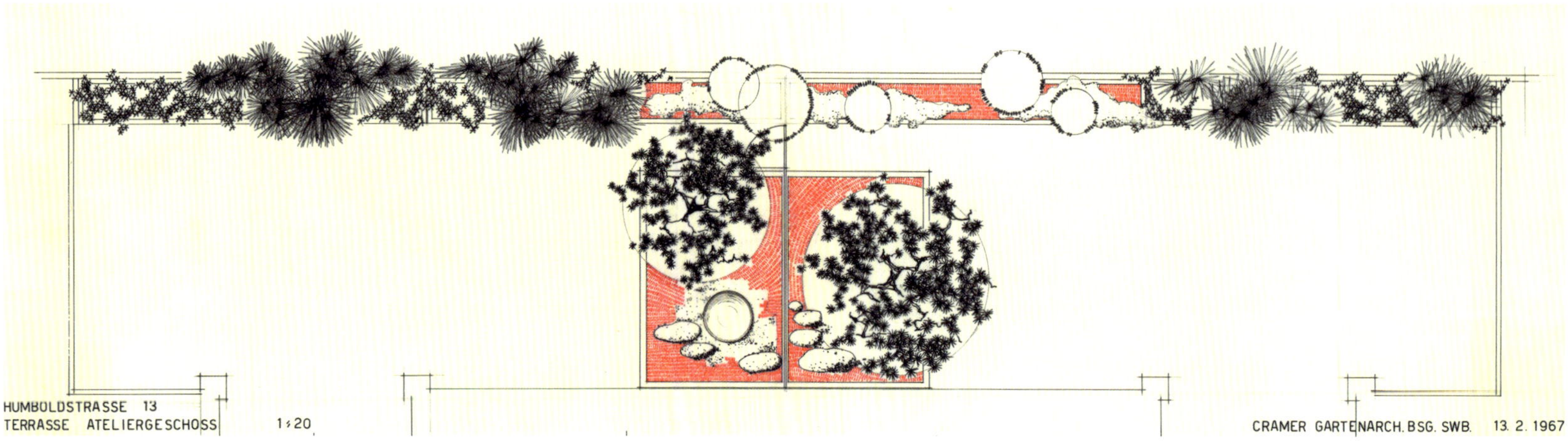

Sketch of a roof garden on Humboldtstrasse in Bern, 1967.

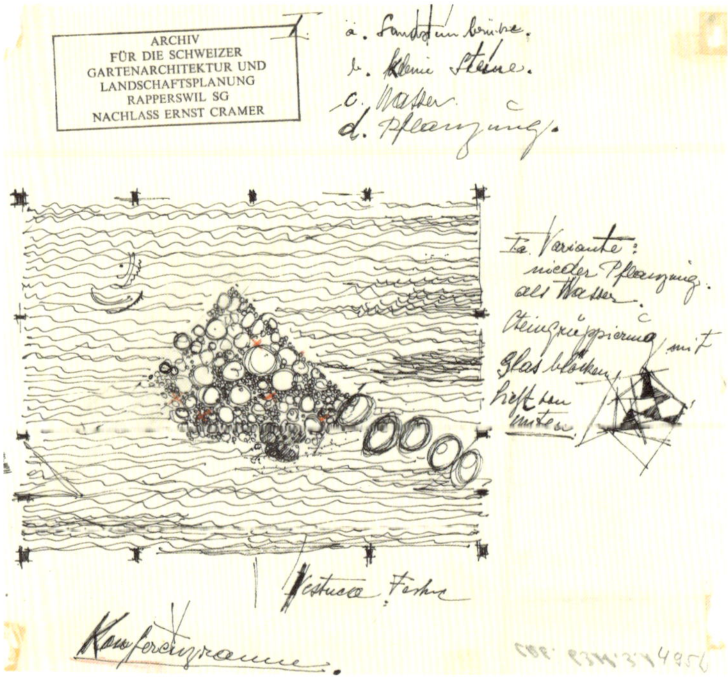

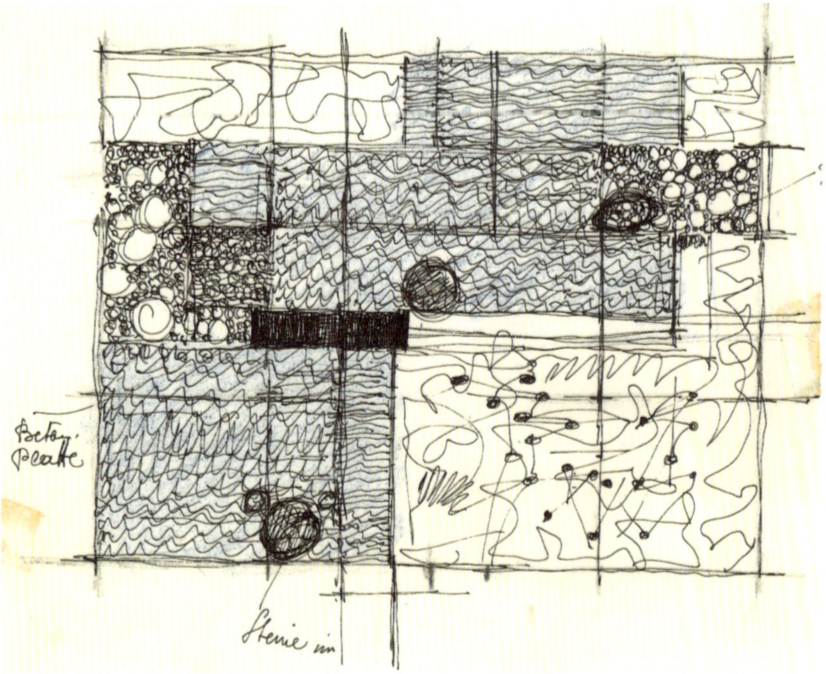

Plan and photograph of one of the two inner courtyards of the Wartmann & Cie. headquarters in Brugg, 1962, exhibiting a geometric adaptation of Japanese garden elements. In an accompanying sketch Cramer labelled these "sandstone bridge", "small stones", "water", and "planting".

Photograph of a detail of an unspecified
Japanese-style garden detail.

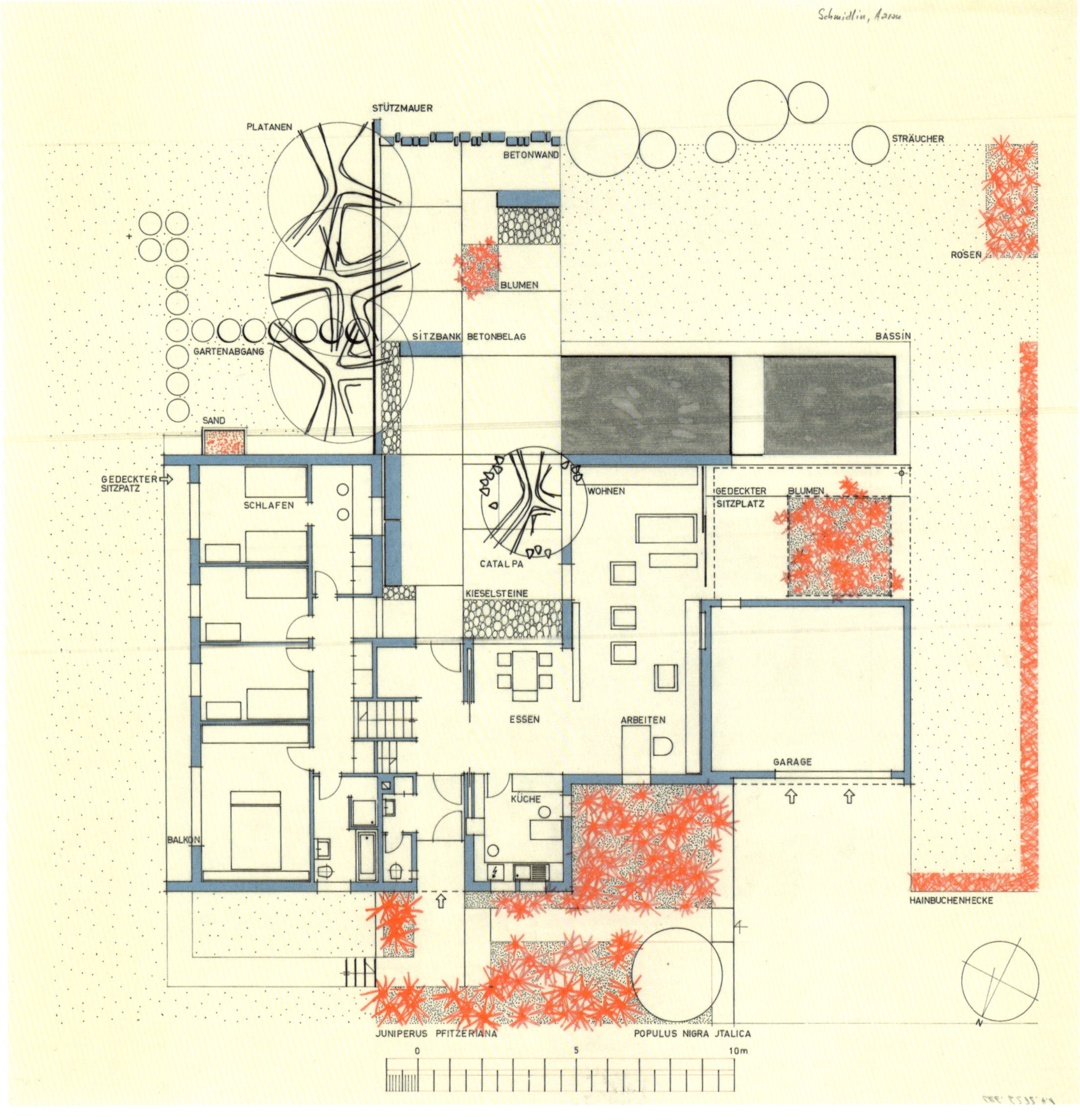

The *Catalpa* is the central point of the Schmidlin house around which the garden hinges.

→ The relationship between inside and outside in the Schmidlin House also extends to the design of the floor, which continues the pebble motif on the inside.

↓ The frame structure of the atrium of the Schmidlin House from 1961 looks like a movable canopy, as if the interior could be pushed outwards during the day and retracted again in the evening.

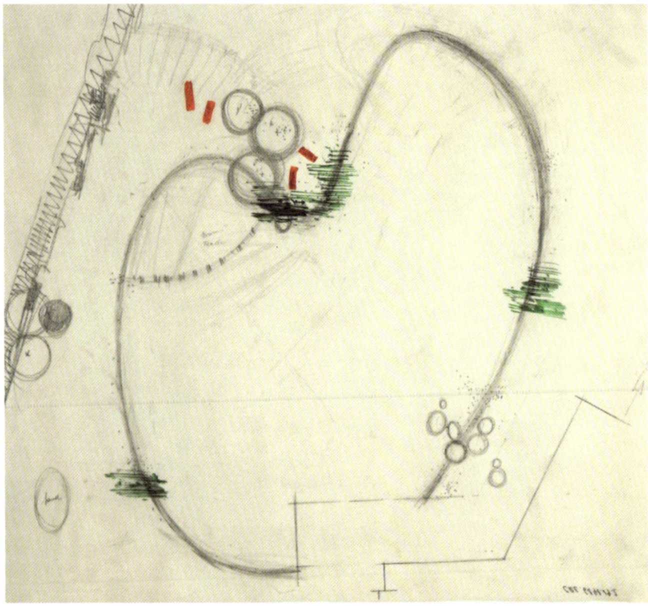

↑ Two sketches of the pond in the "Spirig Garden" with a heart-shaped ground plan – echoing the Japanese custom of shaping bodies of water according to the Chinese character for "heart".

↓ The only surviving photograph of the expansive, navigable pond in the "Spirig Garden" with a possible depth of 70–80 cm – that can therefore be read as a chisen shuyu-shiki-teien (池泉周遊式庭園, pond and boating garden) – dates from 1975.

↑ A photograph of the Prina house from soon after its completion makes it appear to be floating in the pond.

→ In the plan for the Prina House in Dulliken, the biomorphic shape of the pond contrasts with the square ground plan of the house.

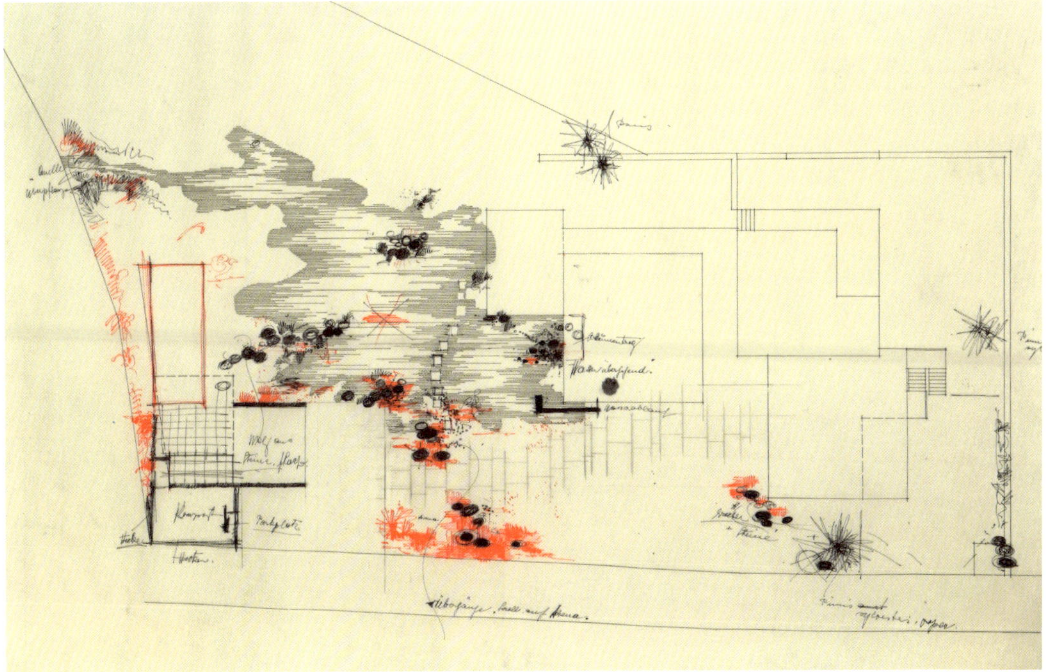

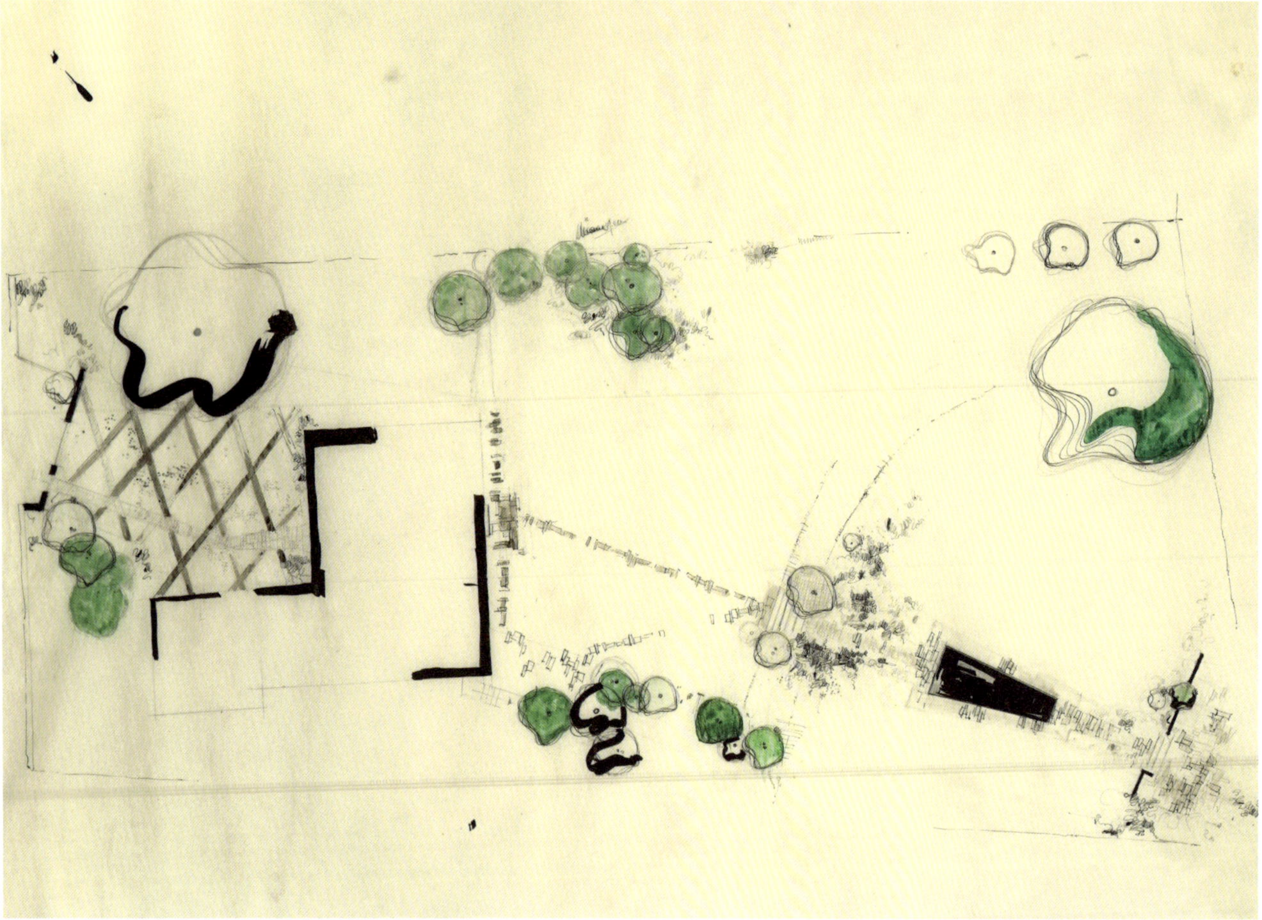

↑ A sketch of the garden of the Angst House in Herrliberg showing sections drawn with an ink brush in the Japanese manner.

→ Japanese stylistic devices are evident in the depiction of the layout of the Gränichen cemetery in the mid-1960s.

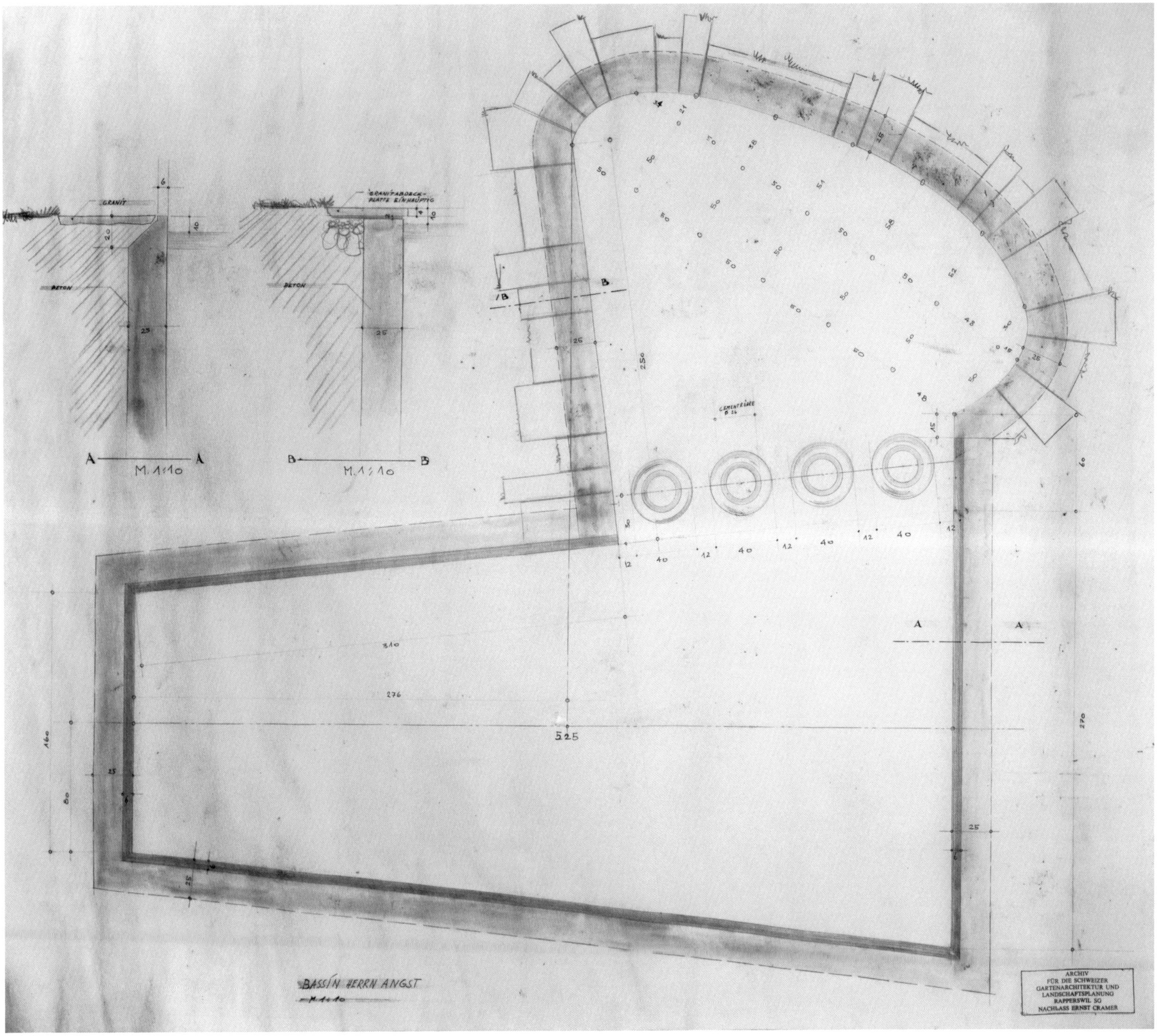

The pool in the Angst Garden is traversed by iconic circular stepping stones.

Sketches for the park design of Meienberg Castle in Jona (St. Gallen), 15 November 1976: Cramer lined up the elements – including an oval inscribed with the yin-yang symbol – as if on a Japanese picture scroll (emakimono).

The undated plan of an unspecified design in an unidentified location shows an unreservedly Japanese garden including a sprawling pond, island, the suggestion of a delta as in the "Katsura Garden" and stone arrangements.

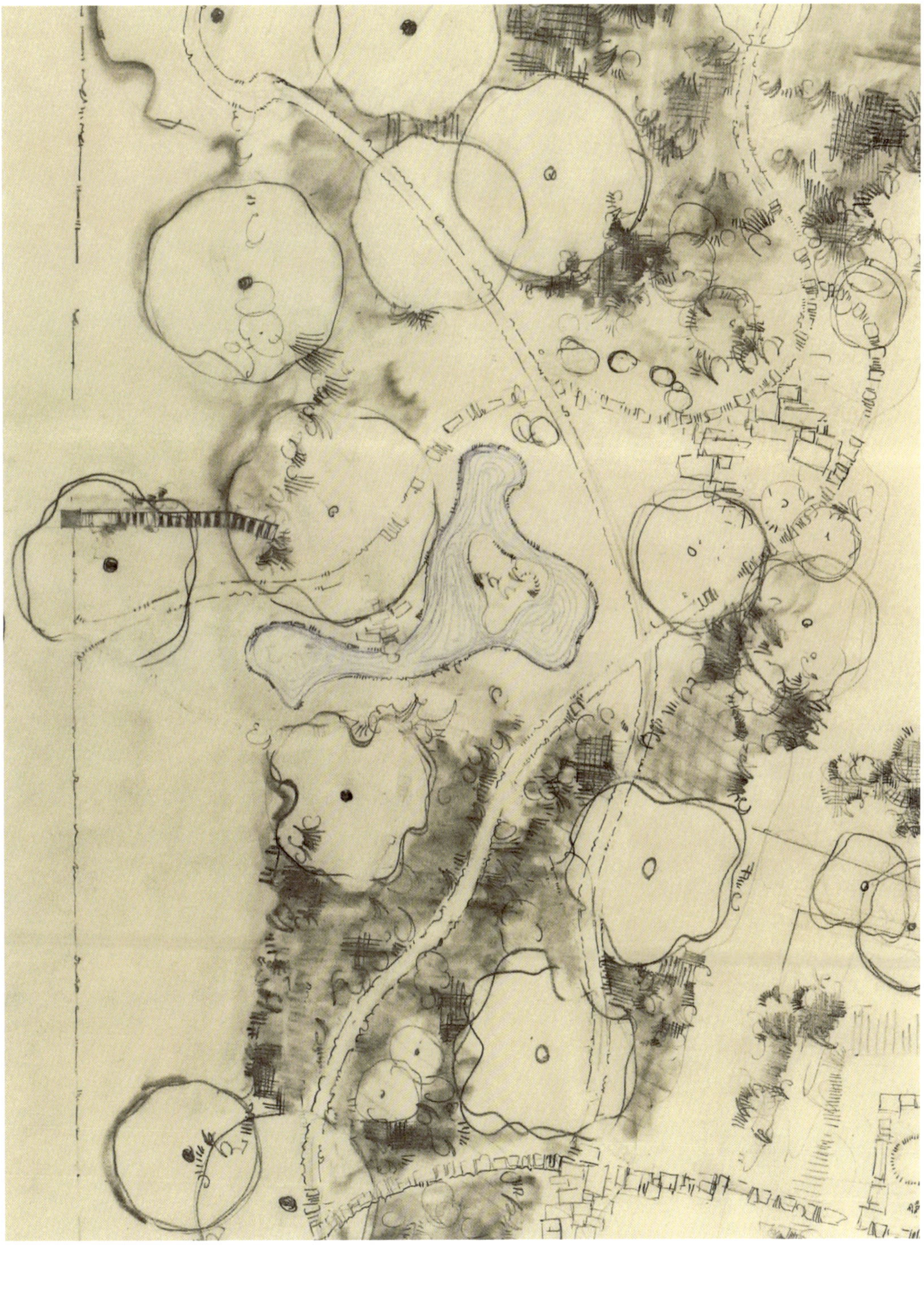

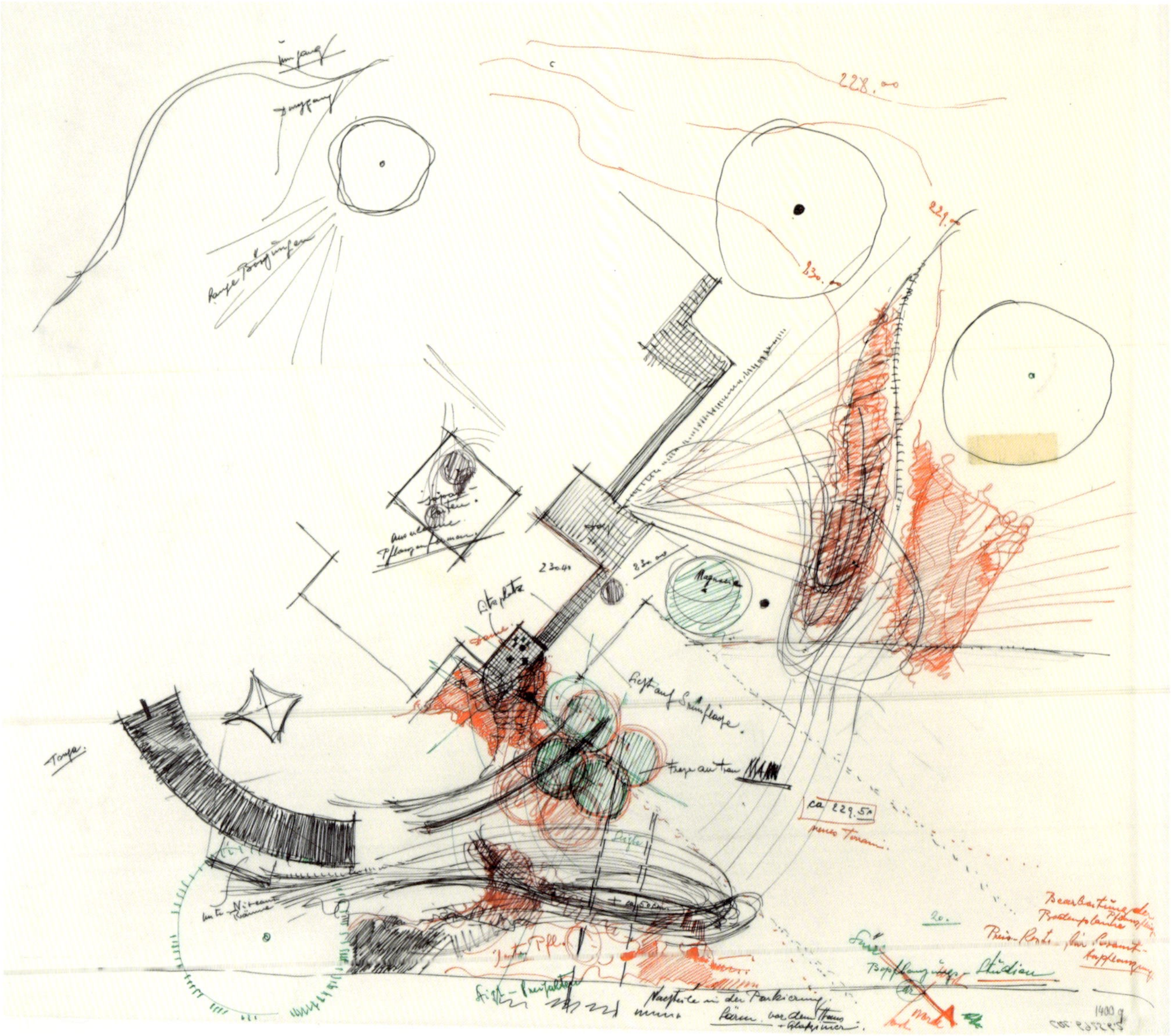

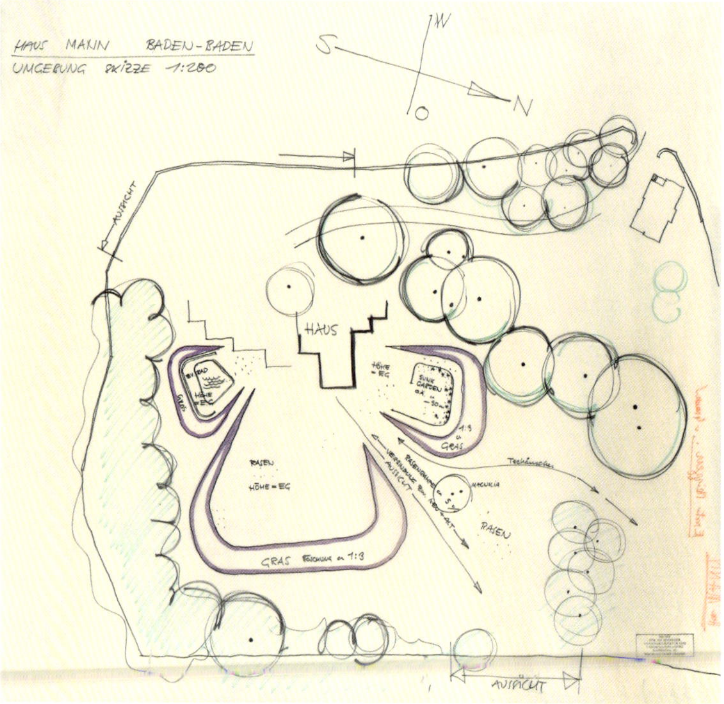

↑ Sketch of a park design for Hugo Mann in Baden-Baden, in the centre of which is an atrium with Japanese Garden and a magnolia as a solitary plant.

→ Alongside a sunken garden, Cramer also envisaged a teahouse in the park for Hugo Mann.

The definitive plan for Hugo Mann's estate includes all the Japanese elements previously sketched: the Japanese-designed atrium, the magnolia, and the "teahouse".

In 1963, Cramer combined the Japanese principle of stone settings with a wall design reminiscent of Mondrian for a single-family house commissioned by the real estate management company Frawa AG.

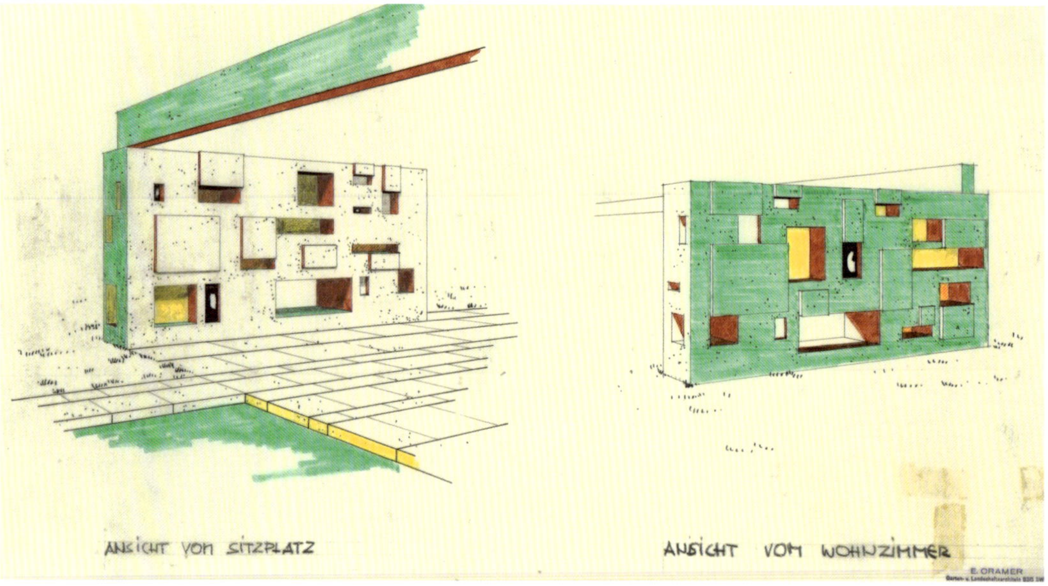

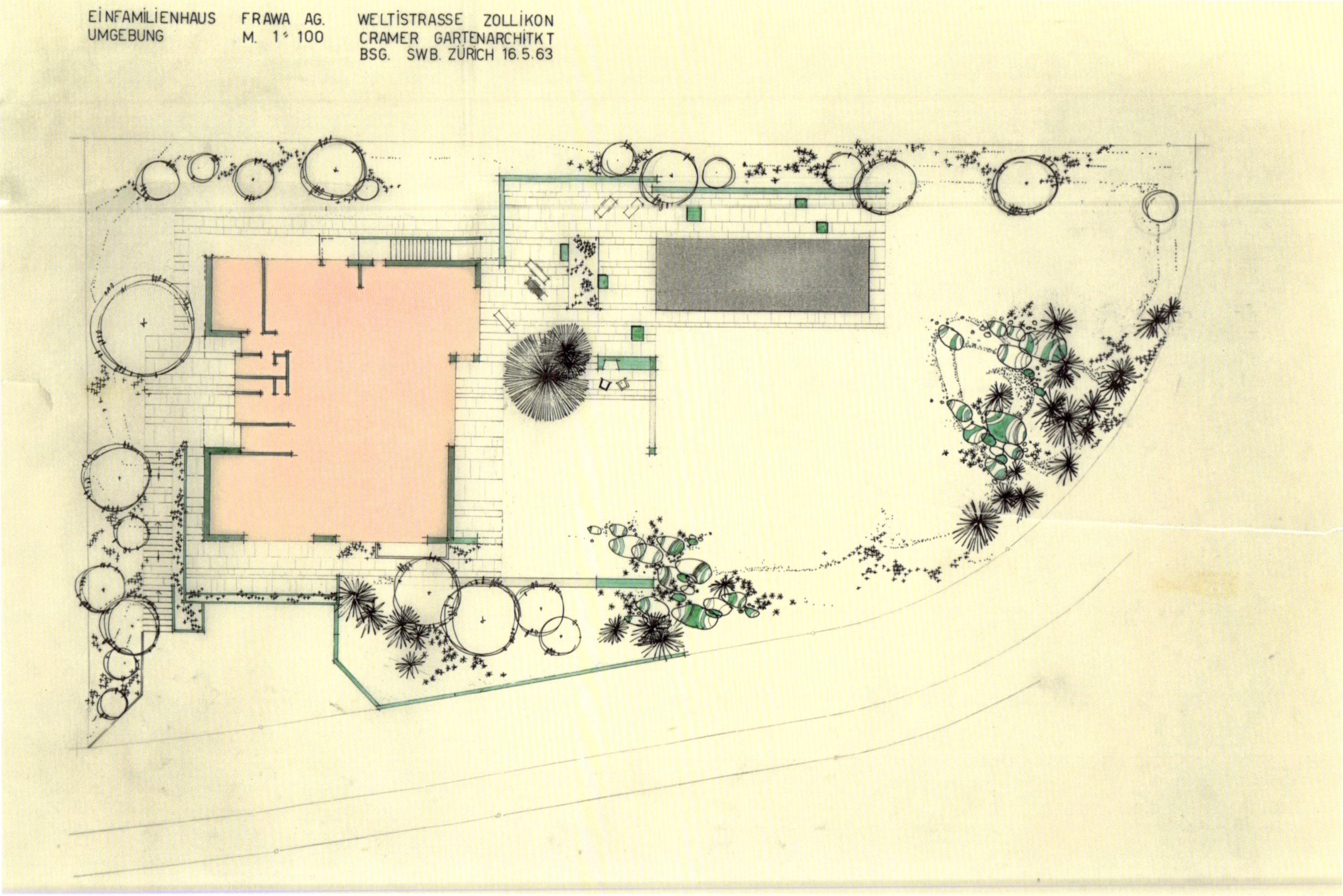

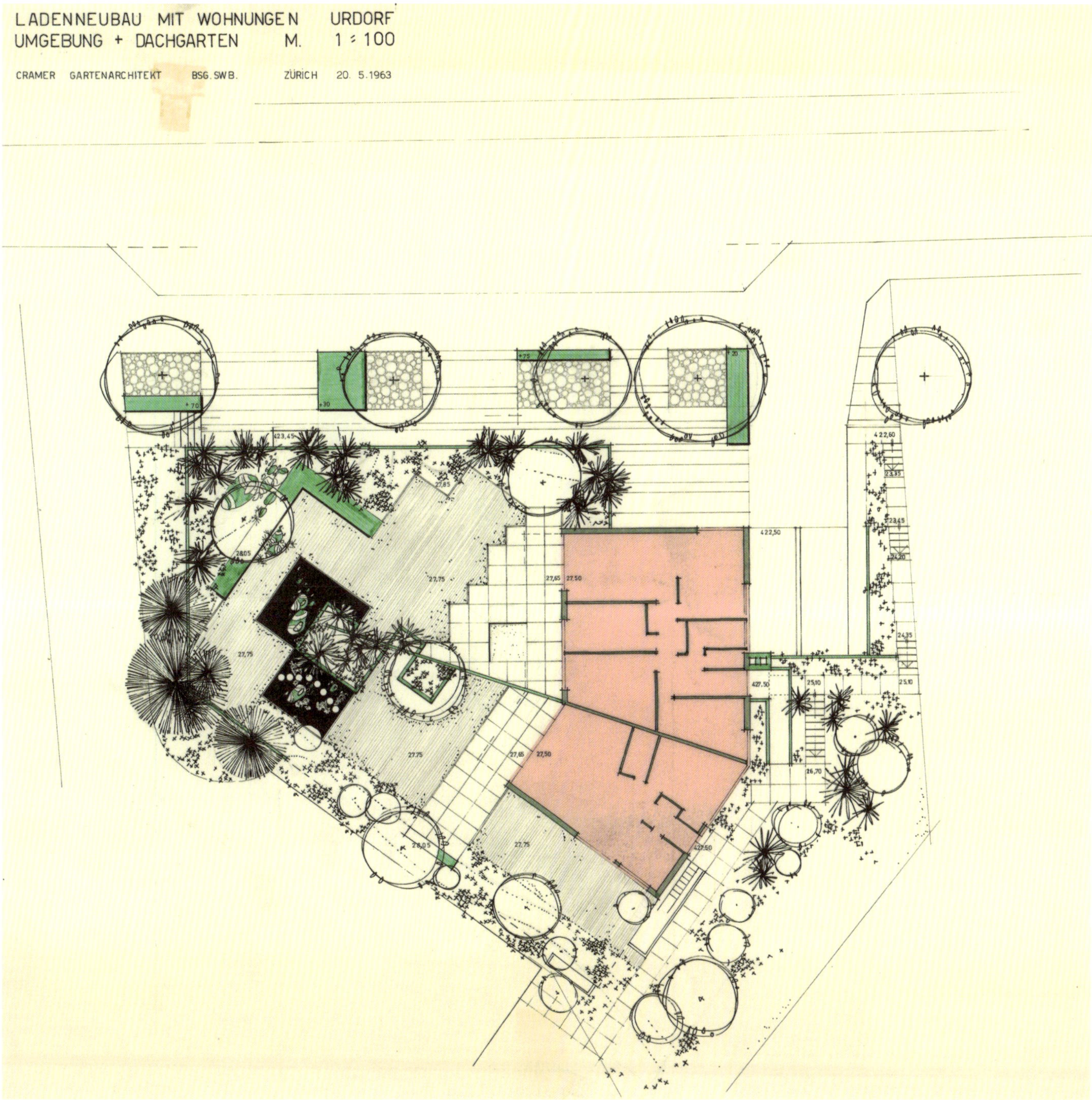

Drawing of the roof garden of the retailer Migros in Urdorf (Zurich), 1963. The angular basin has Japanese connotations with stone settings and circular stepping stones as seen in the G59.

Hans Graf

1919–2014

As in the garden of the Tenryū-ji, where they refer to the mythical Hōrai as Hōrai-ishigumi, seven stones "swim" as islands in the pond of the "Herrmann Garden" in Biel. The stone formation in the background forms a waterfall, possibly a dry karedaki in allusion to the Daisen-in.

Water is the basis of all that lives and therefore worthy of our veneration as seen in the pictures of Japanese gardens. May we all realise this before it is too late.[1]

Amazingly, the *Anthos* editor responsible for these evocative lines on this most precious of liquids took the unusual step of illustrating them with an image of the dry landscape of Daisen-in and the caption "Symbolic mountain scene with torrents, stones and sand"[2]. In Hans Graf's garden designs, by contrast, the element of water was rarely absent, whether he was designing an artfully constructed circular basin with fountain or a biotope as "nature reserve"[3]. He championed "artistic sensibility" but also argued that one should have the courage to let a garden "not be perfect or unchanging" as it is "through change over time" that it lives.[4]

He trained both these aspects, first in 1935–1938 during an apprenticeship as a gardener in Stäfa, then at the Trade School in Wetzikon and in the office of Walter Leder, who taught there. Two winter courses at Oeschberg Cantonal Horticultural School then provided the impetus to settle in the canton of Bern, where he founded his own company for horticulture and garden design in Bolligen in 1947.[5]

In 1937, while still an apprentice, Graf drew a Japanese-inspired garden plan with a two-part lily pond, the banks of which he lined with Japanese swamp irises (*Iris kaempferi*), Chinese reed (*Eulalia*), and hostas (*Hosta*), among other plants. In addition, there were Chinese autumn anemone (*Anemone hupehensis var. Japonica*), Japanese flowering cherry (*Prunus serrulata*), summer lilac (*Buddleja davidii*), Japanese kerria (*Kerria japonica*), Asian flowering dogwood (*Cornus kousa*), threadleaf Japanese maple (*Acer palmatum 'Dissectum'*), and the silver variegated ash maple (*Acer negundo 'Variegatum'*), which, although not native to East Asia, has a similarly picturesque character.

Graf attempted to mediate between "the balancing of proportions" and simplification in order to "unravel confusion" and avoid "overloading", on the one hand, and a desire to preserve the "boundlessness" of the garden in order to give space to the "different qualities of the seasons" and to "diverse activities".[6] He would experiment with baroque set pieces, while also having his eye on the natural garden. The Chinese and Japanese gardens tipped the scales. Graf, fascinated by both – they were "always present"[7] – assembled a sizeable collection of books on East Asian garden culture,[8] and visited China twice, in 1982 and 1997.[9]

In most of his Japanese-inspired gardens, Graf would employ characteristic elements modelled on Japanese precursors in a *chambre separée*, an intimately placed "jewel" to be enjoyed contemplatively from the garden's seating area or an arbour set back from the house, possibly sheltered by a canopy. This can be seen in his designs for the gardens "Diemant" in Zollikofen, 1977;

1 Anon. 1962b, p. 2.
2 Anon. 1962b, p. 2, ill. p. 3.
3 Schwarz/Graf/Mathys 1974, p. 32.
4 Graf 1959, p. 130.
5 See Schubert B 2009.
6 Graf 1959, pp. 129–130.
7 Hans Graf, Jr, conversation with the author, 16 Sept 2021.
8 See Bärtels 1987; Beuchert 1983; Keswick 1978; Lancaster 1989; Nitschke 1993; Schaarschmidt-Richter/Mori 1979; Takama 1983; Seike/Kudō/Schmidt 1983; Weiss 1996; Qian 1982; Zhu 1992.
9 Hans Graf, Jr., conversation with the author, 16 Sept 2021.

Twenty years after sketching the first plans, Graf produced a watercolour of the water lilies in the pond of the "Diemant Garden" in Zollikofen, created in 1977.

"Eicher" in Oberlindach, 1982 [→ p. 195]; "Guggi" in Niederglatt, 1977; "Kiener" in Herrenschwanden, 1974; "Klossner" in Oberlindbach, 1979; "Kohler" in Burgdorf, 1978; "Linder" in Langnau, 1990; "Luder" in Steffisburg, 1981; "Seewer" in Sinneringen, 1995; "Schweizer" in Bern, 1976; "Somm" in Herrenschwanden, 1978; "Suter" in Stettlen, 1979 [→ p. 193, top]; "Stucki" in Rüfenacht, 1972; "Tschan" in Bolligen, 1978; "Wüthrich" in Bremgarten, 1982 [→ p. 189]; and the undated "Schütz" project in Münchenbuchsee. Typically they comprised a lily pond, sometimes with islands, bridged by stepping stones and formed or accompanied by stone arrangements, often nestled at the foot of a hill and surrounded by selected Japanese plants such as bamboo, Japanese maple (*Acer palmatum*), maiden pine (*Pinus parviflora*), pagoda dogwood (*Cornus controversa*), Japanese kerria (*Kerria japonica*), azaleas, and rhododendrons. It is noticeable that Graf, who was not given to copiously labelling planting on the plans, gave preference to the Japanese varieties.

More extensive Japanese-style pond landscapes of the kind intended for strolling through were also planned by Graf for the "Botanical Garden" in Berne, 1971; for the Föhrenweg development in Lyss, 1973; and Breitfeld II in Rüfenacht, 1977; for a nurses' garden at the health resort in Oberägeri, 1986; for a retirement home in Schüpfen, 1988; and – as the only private garden – for Walter Baeggli in Frauenkappelen in 1982.

His most unusual creations, among them the "Luder Garden" in Steffisburg [→ p. 190], "Häni Garden" in Zuzwil, "Hehl Garden" in Hinterkappelen, and "Schindler Garden" in Gümligen [→ p. 191, top], were hybrids of the kind mentioned earlier: they included a baroque designed parterre around a fountain as well as a "nature garden" with a pond and Japanese-style stone settings.

In Steffisburg, he also planned a wall opening that appears to be modelled on the moon gates in the "Humble Administrator's Garden" (拙政园, *Zhuōzhèng yuán*) in Suzhou, China. In perspective, it looks like an adaptation of his circular fountains rotated into the vertical plane.[10] In the other direction, when seen from a semicircular hedge beyond the wall, the round opening acts as a picture window framing the fountain. While the moon gate analogy was not realised, the fountain and the parterre were, but in place of a boxwood border and roses, he used cotoneasters.

In 1997, in the "Schindler Garden", he integrated the architectural garden into a seating area framed by the walls of the house which stood at right angles. He connected the parterre, which was laid out around the fountain as a "small historical garden", with a near-natural water lily pond via a channel. Three steps formed a small waterfall. An oval fountain integrated into the channel served as a link between the architectural and dwelling garden, between artistically sublimated nature and near-natural art, and between past and present.

In the "Schneider Garden" in Münchenbuchsee, designed in 1990, Graf varied the mix of elements. Alongside the fountain, he added a dry garden with "gravel or sand" as well as stones, a bamboo wall, and a rose bed bordered by a serpentine of lavender.[11] Near the house, he placed a pagoda dogwood (*Cornus controversa*) as a solitary feature and, as a distant eye-catcher in the southern corner of the property, a "red prunus" – probably a cherry plum (*Prunus cerasifera*).

These hybrid designs are probably more typical of Graf's late work, but this interpretation is still too simplistic. In his last garden with an East Asian character from 1997, Graf attempted an adaptation of the Ryōan-ji, which is perceived as the quintessence of Japanese garden design.[12]

Instead of gravel, he used lavender covering an organically shaped area to symbolise water. Deviating from the composition of the model, he reduced the number of stone groups embedded in sand banks to three but made their concentration more pronounced.[13] Ivy and bamboo lined the long sides of the rectangular area to the east and west respectively. From the south-facing windows of the house, one had an unobstructed view of the dry landscape, along with a panorama of the mountain massif of the Stockhorn looming in the distance – almost as if Graf had inspired the words of his American colleague A. Keir Davidson: "Maybe a distant line of hills, or a prominent mountain [...] would be easily visible from a garden, and, [...], this feature could be included in the garden itself."[14]

Graf sketched a Japanese miniature for the "Seewer Garden" in Sinneringen in 1995 [→ p. 196]. An atrium measuring approx. 4.50 × 4.50 metres in front of the bedroom contained a small pond flanked by bamboo and grasses on one side and "jap. iris", that is, Japanese swamp irises, on the other, and enclosed a "pinus tree island" 2 metres in diameter.

Finally, in 1980, Graf realised a project for Mr. and Mrs. Herrmann in Biel that was expressly declared a "Japanese garden" and was unique in his oeuvre – equipped with the typical ingredients of ponds, islands, stone settings, two hills, a waterfall, two stone lanterns, bamboo, Japanese swamp irises, Japanese maple, and azaleas.

10 There are three works in Graf's library on the gardens of Suzhou alone, one of which is on the "Humble Administrator's Garden". See Chen 1992; Zhou 1994; Zhaojun/Zhou 1994.
11 "Mr. and Mrs. H. Schneider, Münchenbuchsee, garden redesign", ground plan 1:50, 29 Oct. 1990, ASLA Graf, no rec.
12 Graf also spent time expressly researching Zen gardens. See Davidson 1982.
13 In plan, one can see two sets of six and one of eight blocks, in perspective two lots of five and one of eleven. "Mr. and Mrs. p. and E. Schmid Steffisburg, rock garden west side", ground plan and perspective 1:50, 24 April 1997; ASLA Graf, no rec.
14 See Davidson 1982, p. 28.

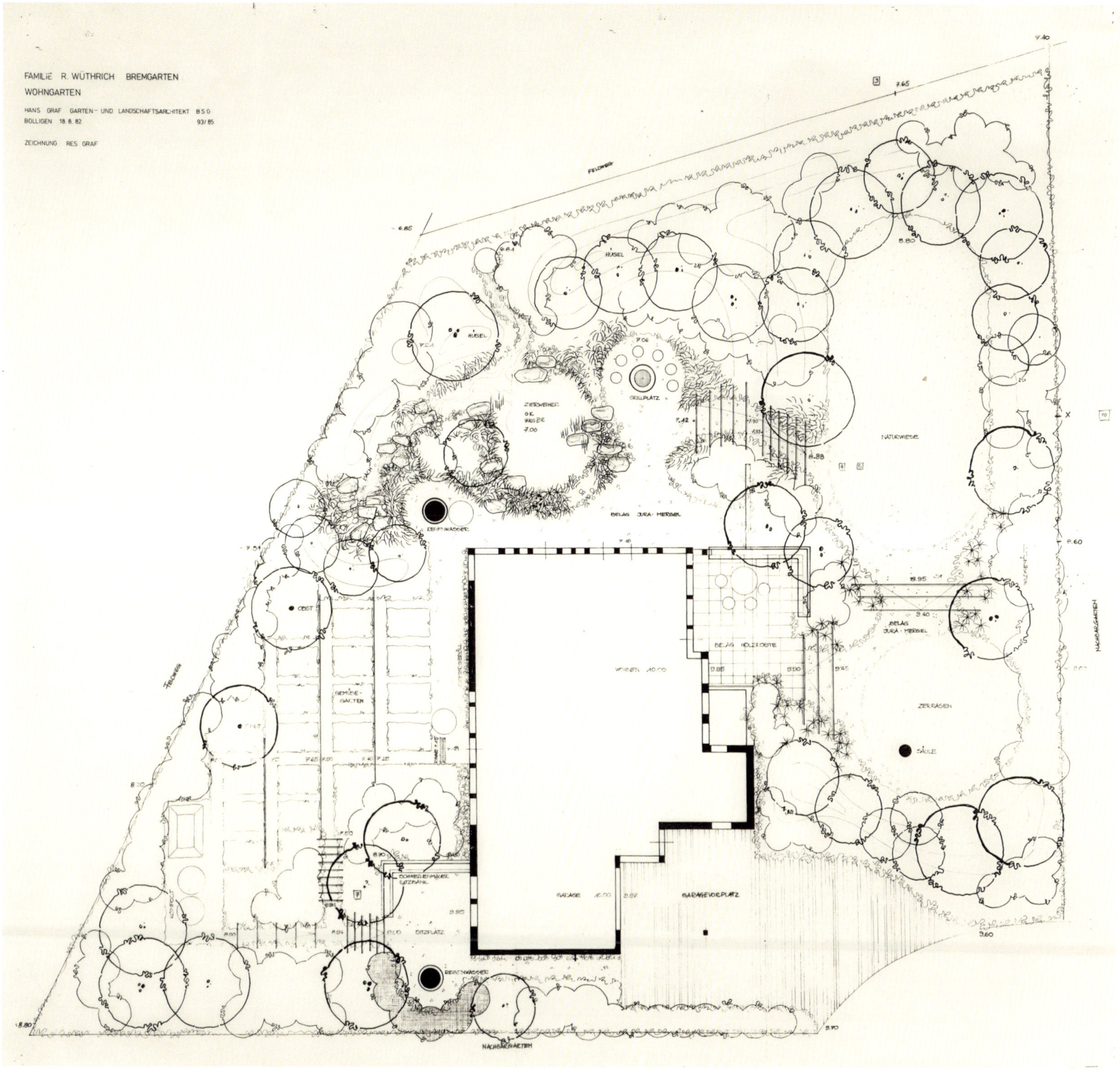

Artificially shaped hills create an intimate atmosphere around the ornamental pond in the "Wüthrich Garden" in Bremgarten, 1982, with stone settings accentuating its Japanese character.

Floor plan and perspectives of the garden plan for the property – a doctor's practice and residence – of Annamarie and Paul Luder in Steffisburg, 13 May 1981: A "small natural garden" in front of the practice is laid out as a *chambre separée*. In an alternative design dated 4 November 1981, Graf proposed a biotope next to the swimming pool in the garden of the house.

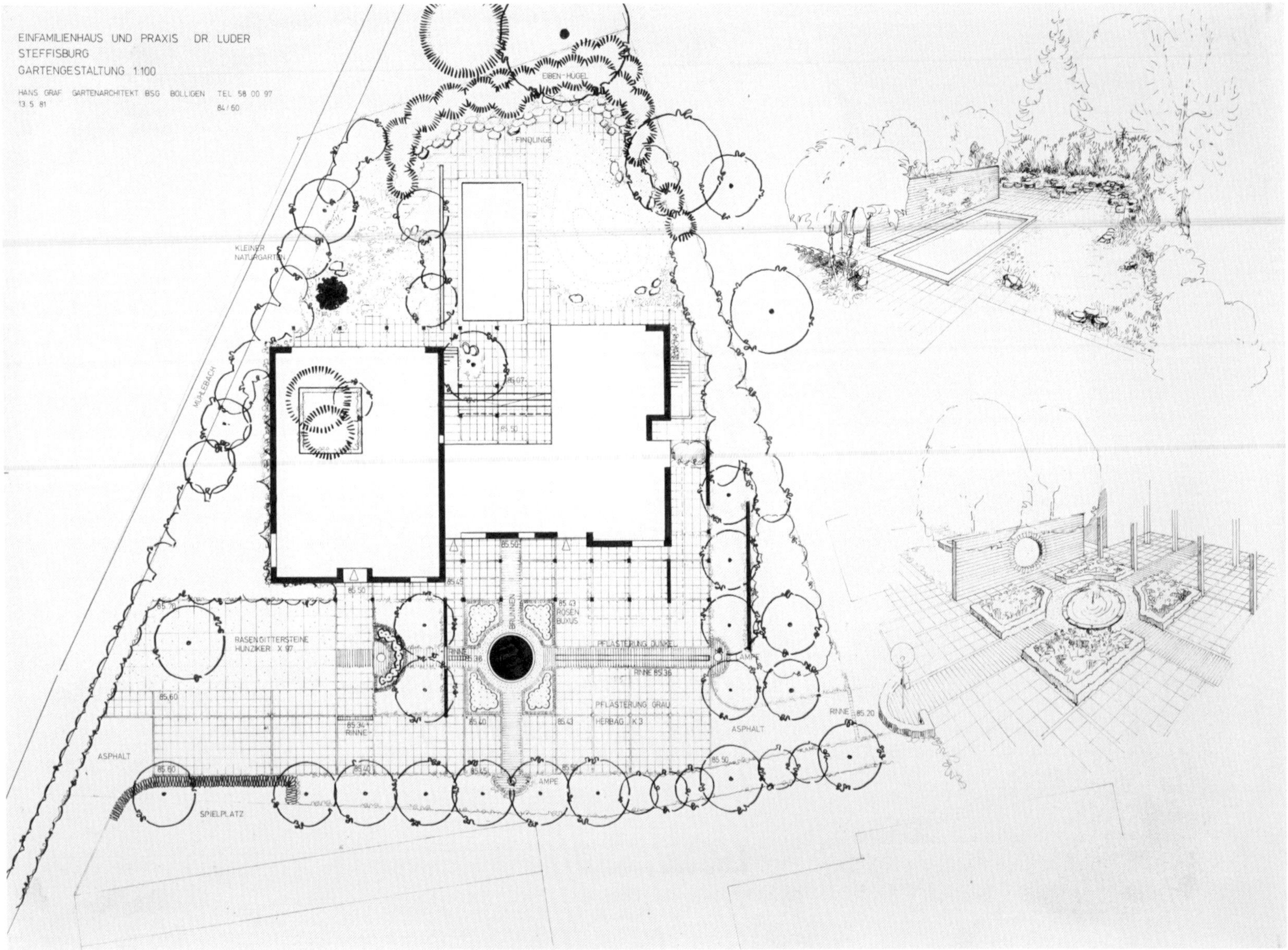

↑ Garden plan for Kurt Schindler in Gümligen, 1997: the parterre is laid out as a “small historical garden” and combined with a Japanese biotope.

→ The combination of Baroque and Asian elements such as the Moon Gate in the park of Schloss Halbau was shown in *Die Gartenkunst* in its 12th edition in 1910.

→ Graf combined water and dry landscape in his project for the "Suter Garden" in Stettlen in 1979, transforming conifers from nature into art.

→ Stone garden inspired by Ryōan-ji for P. and E. Schmid in Steffisburg, 1997. Lavender takes over the function of the watercourse. Flanked by ivy, bamboo, and evergreens, it conforms to the Japanese principle of restricting the range of colours. In the perspective drawing of the dry garden, Graf shows the outline of a silhouette that hints at the mountain range around the Stockhorn.

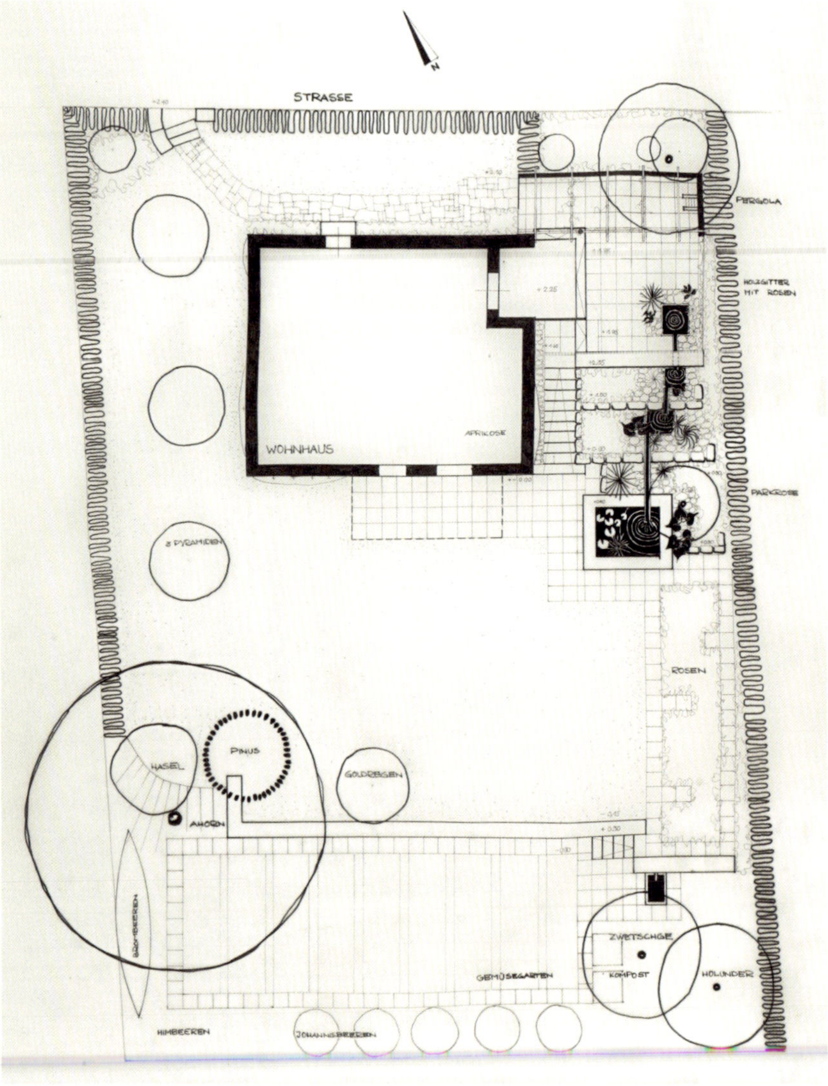

One of Graf's rare geometrically conceived water installations with Japanese connotations for the "Gerber Garden" in Ittigen, 1970. The waterfall runs over an artificial channel, as depicted by Kinkichirō Honda on plate 29. The feed and outlet are offset from one another to simulate natural meandering.

HERRN O. SUTER STETTLEN
GARTENÄNDERUNG 1:50
HANS GRAF GARTENARCHITEKT BSG
3065 BOLLIGEN 28.11.79
HOLZBANK
HALLENBAD
SITZPLATZ
RASEN
ROSEN
HOLZBANK AUF BETONSOCKELI
FAHNE

HERRN UND FRAU P. UND E. SCHMID STEFFISBURG
STEINGARTEN WESTSEITE MASSTAB 1:50
HANS GRAF SEN. 3065 BOLLIGEN 24.4.97
RAMPE
TOR
SAPONARIA
ROSEN (RUGOSA)
EIBEN
SASA
TÜRLI
FÖHRE
IMMERGRÜN
ROSA RUGOSA
BODENBEDECKUNG
KAPPELENWEG

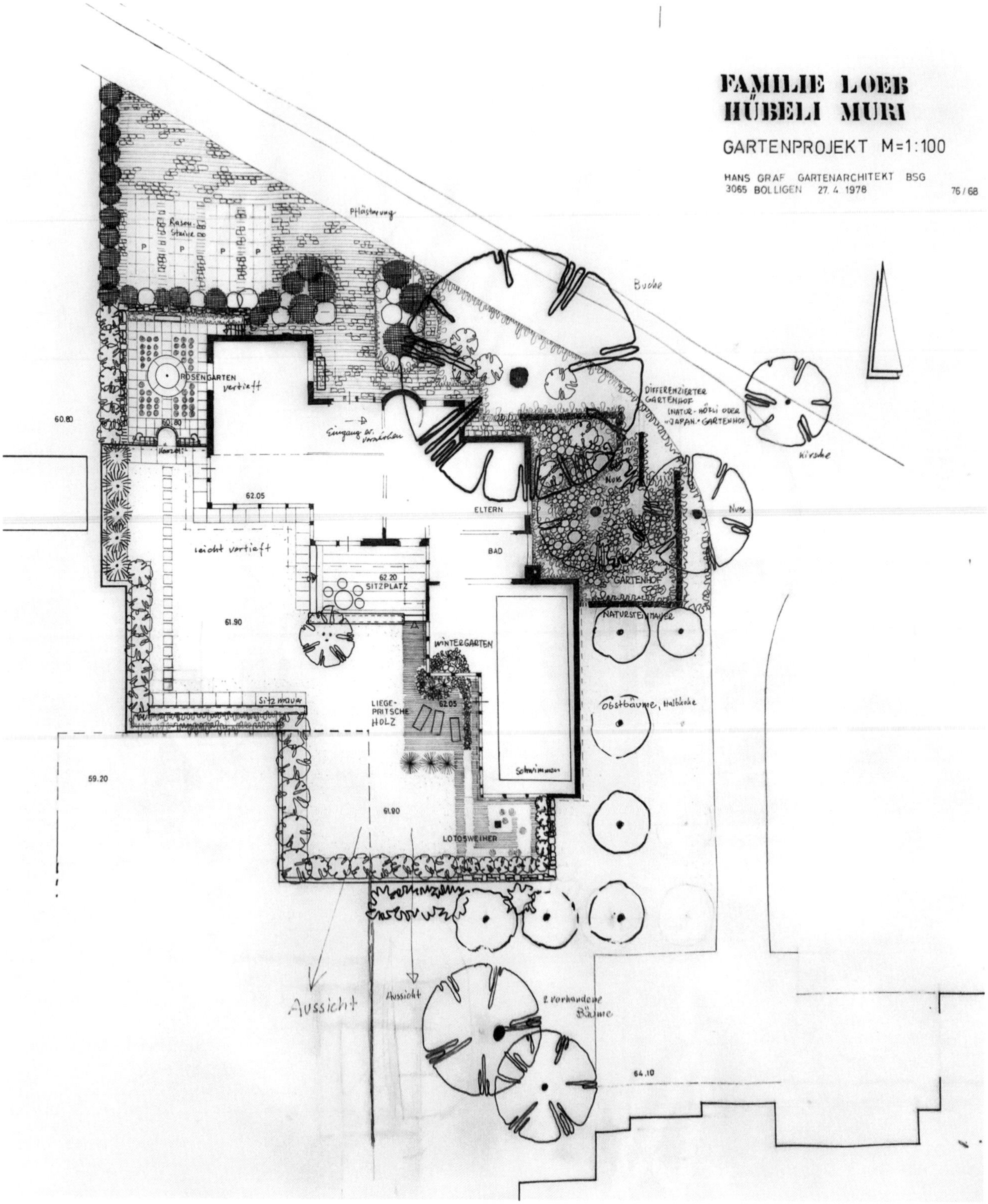

A “lotus pond” and “differentiated garden courtyard (a ‘small nature courtyard’ or ‘Japanese garden courtyard’)” signal the inspiration of Japanese sources.

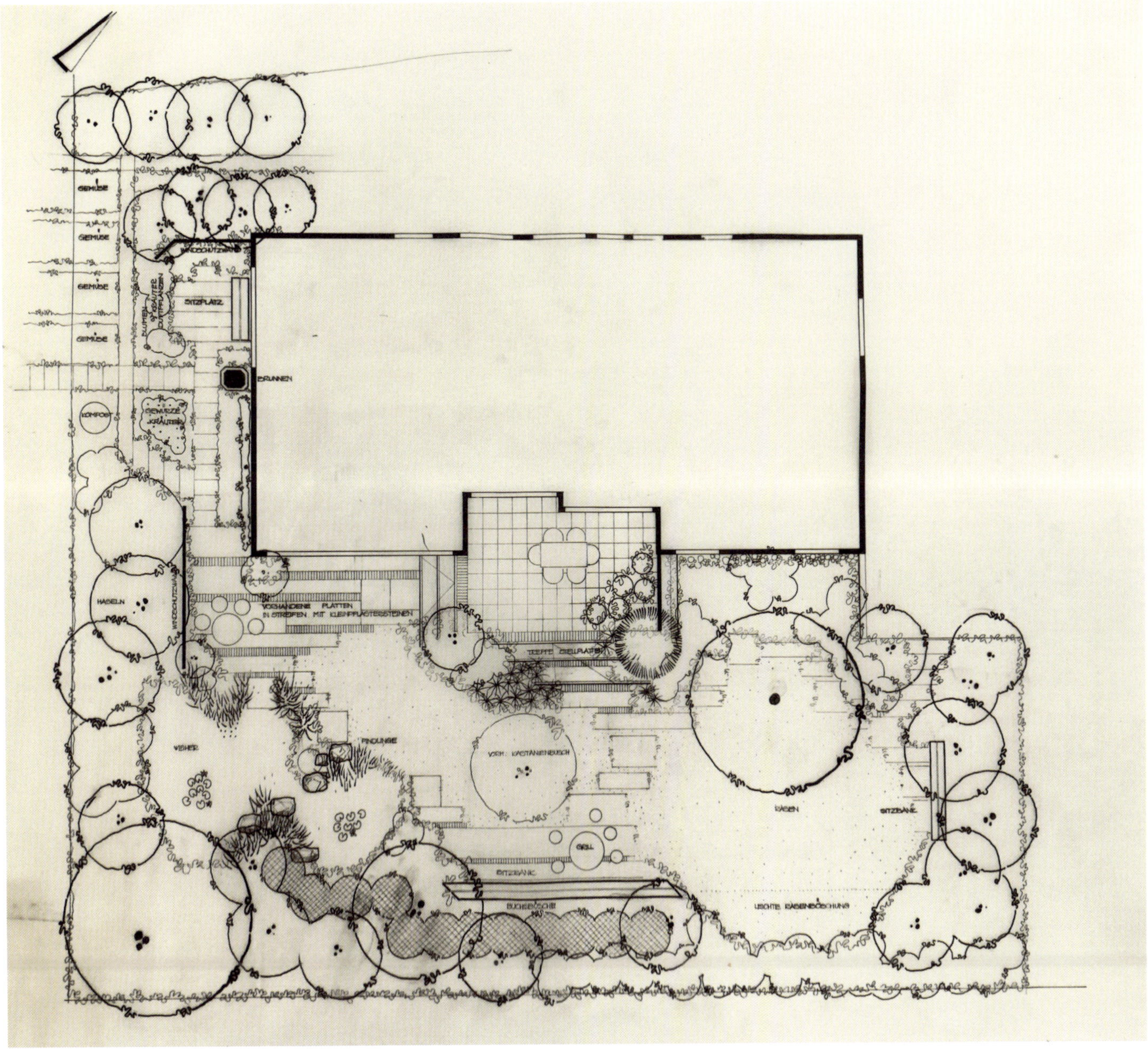

An intimately designed space in the "Eicher Garden" in Oberlindach features a water lily pool, additionally constrained by boulders.

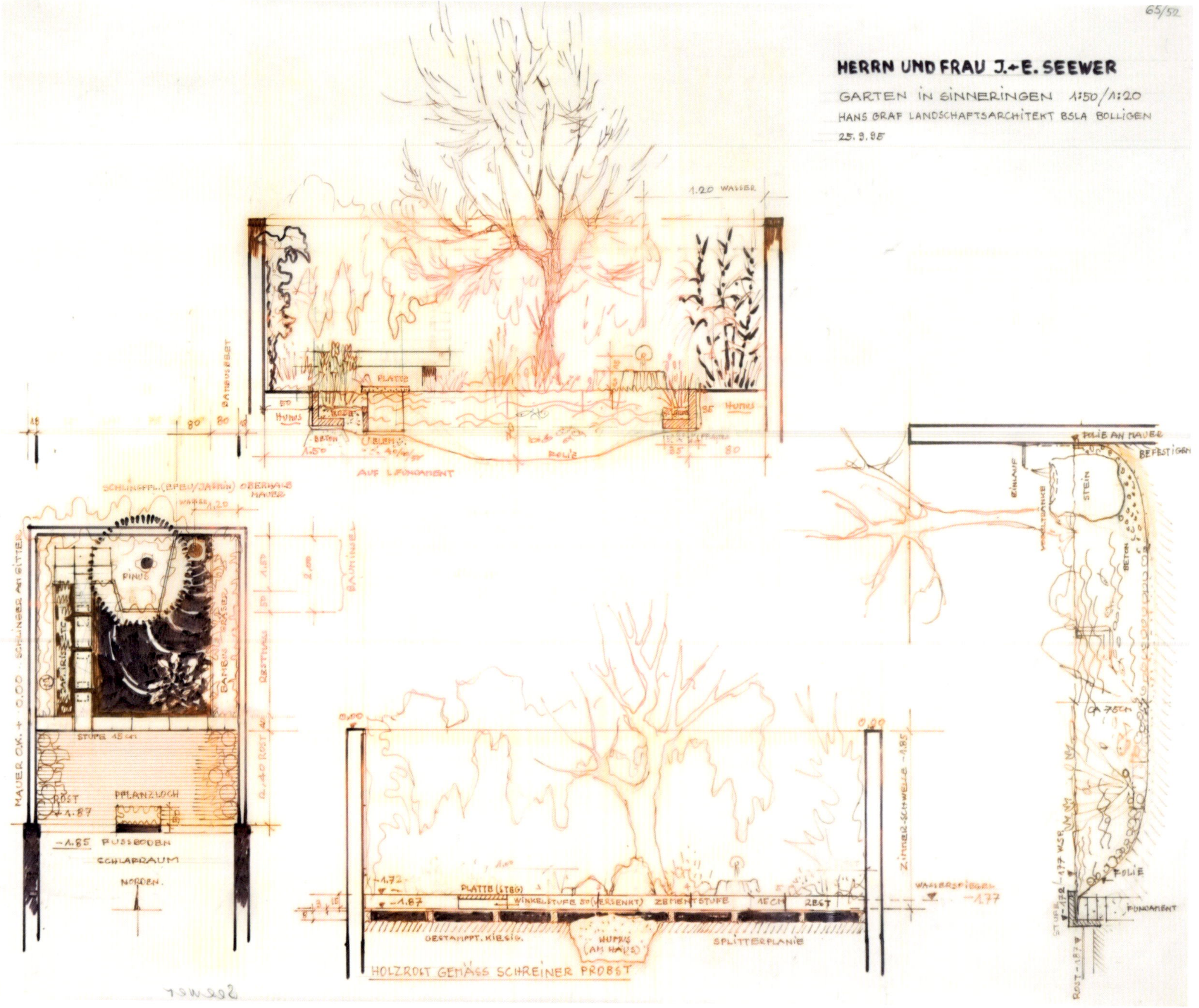

↑ **Graf created a Japanese miniature for the "Seewer Garden" in Sinneringen in 1995.**

→ **The delta-like projection of the bank, and the placement of a lantern at its end in the Japanese garden designed in 1980 for Mr. and Mrs. Herrmann in Biel are reminiscent of the same design element in the garden of the Katsura villa.**

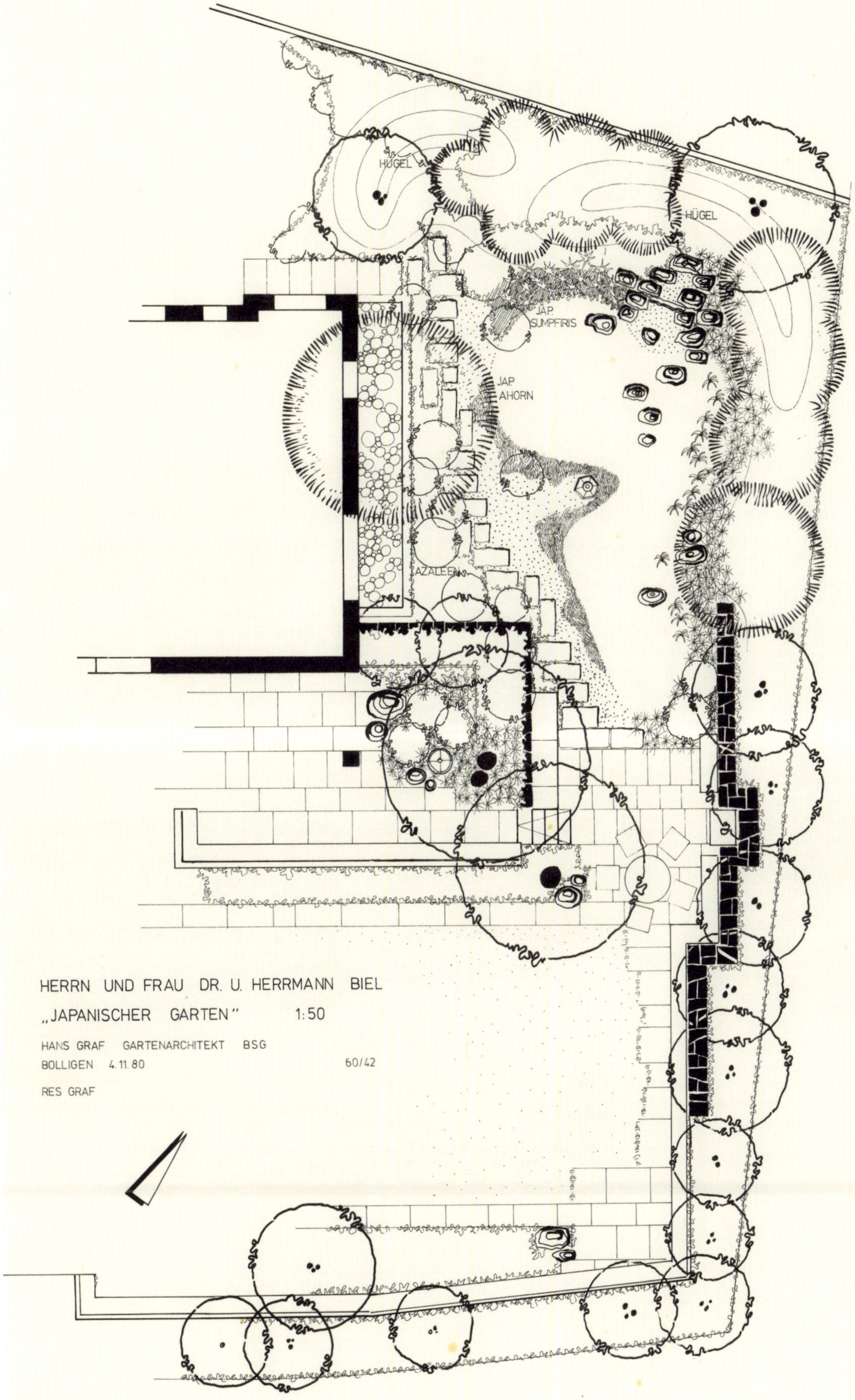
HÜGEL
HÜGEL
JAP. SUMPFIRIS
JAP AHORN
AZALEEN
HERRN UND FRAU DR. U. HERRMANN BIEL
„JAPANISCHER GARTEN" 1:50
HANS GRAF GARTENARCHITEKT BSG
BOLLIGEN 4.11.80 60/42
RES GRAF

Fritz Klauser 1885–1950
Fredy Klauser 1921–2007

This photograph of Fritz Klauser's garden house with crowning pagoda turned up in the estate of the Mertens brothers.

Fritz Klauser's sketch of a garden house topped by a pagoda-like dovecote for the GASGA horticultural show has the character of an English folly.

For a horticultural exhibition in St. Gallen, he designed a 'crazy' garden. It was a sunken garden with a zigzag slope and a prominent garden house that was yellow on the inside and bright blue on the outside. On the roof was a pagoda-like dovecote with white peacock pigeons.[1]

Fritz Klauser's garden house, which his son Fredy described in an interview in 2004, was most probably designed for the Gartenbau-Ausstellung St. Gallen (St. Gallen Horticultural Exhibition, GASGA) in 1938, which focused on the theme of the "Gardens in Art".[2] It looked like an English folly, one of those sensational but bizarre creations so popular in Britain [→ pp. 198, 199].

This geographical reference has a biographical background. Klauser's father spent five years of his apprenticeship in Colchester in the English county of Essex, north-east of London[3], after working in Switzerland – Neu-Allschwil (Basel City), Brodoghera (Ticino), Châtelaine (Genève) – and Germany. His teacher was Robert Whistler Wallace (1867–1955), son of Alexander Wallace (1829–1899). The latter was a botanist and entomologist who ran the New Plant and Bulb Company in Colchester[4], introduced Japanese plants, experimented with silkworm breeding,[5] and also imported Japanese works of art.[6] Alexander's pioneering work was taken over by Robert, who renamed the nursery in Kinfield Gardens, Colchester, to R. Wallace & Co. and expanded its specialisation to encompass orchids, azaleas and irises.

This was not, however, Colchester's first exposure to Japan: in 1885, the year of Klauser's birth, a Japanese village was erected in the town's Public Hall – modelled on a similar well-received project by Tannaker Buhicrosan at Humphreys' Hall, Knightsbridge, London, and opened on 10 January 1885.[7] The impetus came from the local art dealer Charles Clark, who imported the attraction from Japan in the same year he took over management of the hall.[8]

Apart from the fact that he worked in an environment open to Japanese influences, Klauser, who was born on the same day as Gustav Ammann[9], is unlikely to have missed the Japan-British Exhibition in London in 1910.

The alpine gardens, which Klauser "as a specialist"[10] was often called upon to design, were presumably also inspired by Japan. However, after his "forced" return to Rorschach in 1914 due to the outbreak of the First World War and the opening of his own firm, he avoided producing designs for rock gardens.[11]

The end of the Second World War marked a turning point for his son Fredy. Although convinced to follow his father's path from childhood on, he had to step into his father's shoes when only 24 years old. He acquired a wealth of plant knowledge during his schooling at Arnold Vogt's perennial nursery in Erlenbach, which at that time had made a name for itself with a number of Japanese-inspired show gardens. Unlike his father, however, Fredy did not have the opportunity to travel or study abroad.[12]

In a conversation with Annemarie Bucher Bauer and Beatrice Nater, Fredy recounts that his father had been influenced in England by the "austere, architectural gardens" as well as by the "country gardens" depicting an idealised nature, and he "favoured the idea of the extended home, the dwelling garden".[13] The plans in Klauser's estate can thus be divided into these three categories: geometric-axial layouts contrast with residential garden concepts, while the two are used together to produce hybrid variants in which ornamental parterres are combined with landscaped sections.

The garden plan drawn in 1924 for the textile manufacturer Paul Stierlin-Naumann in Wängi, Thurgau, shows a finely attuned arrangement. Klauser divided the garden space into five different typologies: a vegetable garden divided into linear beds, a perennials section with curved paths, a rose parterre with a sundial

1 Bucher Bauer/Nater 2004, p. 15.
2 Egli 1939, p. [97].
3 Cf. H. B. 1945b.
4 Papini 2021, p. 211.
5 See Wallace 1869; cf. anon. 1869, p. 91.
6 Papini 2021, p. 211.
7 Cf. anon. 1885.
8 Cf. Schöppler 2020, pp. 117, 118; "Colchester's Sleeping Beauty – St Georges Hall", https://colchesteruncovered.wordpress.com/2015/09/25/colchesters-sleeping-beauty-st-georges-hall/ (retrieved 15 July 2023). "The Colchester public are fortunate in being provided, as far as possible, with the greatest of various 'London Successes'. Colchester is certainly one of the very first provincial towns in England to follow the example of the Metropolis in having a Japanese Village." anon. 1885, p. 5.
9 Ammann also kept a copy of the lines for Klauser's 60th birthday, which appeared at the same time as those for his own birthday on 9 July. Cf. H. B. 1945b; Gustav Ammann Belegbuch IV, 1942-50, p. 43; NSL 2-S-8-4.
10 Bucher Bauer/Nater 2004, p. 15.
11 Ibid.
12 Cf. Roelly 2007.
13 Bucher Bauer/Nater 2004, p. 15.

as focal element, a lawn bordered by a small wood and a group of rhododendrons, and a kidney-shaped meadow. Its attraction was a pond landscape in a pleasingly Japanese character. Klauser designed it as a garden for strolling in the spirit of the Chisen-kaiyū-shiki-teien. Stepping stones of well-proportioned but irregular shape form a path around the pond that passes through the waterside vegetation, crosses the pond, and traverses the stream that feeds it. A hanging willow, or given its size more likely an East Asian weeping willow (*Salix babylonica*), lends the scene a melancholy accent, for which the tree is apparently prized in Japan.[14] In his design for the "Henz Garden" in Aarau in 1948, a hanging beech occupies this place at the pond's edge, giving it a distinctly Japanese atmosphere.

Klauser was less liberal than his colleagues in labelling his designs with plant names, and there are very few actual plant lists. Often he seemed content to use only summary terms such as "shrubs", "hedges", "flowering shrubs", "flowers". In the plan for the "Heitz Garden" in Kreuzlingen in 1925, he only noted willow, walnut, copper beech, and red maple. Conversely, in the plan for a "Garden in Ticino" in 1919, he named the plants of mainly Asian origin in comparative detail [→ p. 201]. In a subsequent revision, this time titled as a project for the gardens of the Villa "Felice" in Minusio, he added the camphor tree (*Cinnamomum camphora*) native to Taiwan, China, and Japan – a rarity. Not only is this the only time this specimen is found in the Japanese-influenced plans of the 13 landscape architects collected in this book, but Sei Shōnagon remarks in her "Pillow Book" that it is rarely planted in Japanese gardens because it resembles "rampant wilderness".[15]

For one of the two Müller residences in Rheinfelden, Klauser again grouped various garden areas around the house but segregated them from each other less strictly. On the plan dated 1946 [→ p. 204], an extensive lawn ties the sections together. A Japanese-style "water channel" meandering through the lawn connects a new pool near the house with an existing pond of amoeba-like shape at the intersection of the two plots.

Klauser added another Japanese element – an island – to the plan for the Keller Garden in Rorschach, which he designed with his son Fredy in 1950, the year of his death.

When Fredy Klauser took over, the clientele gradually changed. Fritz's clients included numerous owners of venerable estates – castles such as Eugensberg and Wolfsberg in Ermatingen, Wiggen in Rorschach, Hahnberg in Berg in St. Gallen and extensive estates such as for the Villa Grüthof in Herrliberg and Villa Boller-Winkler in Turbenthal or the Bellevue health resort in Kreuzlingen. Fredy began to increasingly design gardens for more modern houses. The highlight of this phase was undoubtedly the Tugendhat atrium house in St. Gallen, built by local architects Danzeisen + Voser in 1956–1958 for Fritz Tugendhat, owner of the legendary villa built in 1929–1930 by Ludwig Mies van der Rohe in Brno. In April 1957, Fredy Klauser designed an inner courtyard as a Japanese miniature with a bird bath carved out of a natural stone and a small rectangular basin adorned with marsh plants [→ p. 206]. He planted the courtyard with plants almost exclusively of East Asian origin: threadleaf maple (*Acer palmatum 'Dissectum Atropurpureum'*), climbing hydrangea (*Hydrangea petiolaris*), Hinoki cypress (*Cupressus obtusa 'Nana'*), Chinese plumbago (*Plumbago larpentiae/Ceratostigma larpentiae*), cushion snowball (*Viburnum davidii*), Boston ivy (*Ampelopsis veitchii*), Zabel cherry laurel (*Prunus laurocerasus 'Zabeliana'*), small-leaved cotoneaster (*Cotoneaster microphyllus*), bearberry cotoneaster (Cotoneaster dammeri), mountain clematis (*Clematis montana 'Rubens'*), and firethorn (*Pyracantha 'Kazan'*).

As a counterpart to the Japanese-style courtyard, Fredy Klauser placed a rock garden in the southern corner of the property. The St. Gallen Botanical Garden may well have served here as inspiration, as the Japanese Garden and Alpine Garden had featured shortly beforehand in 1954 in the *Schweizerische Gartenbau-Blatt*.[16] The ground plan reveals the permeability of the design: the inner courtyard communicated through the floor-to-ceiling glazing with the outer seating area, which was aligned with rock garden. The flowing transition is also indicated by the "suggestion" of the stone slabs.

The elder Klauser had also experimented with the aspect of transparency, sporadically transforming a dwelling garden into a garden flat. This positive-negative inversion can be seen in the design of the "Häuptli Garden" in Aarau 1935–1936, which appears as both an exterior and an interior space. The pivotal point is the garden pavilion, whose architecture bears the unmistakable influence of Neues Bauen. Fritz drew two variants [→ p. 202]. The first shows a winged building with a mullioned window in place of the back wall. To the left is an arcade with two full height cut-out openings that frame the landscape like a painting and articulate the theme of the "window as picture".

In the other version, the view is more directed, wandering first over the elongated oval water basin before focusing on – actually literally being drawn to – the rear window onto which the landscape appears to be projected. In this variant, the garden house functions like a camera lens in whose aperture the image is reflected.[17] By the same token, the rear window can also be read as a backdrop, or even – an impression heightened by its division into vertical panes – as a Japanese painted screen (屏風, Byōbu).

14 "Willows (especially weeping willows), planted near water, at the pond edge, or on an island, [...]." Takei/Keane 2004, p. 38
15 Cf. Shōnagon 2017, pp. 53–54.
16 Anliker 1954, p. 667, ill. p. 668.
17 Michael Fowler described the Japanese strolling garden as a sequence of cinematically orchestrated images: "Numerous paths circumnavigate shinjiike [shinji ike 心字池, heart character pond] and through the technique of miegakure [hide and reveal] a visitor's movements through the garden is carefully orchestrated so as to both reveal and focus awareness on specific aspects, features, or sounds within the garden at particular points." Fowler 2013, p. 106. Beatriz Colomina analysed the phenomenon of the house as a camera: "If the window is a lens, the house itself is a camera pointed at nature." Colomina 1996, p. 203.

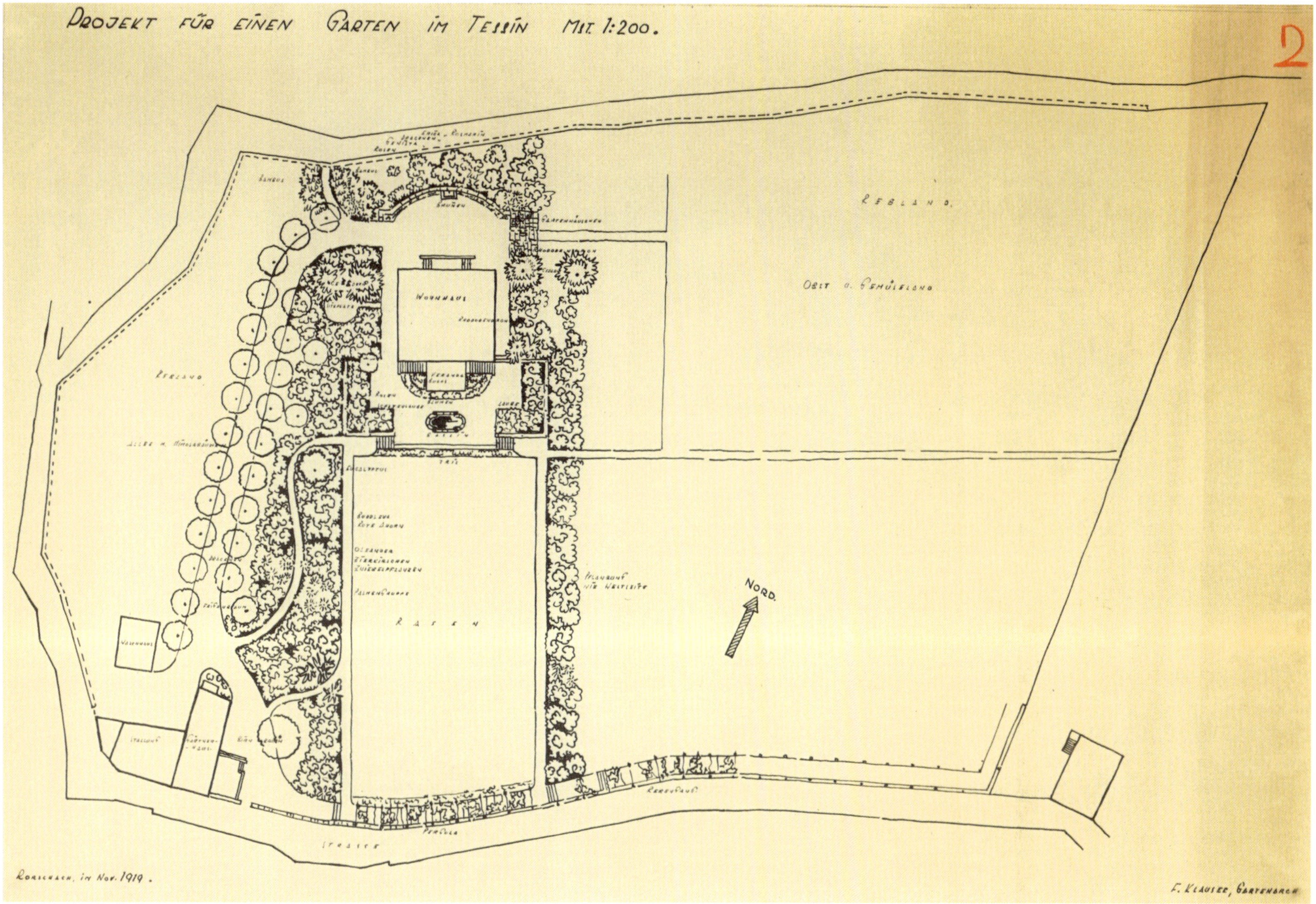

Klauser's plan for a "Garden in Ticino" from 1919 includes the names of shrubs of Asian origin: summer lilac, ornamental cherry, rhododendron, bamboo, and probably Japanese maple.

This variant of the pavilion in the "Häuptli Garden" in Aarau, drawn in 1935, is designed as a building with wings that capture the landscape as a gallery of paintings.

In another variant of the pavilion in the "Häuptli Garden" in Aarau with a circular ground plan, the landscape beyond the garden – "rhododendrons, forest meadow, trees and shrubs, sculpture" – is focussed onto a single point, as if in a magnifying glass.

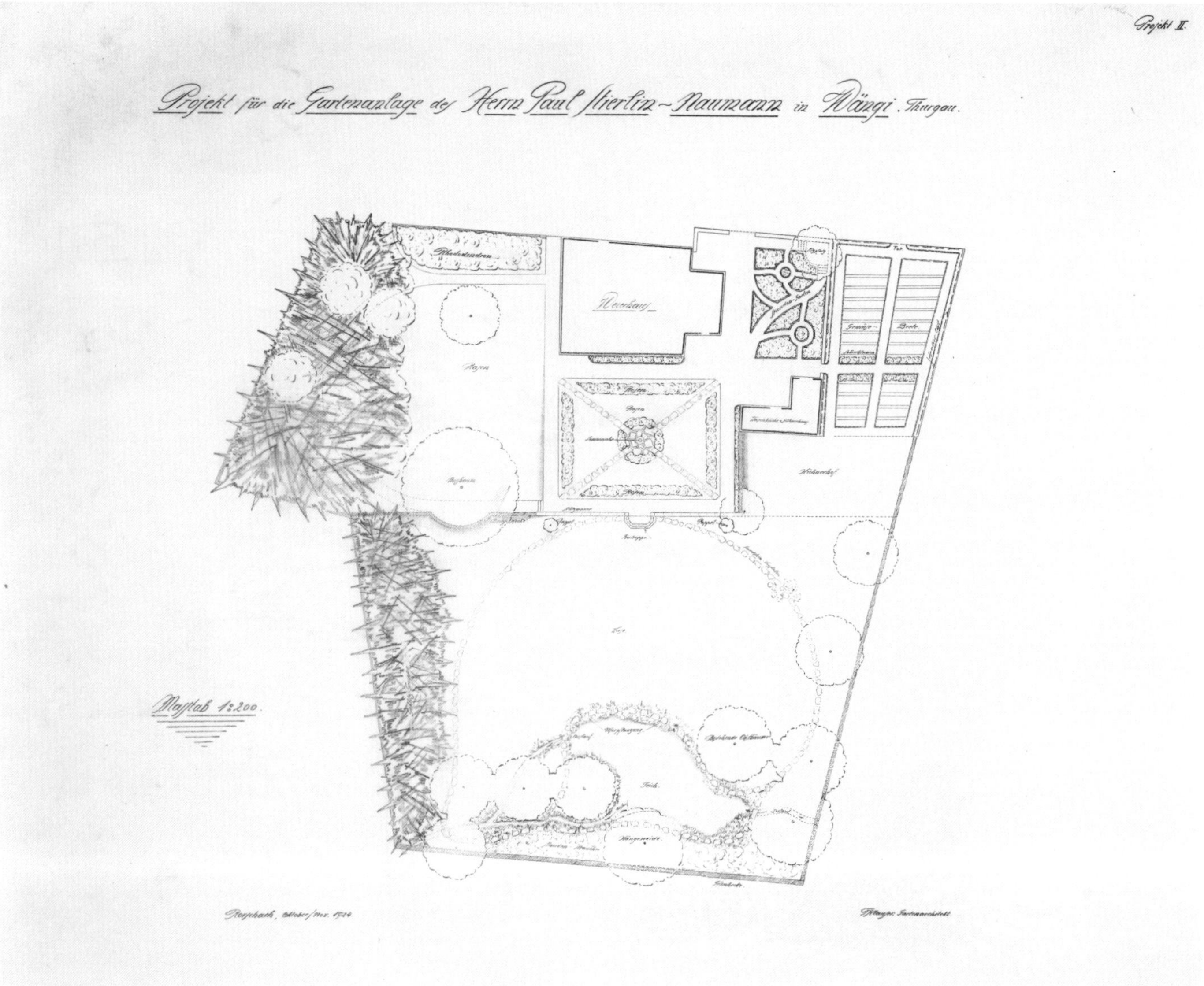

↑ In 1924, Klauser drew a scenery with the ingredients of a Japanese garden – a pond, a stream, stepping stones, and a weeping willow – for Paul Stierlin-Naumann in Wängi, Thurgau.

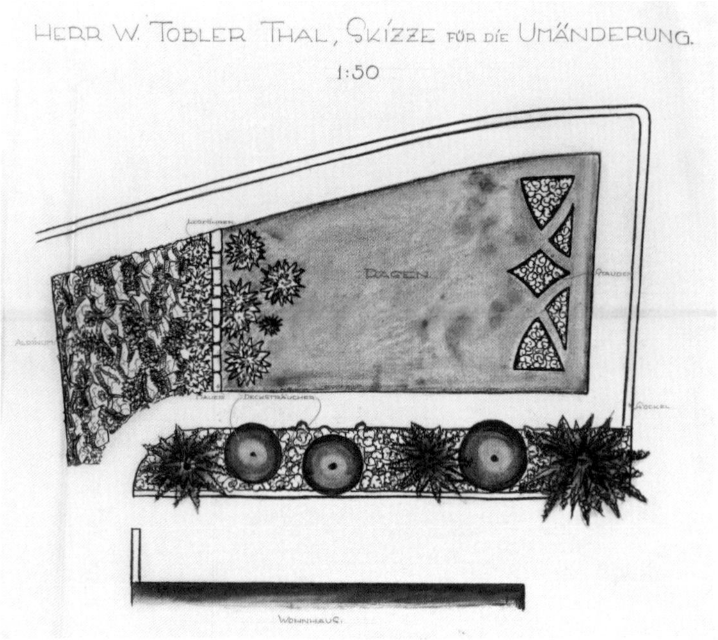

→ A plan for an alpine garden in the Tobler Garden in Thal, St. Gallen, drawn by Klauser in 1930. Its rather schematic representation appears to confirm Fredy Klauser's opinion that his father only begrudgingly undertook designs for rock gardens in Switzerland.

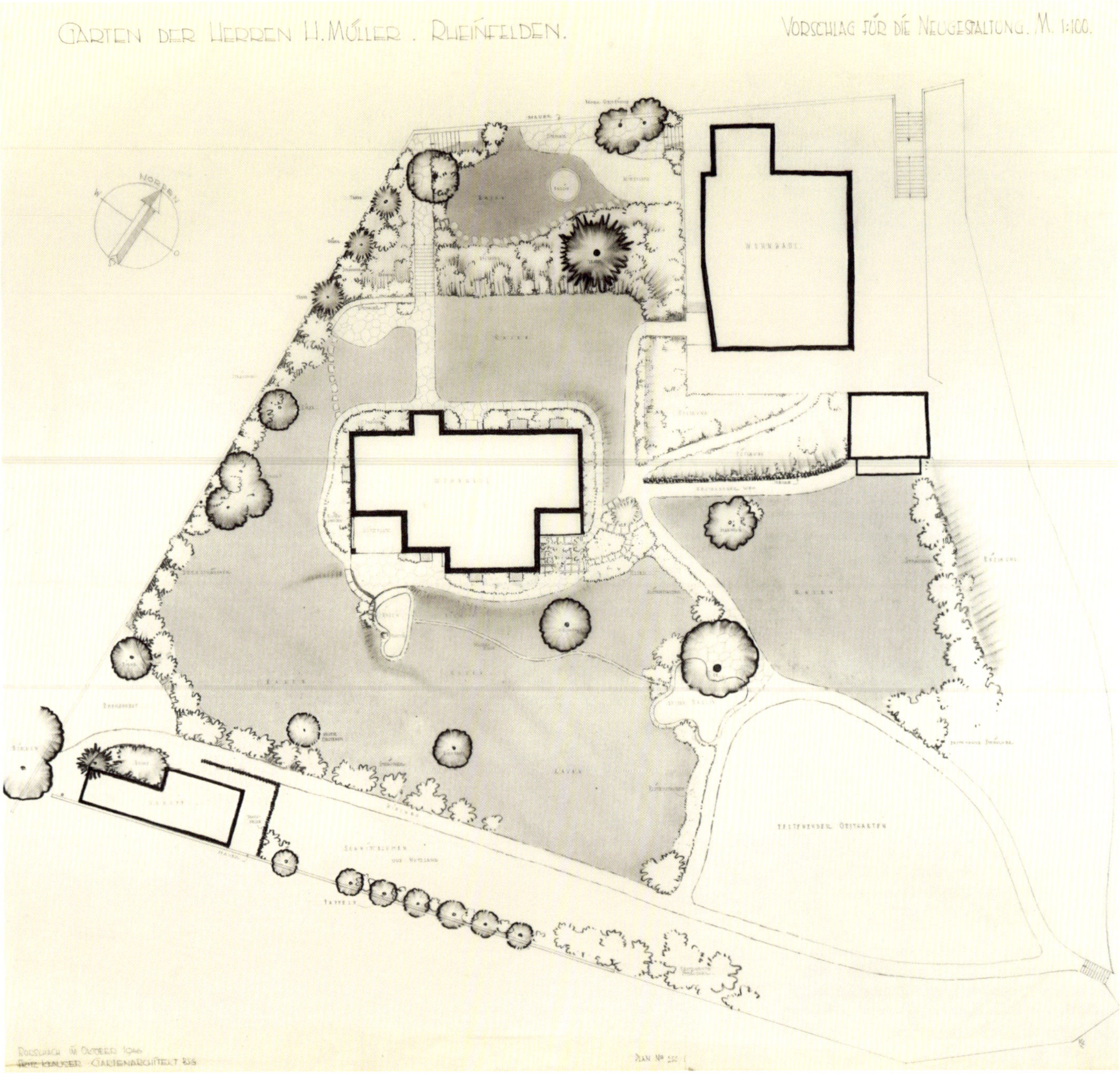

In his 1946 plan for the "Müller Garden" in Rheinfelden, a cherry tree accentuates the "water channel" mid-way between the two basins. This and a magnolia are the only conspicuous flowering trees that Klauser mentions by name.

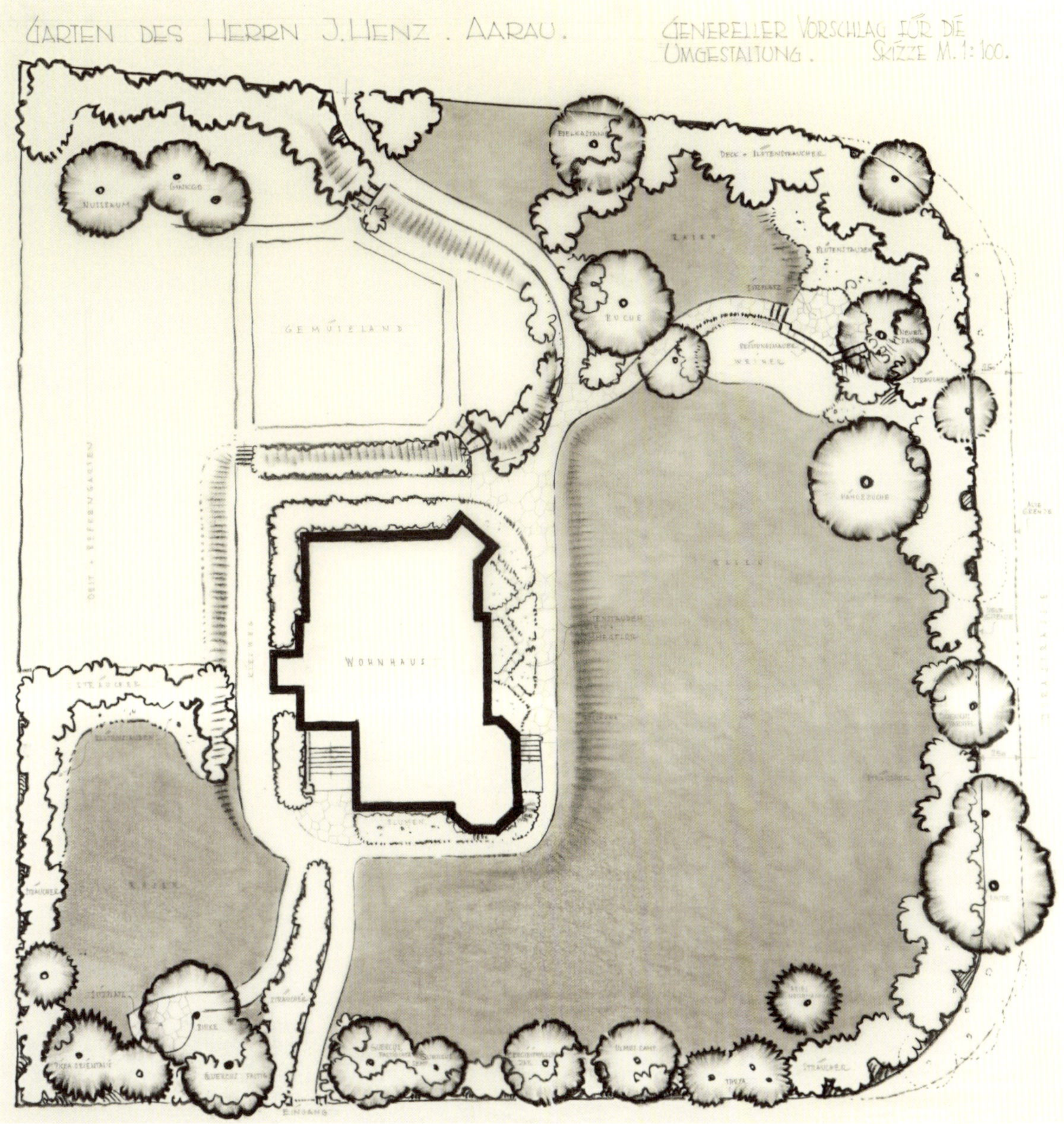

In the plan for the “Henz Garden” in Aarau, 1948, Klauser specifies a katsura tree, a species native to Japan whose bright autumn colouring sets it apart from the neighbouring woody plants.

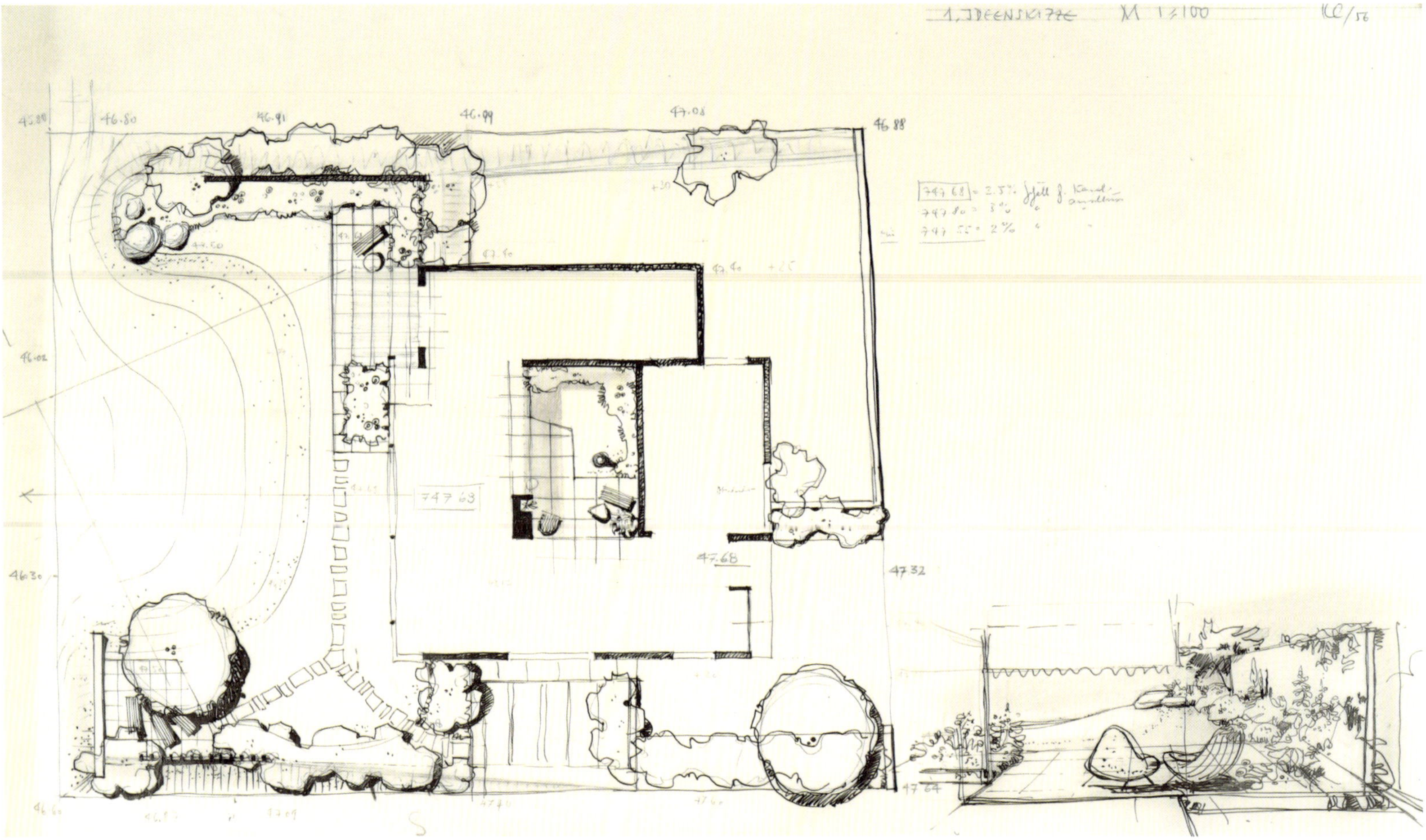

Fredy Klauser's perspective shows how the wider outdoor surroundings form an ever-present backdrop to an intimate garden room in the garden of the Villa Tugendhat in St. Gallen.

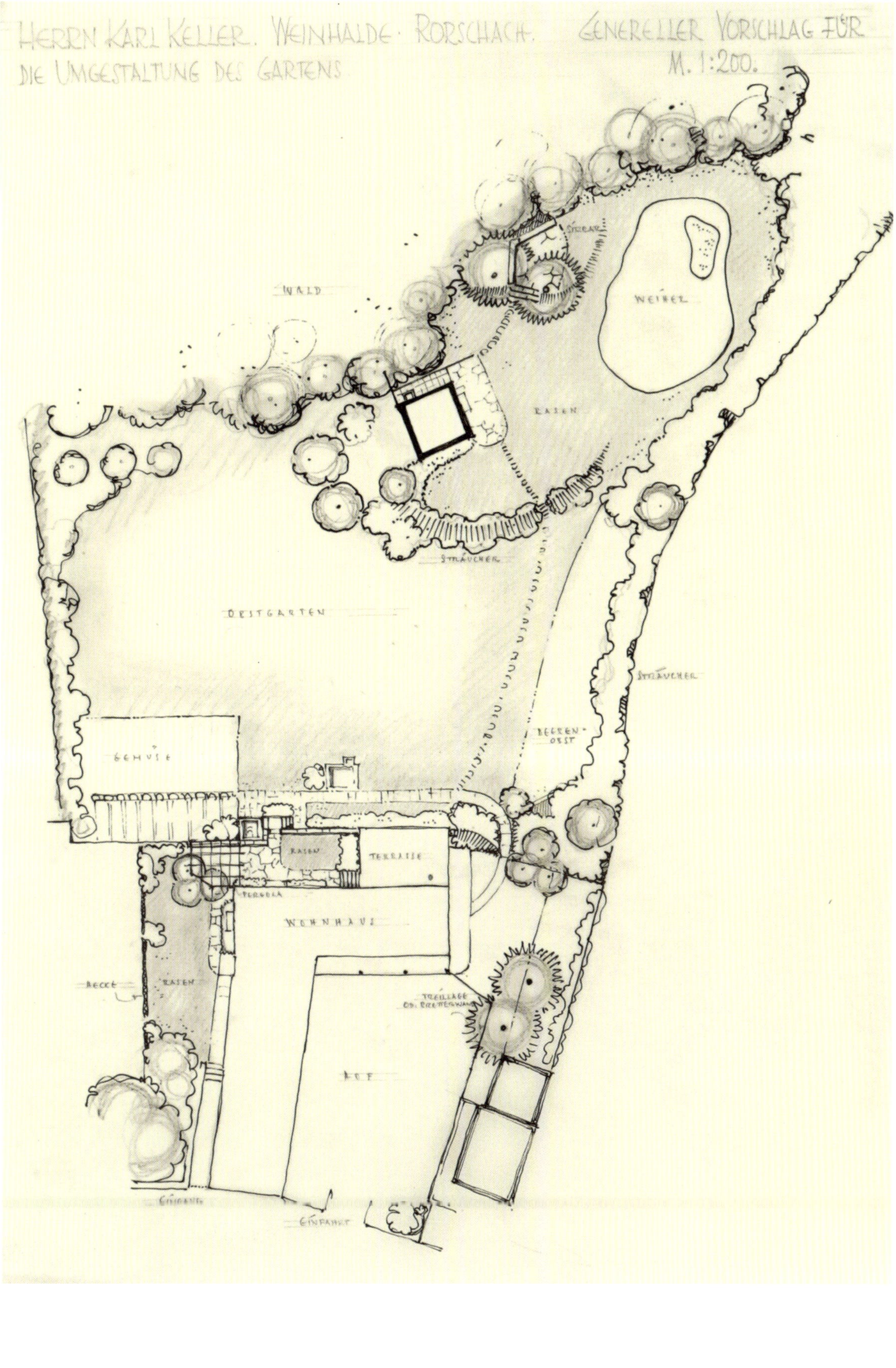

In a plan from 1950, Klauser adds a Japanese supplement at the end of the Keller Garden in Rorschach in the form of an island in the pond.

← In his plan for the Kopp Garden in Rorschacherberg from 1955, Fredy Klauser conveys a sense of Japanese fluidity in the form of a "little stream" bestrewn with a loose arrangement of stones.

↓ For the design of the pavilion in the "Boller Garden" in Turbenthal in 1928–1930, Fritz Klauser combines classical column forms with a Japanese roof shape and fronts it with an ornamental pond that, although rectangular in shape, appears biomorphic due to its planting. He softens the accuracy of the steps down to the pond by interspersing them with clustered tufts of grass.

→ Fredy Klauser places the bird bath in the "Bütler Garden" in Kradolf, the design of which is modelled on a Japanese pond, amid a planting of red-flowering Japanese cherry (*Prunus serrulata 'Hisakura'*), fountain grass (*Gymnotrix japonica*), beauty bush (*Kolkwitzia amabilis*), David viburnum (*Viburnum davidii*), false holly (*Osmanthus ilicifolius*), silver blush (*Cimicifuga japonica*), and Christmas rose (*Helleborus niger*).

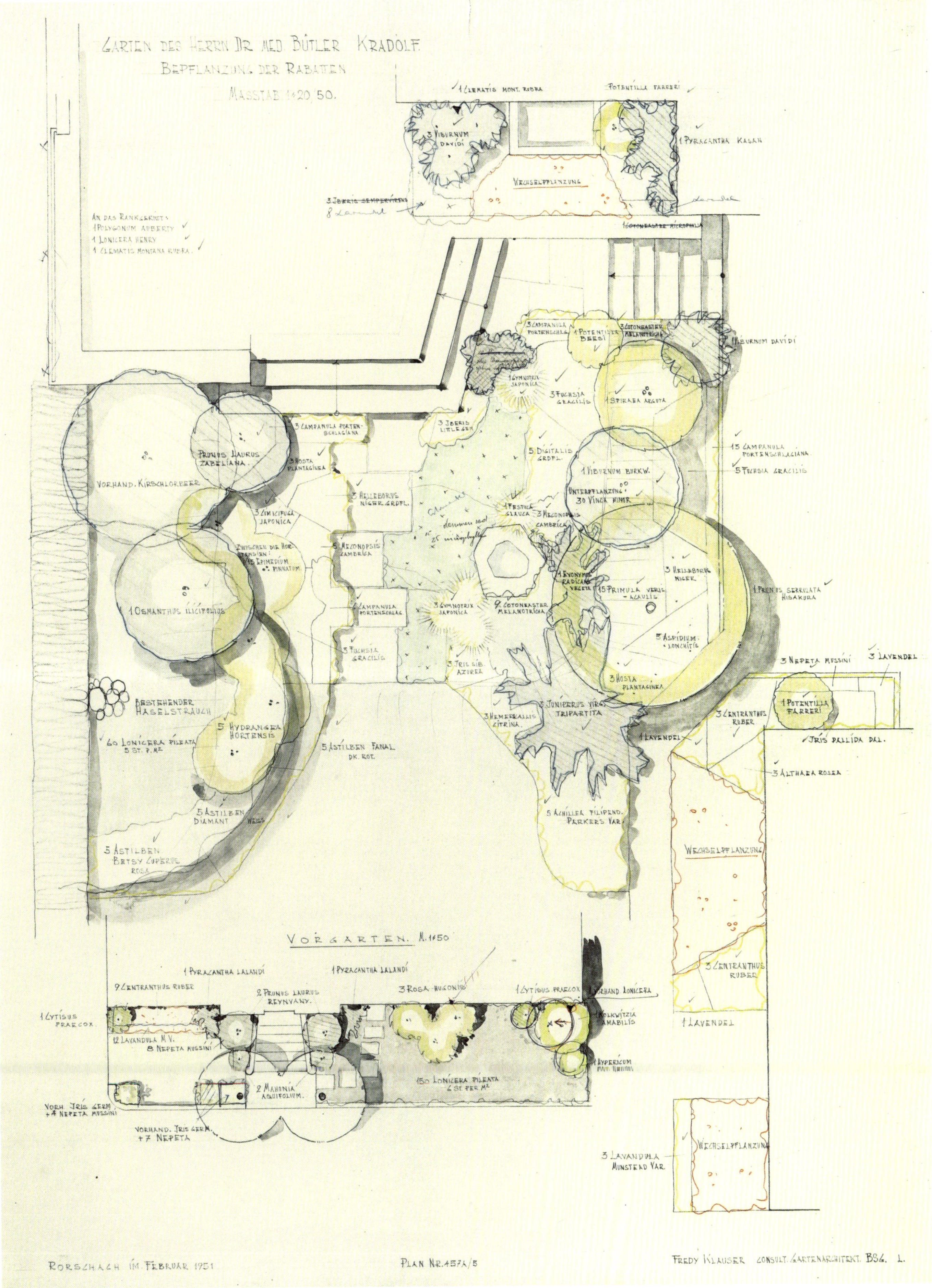

Garten des Herrn Dr. med. Bütler Kradolf
Bepflanzung der Rabatten
Masstab 1:20 50.
1 Clematis mont. rubra
Potentilla farreri
3 Viburnum davidi
1 Pyracantha Kasan
Wechselpflanzung
3 Iberis sempervirens
8 Lavendel
An das Rankgerüst:
1 Polygonum auberty
1 Lonicera henry
1 Clematis montana rubra
1 Cotoneaster microphylla
3 Campanula portenschlagiana
1 Potentilla beesi
3 Cotoneaster melanotricha
1 Viburnum davidi
1 Gymnotrix japonica
3 Fuchsia gracilis
1 Spiraea arguta
3 Iberis little gem
3 Campanula portenschlagiana
5 Digitalis grdfl.
15 Campanula portenschlagiana
5 Fuchsia gracilis
Prunus laurus zabeliana
3 Hosta plantaginea
Vorhand. Kirschlorbeer
3 Helleborus niger grdfl.
1 Viburnum burkw.
Unterpflanzung 30 Vinca minor
3 Cimicifuga japonica
1 Festuca glauca
3 Meconopsis cambrica
Zwischen die Hortensien 15 Epimedium pinnatum
5 Meconopsis cambrica
1 Evonymus radicans vegeta
3 Helleborus niger
15 Primula veris acaulis
1 Prunus serrulata Hisakura
1 Osmanthus ilicifolius
Campanula portenschlag.
3 Gymnotrix japonica
9 Cotoneaster melanotricha
5 Aspidium lonchitis
3 Fuchsia gracilis
3 Iris sib. azurea
3 Hosta plantaginea
3 Nepeta mussini
3 Lavendel
Bestehender Haselstrauch
1 Potentilla farreri
3 Hemerocallis citrina
3 Juniperus virg. tripartita
3 Centranthus ruber
5 Hydrangea hortensis
1 Lavendel
Iris pallida dal.
60 Lonicera pileata 5 St. p. m²
5 Astilben Fanal dk. rot.
3 Althaea rosea
5 Astilben Diamant weiss
5 Achillea filipend. Parker's Var.
Wechselpflanzung
5 Astilben Betsy Cuperus rosa
Vorgarten M. 1:50
1 Pyracantha lalandi
1 Pyracantha lalandi
9 Centranthus ruber
2 Prunus laurus reynvany.
3 Rosa hugonis
1 Cytisus praecox
1 Lonicera
3 Centranthus ruber
1 Cytisus praecox
1 Kolkwitzia amabilis
12 Lavandula M. V.
8 Nepeta mussini
1 Lavendel
1 Hypericum
2 Mahonia aquifolium.
150 Lonicera pileata 6 St. per m²
Vorh. Iris germ. + 4 Nepeta mussini
Vorhand. Iris germ. + 7 Nepeta
Wechselpflanzung
3 Lavandula Munstead Var.
Rorschach im Februar 1951
Plan Nr. 457a/5
Fredy Klauser consult. Gartenarchitekt BSG. L.

Walter Leder 1892–1985

The degree to which Leder strove to recreate an original situation in the "Weber Garden" near Bern in 1944 can be seen in the coloured perspective of the garden section with the stone lantern. The triad of blue, red, and yellow of modernist art recurs repeatedly in his designs.

Japan at heart

> Though our visit in Japan was short (5 days Tokyo, 9 days excursions) we received anyway a good impression from the country, the urban and rural, the industrial and natural landscape. We saw the town planning, the problems with traffic and green spaces, the gardens, parks, playgrounds and recreational landscapes. We saw historical and modern architecture, modern hotels in Tokyo, Osaka, Kyoto, Nagoya and Hakone. We were invited to a Japanese home, to Geisha restaurants, private parks, official receptions, performances and ceremonies. In rural areas, we saw horticulture – agriculture – and forestry – work sites, country houses, villages.[1]

Walter Leder had a deep respect for the East Asian culture. The many journal articles on Japanese gardens that he kept, his correspondence with his Japanese counterparts as part of the 9th IFLA Congress in Tōykō in 1964, the Japanese-infused Swiss garden design there, not to mention the contacts he cultivated over many years testify to this as much as his numerous own garden designs. Born in Genoa and raised in Barcelona and Naples, Leder learned to love travelling and be open-minded towards foreign cultures from an early age, which may have influenced his choice of profession. After school in Bern, Wimmis (Bern), and Les Verrières (Neuchâtel), he became a postal clerk in Cham (Zug) and quickly grew familiar with international export routes.[2]

His global horizon stood him in good stead when he was forced to change jobs as a result of the detrimental hygienic conditions of his workplace. In 1909, he enrolled in the École cantonale d'horticulture de Châtelaine (Châtelaine School of Horticulture) in Geneva, which exposed him to the world and ideas of French garden design. From 1912–1915, he worked for the Mertens brothers before leaving for Germany to study at the Gartentechnikum in Köstritz. In 1916, he found employment with Ludwig Lesser in Berlin, and in 1918 with Leberecht Migge in Hamburg. He continued to travel internationally after his return to Switzerland in 1920. No other landscape architect discussed in this book worked in as many cantons as Leder.

In Germany, he undoubtedly followed the debates conducted in the garden journals – not least those on Japanese gardens – for example, in *Die Gartenkunst*, in which Leberecht Migge had published his article "Die gute Garten-Wohnung" (The Good Garden Dwelling) praising the virtues of a Japanese garden (see "Inspiration, imitation, integration").[3]

Just how formative Leder's period of employ with Leberecht Migge was is shown by the fact that he took a compendium of his teacher's plans from the years 1910–1920 with him to Switzerland and – having gained a secure position as a teacher at the Gewerbeschule in Zurich – took up independent work as a freelance garden

1 Leder wrote these lines in English. Leder n. d. ASLA, p. 1.
2 "[…] and as a result of intensive export traffic by the Anglo-Swiss Milk Company [he learned] the fastest postal routes all over the world almost by heart, further fostering his interest in international connections." Leder/Mathys 1985, p. 44.
3 Migge 1927a, p. 10.

Leder's garden pavilion for the Zurich Horticultural Exhibition (Züga) in 1933 is reduced to a minimum, making the garden the most important protagonist.

The notion of the garden as an image is accentuated in Leder's 1944 design for the "Weber Garden" near Bern by framing it with a tokobashira – a one-sided post that rests on an unhewn stone.

architect.[4] What also linked them was their shared self-conception as architects. "I am beginning to make architecture"[5], Migge wrote in 1918, and Leder also signed the plan label on Migge's plan of the "Pergola" and the "Light and Air Bath" as "Architect and Teacher" (see "Inspiration, imitation, integration").[6]

Despite working on garden designs for innumerable Wilhelminian villas, his own architectural design approach was modern [→ p. 215]. Leder's garden pavilion for the Zurich Horticultural Exhibition (Züga) in 1933 was transparent to the point of being almost fleeting and dematerialised, and broke the architectural mould, just as Cramer did later, in 1947, at the Züka (see "Ernst Cramer – Japanese tempered"). As a "place for living in light, air and sun"[7], as Valentien had postulated, Leder's creation translated the maxims of Neues Bauen to the context of garden design [→ p. 211]. Its filigree radicalism was still remembered decades later: "At the Zurich horticultural exhibition Züga 1933, he had created a small dwelling garden with a completely modern garden room made of tubular steel and glass in the smallest of spaces."[8]

Leder also shared Migge's opinions on Le Corbusier. His estate contains slides he had made of Le Corbusier's sketches of landscape architectural elements – green room, vegetative branching, artificial hills – for the Capitol in Chandigarh and published in the *Œuvre complète 1950–57*.[9] Further evidence of Leder's affinity with architecture can be seen in the numerous plans for pavilions and garden houses that he designed. For the architects he worked with, he was therefore an equal partner. Otto Rudolf Salvisberg (1882–1940), for example, entrusted him with the design of the garden of his home on Restelbergstrasse in Zurich, which he built in 1932.[10] In the same year, Peter Meyer remarked on "the connection with the garden and the surrounding landscape" at the Salvisberg house, whose terrace had the character of an "outdoor living space".[11]

Leder emancipated the garden from the architecture and interpreted it as a walk-in three-dimensional image, as if "photographed" by the house as a camera. He drew on the visual language of Neues Bauen as liberally as he adapted the traditional Japanese arts. He sketched the view from a roof garden framed by two ribbon windows[12] and in some floor plans often drew the Japanese elements from a slightly oblique viewpoint to reveal their spatial dimension. This method of representation recalls the Japanese practice of parallel projection, which enabled the artists to create the illusion of three-dimensional space on a flat surface, while at the same time allowing the objects in the background, presented as if in plan, to stand out just as clearly as those in the foreground.

In the year of the national exhibition, Leder designed a garden for Dr. F. Meyer-Kaiser in Sursee that virtually encompasses the house. Its plans are suffused with Japanese charm. Two stones rise up out of the curved pond as islands, while natural-looking stepping stones

4 Leder/Mathys 1985, p. 44.
5 Migge 1936. Although printed in 1936, the letter dated from 1918, as noted by the editors of the NZZ: "The recently deceased famous garden architect Leberecht Migge sent the following letter to a Swiss colleague in 1918, in which he set out some basic ideas about horticulture in his inimitable manner".
6 ASLA, mig_4_21-0002, o. D., Im Eisernen Zeit, Zurich.
7 Valentien 1932, p. 33.
8 Leder/Mathys 1985, p. 45.
9 Boesiger 1957, p. 113. The review of the work appeared in the same issue of the journal *Das Werk* as the one on *Der japanische Garten* by Tetsurō Yoshida. See Huber 1957; Theile 1957.
10 See Meyer 1932a; Meyer 1932b. Leder used the example of intimately interlocking house and garden in that project to propose a project two decades later to Alfred Roth's (1903–1998) client Philippe Rossiez. Rossiez, however, seems to have invited offers from competitors. The first choice was probably Leder, whose estate contains the architect's plans – floor plans, sections, views – dated 15 Dec. 1953 and an axonometry of Leder's garden room dated 5 April 1954; ASLA WL_14493_1 – WL_14493_5; WL_14493_6. Gustav Ammann drew a single garden plan, dated 16 Sept. 1954; gtaA NSL 2–0269. On 13 March 1955, Ernst Baumann took over the project; ASLA BAE, "Haus Dr. Rossiez, Küsnacht, Umgebungsgestaltung", no rec.
11 Meyer 1932b, p. 169.

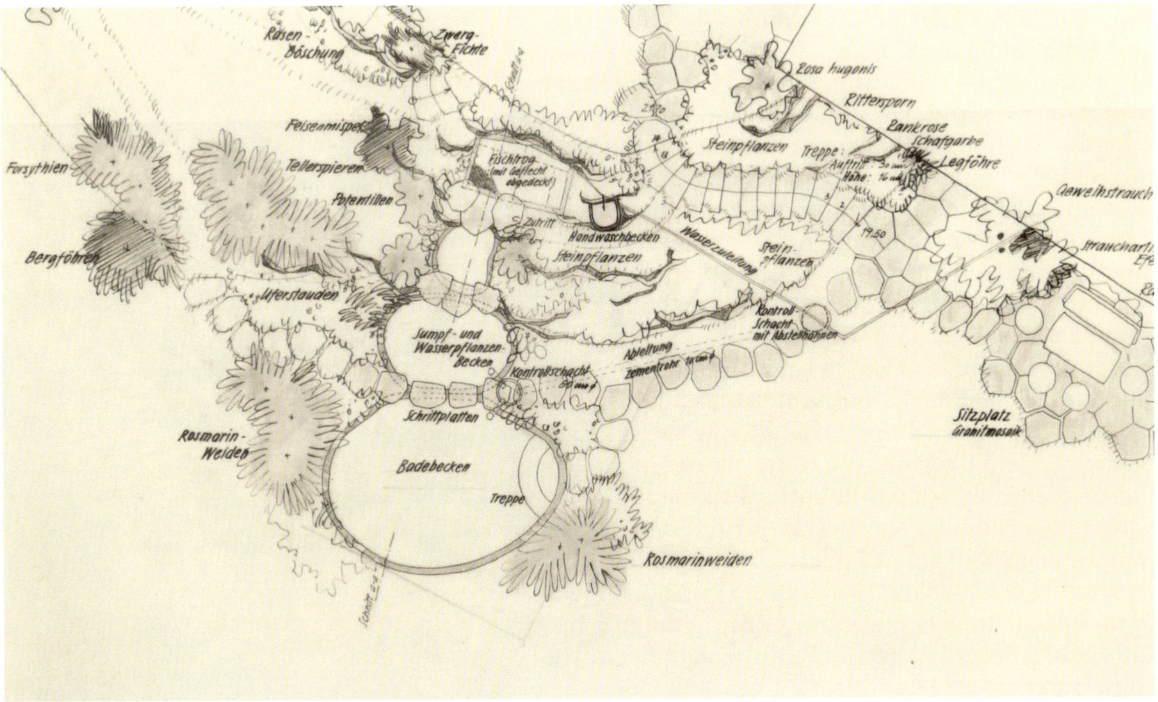

↑ Detail of the cascade in the “Schönenberger Garden” in Uznach.

← A rare surviving photograph of a Japanese-inspired winding stream with a similarly meandering path in the “Hugenin and Allemand Garden”.

allow one to traverse the water's surface, which is partly covered by water lilies, in the Japanese manner. He concealed the inflow with a stone bench so that the water appears to pour into the basin from a virtually invisible source within the small adjacent rock garden. Japanese maple, summer lilac, Japanese snowball, and rhododendron as well as creeping juniper and dwarf spruce provide organic Japanese accents.

Leder used "circular stepping stones" in the design for the "Brandestini Garden" in Küsnacht in the same year that Tetsurō Yoshida depicted the iconic stepping stones in the garden of the Heian shrine in his publication *Der japanische Garten*, which was published in 1957[13]. The shape of the basin, alongside which he planted Japanese maples, resembled the one in the garden of the Hompō-ji, which Yoshida attributed to the painter Ogata Kōrin.[14] In the same year, Leder gave his own garden a Japanese character and again two years later at the G59, where "he was able to realise his idea of the combination of home and garden in the enchanting country house situated there in the midst of the water garden below Belvoirpark" [→ p. 102].[15]

As with numerous projects, he devoted particular attention to the Japanese section of the "Vogt Garden" in Reinach in 1944, sketching several variant studies. Here the stream ripples over conical half-shells, a device he would use again later in his own garden.

"An intimate friend of mine"

Leder had been personally acquainted with Akira Satō, the president of the Japanese Institute of Landscape Architecture (JILA), since at least 1954, when at the Vienna meeting of the IFLA, Leder was elected president for two years as the successor to Sir Geoffrey Jellicoe.[16] It must have been a trusting relationship, as Satō recommended friends and companions visit Leder on their trips to Europe. In 1960, for example, he sent an "intimate friend of mine", Kuro Kaneko (see "Inspiration, imitation, integration"), who was to stand in for Satō at the IFLA Congress in Amsterdam[17], as well as two ministry officials Yoshio Sekimori and Hideo Kamura.[18] Satō and Leder obviously had a personal connection, as documented by handwritten greetings to Leder's wife Emmy, to their son Klaus, and numerous greeting cards.[19]

It was certainly thanks to Leder's initiative that Satō, who was in Zurich after the Vienna IFLA Congress, presented Japanese gardens to the members of the Bund Schweizerischer Gartengestalter (Federation of Swiss Garden Designers, BSG) at a "quickly improvised meeting".[20] Margaret Brasch opened proceedings with an introduction to the "manifold Japanese landscape" with the help of "splendid coloured photographs that related the special character and beauty of the country", instantly awakening among those present "a desire to travel" to the country.

She had grown up in Japan as the daughter of Heinz Brasch (1905–1995), who was born in Tōkyō and worked in the Volkart brothers' trading house, which was founded in Winterthur in 1851 and specialised in trading in colonial goods.[21] In addition, he collected Japanese art, translated and wrote books – in particular about Nanga and Haiga.[22]

After the Second World War, Brasch took up residence in Forch, Zurich, in whose garden Leder designed a Japanese section in 1964. He arranged it around a curved pool, which he inscribed in a circle so that the water's surface roughly corresponded to the shape of one-half of the yin-yang symbol, and the planted area the other. A slightly curved bridge traversed the pond and led into a path of rounded stepping stones that at one end bunched together to form a circular seating area and at the other end led on to the pergola. Leder drew the pergola as a kind of living folding screen, "painted" with stones and boulders, against a background of azaleas, rhododendrons, clematis, heather, grasses, and ferns.

This took place in the run-up to the departure for the 9th IFLA Congress in Japan, where Brasch was probably the contact person for the Swiss delegation.[23] Leder, who organised the visit to Japan (see "Inspiration, imitation, integration"), prepared meticulously for the stay, studying traditional architecture as well as garden design.[24] Leder felt particularly honoured by Satō's invitation to design a "small Swiss garden" as part of the national gardens planned in Mukōgaoka Park,[25] and accepted post-haste. He soon had an idea for its concept: "I would like to design the garden as a picture," he wrote to Satō, alluding to the traditionally close relationship between art and garden design in Japan. He harboured a passion "for the connection between garden and open countryside" with a view to bringing "the landscape closer to architecture in the composition of the picture".[26]

12 Roof garden Balhani MD [Balkanyi], Seefeldstr. 152, Zurich, 4 April 1973, with Klaus Leder, ASLA, WL_13943_4.
13 ASLA, WL_13877_6; Yoshida 1957, p. 121.
14 Yoshida 1957, p. 21.
15 Leder/Mathys 1985, p. 46.
16 Ibid. (Leder/Mathys 1985, p. 46.)
17 Leder 1960a ASLA.
18 Yoshio Sekimori, Director of Town Planning Bureau, Ministry of Construction; Hideo Kamura, Chief of Park Section of Town Planning Bureau, Ministry of Construction; Leder 1960b ASLA.
19 Leder kept these evidence of friendly exchanges as well as articles in newspapers and magazines about Japanese architecture, which also document his interest in traditional Nippon architecture. See U. I. 1964; Domenig 1965.
20 Hofmann 1954. It is this newspaper article that the Leders sent to Gustav Ammann.
21 Brasch met Elsy Leuzinger (1910–2010), who was director of the Museum Rietberg from 1956–1972. His friendship with her led him to bequeath a large part of his collection to the museum as well as his library of almost 200 volumes "mainly containing works on Japanese woodblock prints". See Leuzinger 1967, p. 21.
22 See Brasch 1965; Japanese Ink Painting 1962; Colour Woodcuts 1965.

Perspective of a villa project: Leder blurs the boundaries between house and garden by using the extending wall panel as a canvas for a tableau of plants and allowing the water basin to cross over the threshold into the interior.

← Leder arranges the window as a frame to an outdoor picture in the "Zollinger Garden", designed in 1960. Pines and Japanese maple surround the water lily pool with its iconic round stepping stones.

↓ Leder's stone setting combines a standing main stone and two lying stones according to the Buddhist model.

Framed, Japanese-inspired miniature gardens created by Leder for the community centre in Opfikon, Zurich, 1959.

← A picture within a picture – Leder drew the view through an opening into the "Pulver Garden" 1946, which itself features a framed view.

↓ "H. Wolf-Bender Garden": The view of the lake and mountains framed and partially articulated by the terrace door refers just as much to the European topos of the window as an image as it does to Japanese photographically captured views.

The notion of the garden as an image as captured in a photograph by Leder of the view from a house onto a Japanese garden in Kyōto.

"Framed vistas to the water-fountain, to the iris garden in Japanese style, to the large pond, to the bridge, to the pergola, to the sculpture, and to all the nice details of vegetation"[27] envisaged Leder for the Mukōgaoka project, as well as a stream, a peninsula, a hill, and a background covered with trees.[28] On 22 February, Leder delivered the plans. Since he could not supervise the work on site, he gave detailed instructions[29] – including for its planting. His planting list of 17 March 1964 ran to almost a hundred items.[30] Even more exceptional was that Leder knew the Japanese names of almost all the plants, noting them in his catalogue of chosen plants (see plant catalogues).

He had obviously began studying Japanese plant names several years earlier – possibly inspired by Yoshida's publication of 1957. In any case, he used the name of the Japanese ornamental cherry "Hisakura" in that year in one of the plans for a garden for the industrialist Emil Meyerhans (1906–1997) in Weinfelden, Thurgau.[31] Its design was a work in progress that spanned fifteen years – from 1944 to 1959.[32] Leder created a collage of plants in the primary colours red, blue, and yellow as well as white, as Piet Mondrian defined them, and gave the entire composition a Japanese-style with weigela, spier shrub, Japanese maple, ornamental cherry, magnolia, lilac, peony, hydrangea, rhododendron, and azalea. A development from naturalism to abstraction – in the manner of Mondrian – as Heini Mathys diagnosed in Ernst Cramer,[33] is also evident in Leder's work. In 1967, the naturalistic borders in the "Meyerhans Garden" were replaced with a geometric segmentation that bore striking similarity to Mondrian's compositions [→ p. 227]. The centre of the garden, which Leder designed together with his son Klaus in a villa quarter in St. Gallen, was an extensive pond in the shape of a figure of eight that was "conceived not for bathing, only for the eye" and which "magically reflected" the beech woodland in the background.[34] "Two slightly raised sections with a green carpet of sedum and beautiful forms of Japanese ornamental shrubs",[35] formed the visual middle ground. Leder's perhaps most attractive architectural project, which he seems to have drawn without a commission, has red walls, yellow awnings, and a blue water surface and, as such, can be read as a three-dimensional Mondrian [→ p. 215]. Saburō Hasegawa (長谷川 三郎, 1906–1957), for example, testified that Japanese artists associate the buildings in their homeland with Mondrian: "In a sense, we live in and on Mondrian paintings."[36]

His enthusiasm about the trip to Japan with the excursions to historical and contemporary gardens and temples, the demonstration of a tea ceremony, the instruction in ikebana, the insights into pottery and textile crafts and – above all – the encounters with "nice and important people", is evident in the lines Leder wrote on his return.[37] Of all the Swiss participants of the trip to Japan, he brought back the most extensive visual documentation. Not only did he take numerous photographs of both historical and contemporary installations himself, but he also acquired several boxes of slides.[38] And, as he himself wrote "we came back with Japan in our heart[s]".[39]

23 Leder noted the hotel accommodation of the Brasch and his wife – "Hotel Palace Tokyo" – and the business address: "Kaigai Tsusho K.K. [Kabushiki Kaisha], Kanda-Tsukasacho, Chiyoda-ku, Tokyo"; Leder n. d. ASLA, IFLA II. Kaigai Tsū shō is the Japanese translation for overseas trade. The company Übersee-Handel AG (UHAG) was founded in 1927 by Julius Müller, a former employee of the Volkart brothers, in Zurich and Tōkyō. See Brauchlin/Wehrli, p. 222. As head of UHAG, he led a Federal Council delegation through Kyōto in 1967 to "supplement the impression of the rapidly advancing modern country as it had confronted us in the capital with the image of traditional and cultural Japan". Probst 1969/2012, p. 61. The trip had originally been planned to celebrate the 100th anniversary of the first trade and friendship treaty between Switzerland and Japan, concluded on 6 February 1864.
24 Leder 1960b ASLA.
25 Leder 1963a ASLA.
26 Leder/Mathys 1985, as note 2, p. 45.
27 Leder 1964b, p. 2.
28 Leder 1963b ASLA.
29 Leder 1964b.
30 Leder 1964c ASLA, pp. 9–14.
31 ASLA WL_14421, Plan23 [WL_14421_23], 8 May 1957. On Emil Meyerhans see anon. 1998; Etter 1974, pp. 93–95.
32 After 1944, 46, 48, and 49, Leder would revisit plans every year from 1954–1959.
33 See Mathys 1978, p. 32.
34 Anon. 1967, p. 32.
35 Ibid.
36 Johnson/Hart 2019, p. 31.
37 Leder o. D. ASLA, p. 1. Contact with Japanese colleagues continued after Leder's return to Switzerland. Some also visited him, such as Yoshitaka Akui (see "Inspiration, imitation, integration"). The 28 people on Walter Leder's list of participants were joined by Yoshitaka Akui, Mitsuo Yokyoama, and Teitaro Kitamura, who were already in Switzerland. Leder 1972c ASLA. With only one day, 28 August, available for visits, the Swiss programme was comparatively meagre. It included visits to the lakeside gardens: on the right bank, the lakeside path from the Zürichhorn, via Tiefenbrunnen lido to the Utoquai, with a detour to the Töchterschule on the Hohe Promenade; on the left bank, the Arboretum, the Succulent Collection and the Belvoirpark, again with an architectural detour to the Freudenberg Cantonal School; Leder 1972e ASLA.
38 "Japanese Gardens I", "Japanese Gardens II", "Flower Arrangement", "Beautiful Fuji", "Katsura Detached Palace", "The Cherry Blossoms" and "Beautiful Geisha Girl" by SK Slide Co, LTD, Tōkyō (Kanda-Sarugakuchō, Chiyoda-ku, Tōkyō) and "Japanese Garden (1) – Gardens in Kyoto" by Nippon Bunka Film Co. Ltd.
39 "we came back with Japan in our heart" [sic] Leder n. d. ASLA, p. 2.

The Japanese section of the "Weber Garden" in Wabern near Bern from 1944 – with spring, stream, fast-flowing water, bridge, water lily pond crossed by irregularly placed stepping stones, silver willow, and Japanese maple. Leder's drawing has a slightly oblique aspect to make the garden stand out more vividly.

In this undated plan of the “Kaufmann Garden” in Thörishaus, Leder devotes particular attention to the Japanese section of the garden with its spring, stream, fast-flowing sections and water-lily pond.

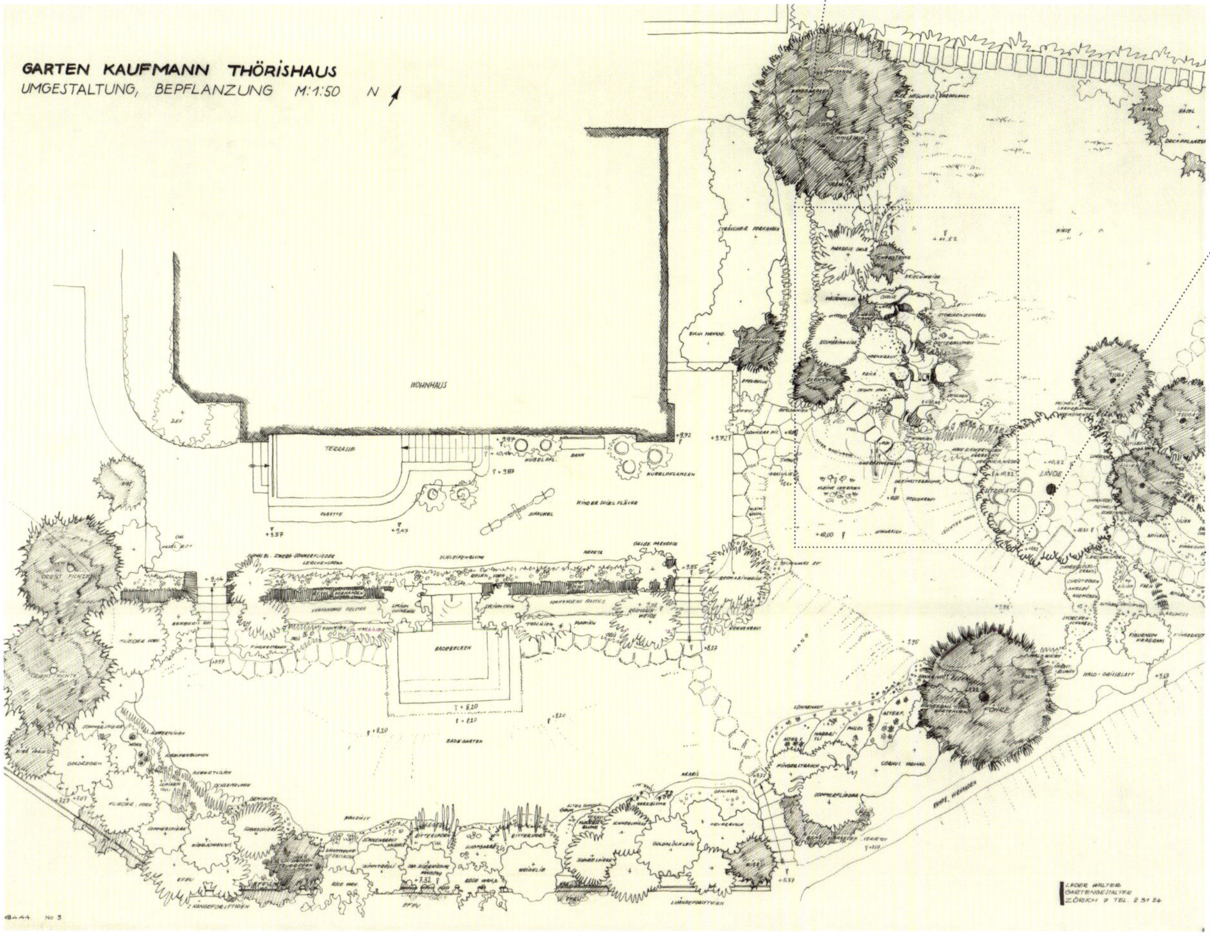

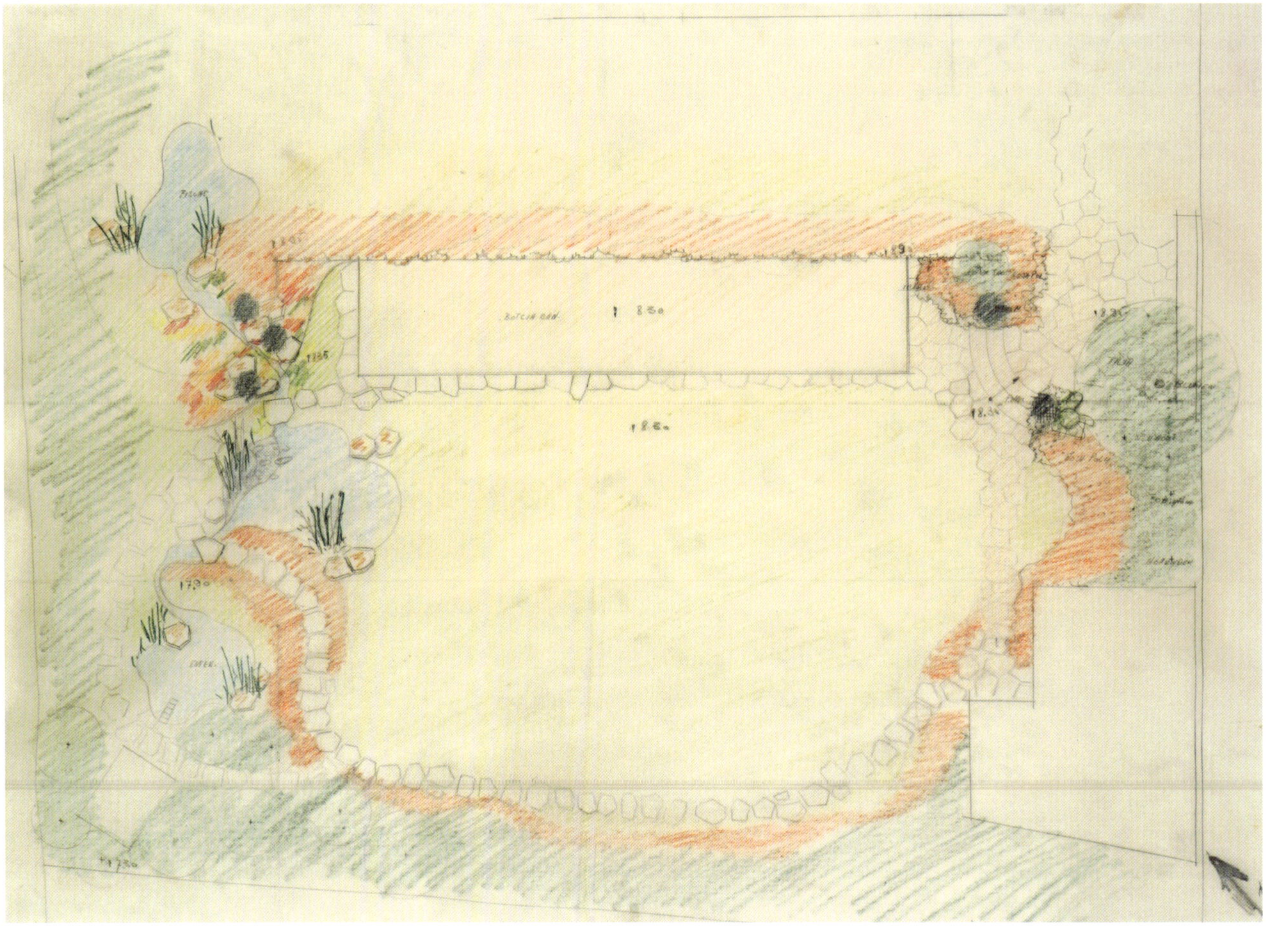

↑ Coloured sketch of the Japanese part of the "Bruppacher Garden" in Uitikon, designed in 1944.

← A perspective of the "Brasch Garden" showing a miniature landscape of pond, arched bridge, round stepping stones as well as irises, water lilies, lotuses, and bulrushes.

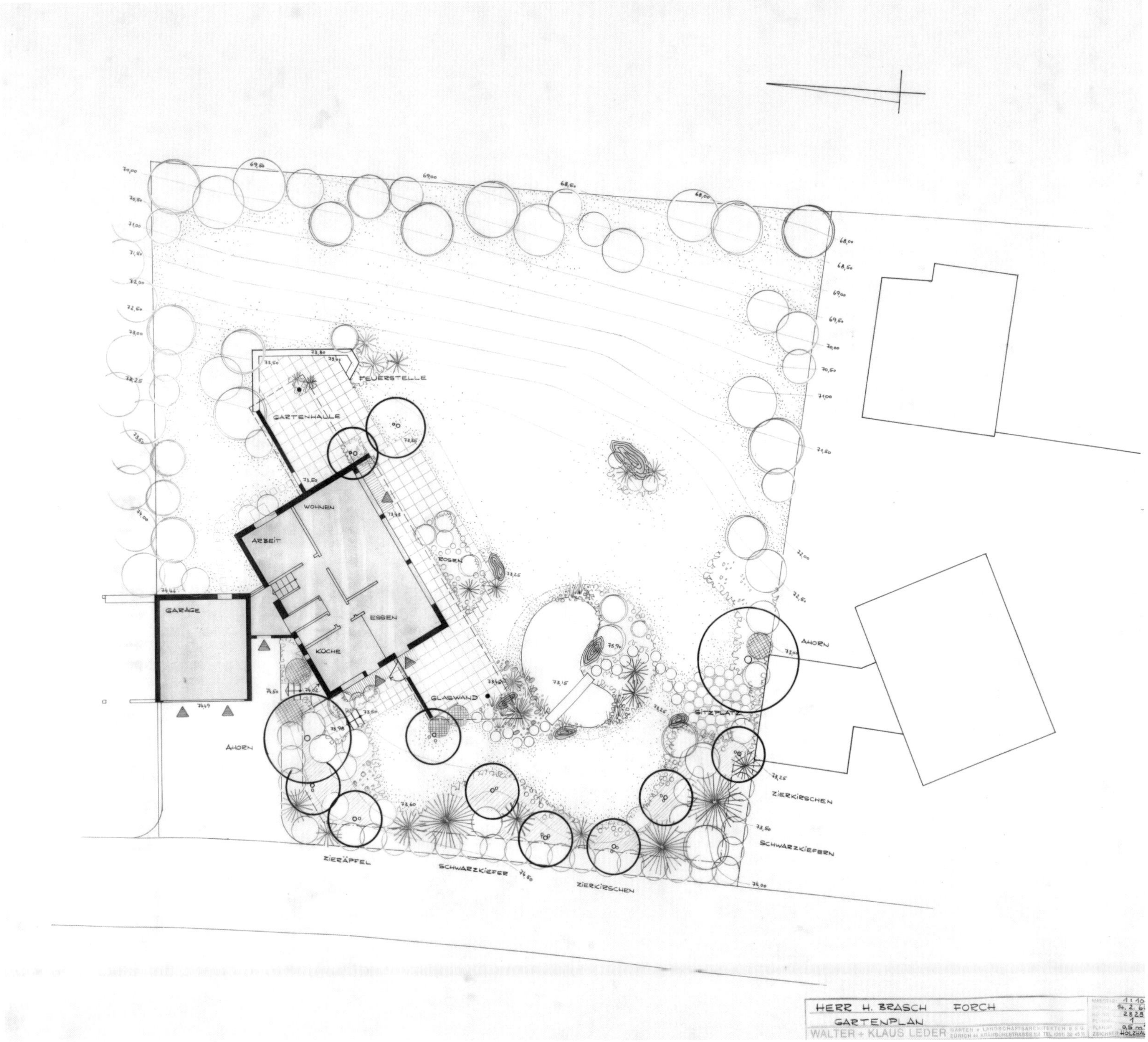

Leder employed stone settings and the planting of birch, silver maple, black and mountain pine, and Japanese ornamental cherry to heighten the Japanese character of the pond landscape in the definitive plan of the "Brasch Garden" dated 24 April.

↑ A stream rippling over stones and spanned by arched bridges, flows into a circular basin in the “Dierauer Garden”.

← For the undated “Schönenberger Garden” in Uznach, Leder created an enchanting sketch of a stream that splashes as a waterfall over several steps and flows into a natural pond. The stepping stones crossing the pond once again draw on the Japanese model.

The ground plan of the garden layout for Meyer-Kaiser in Sursee from 1939 reveals how the house and garden interlock. Two stones figure as islands in a pond with a curving outline.

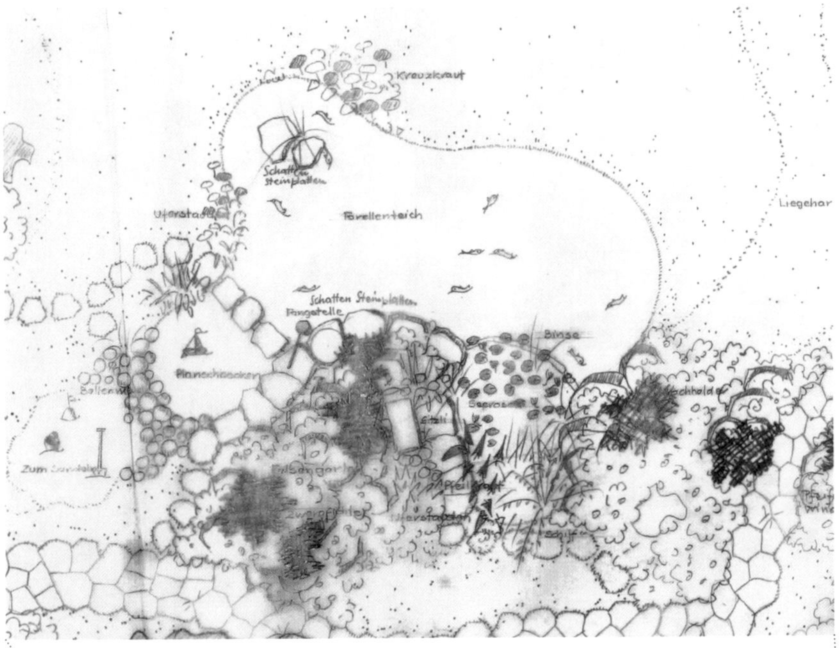

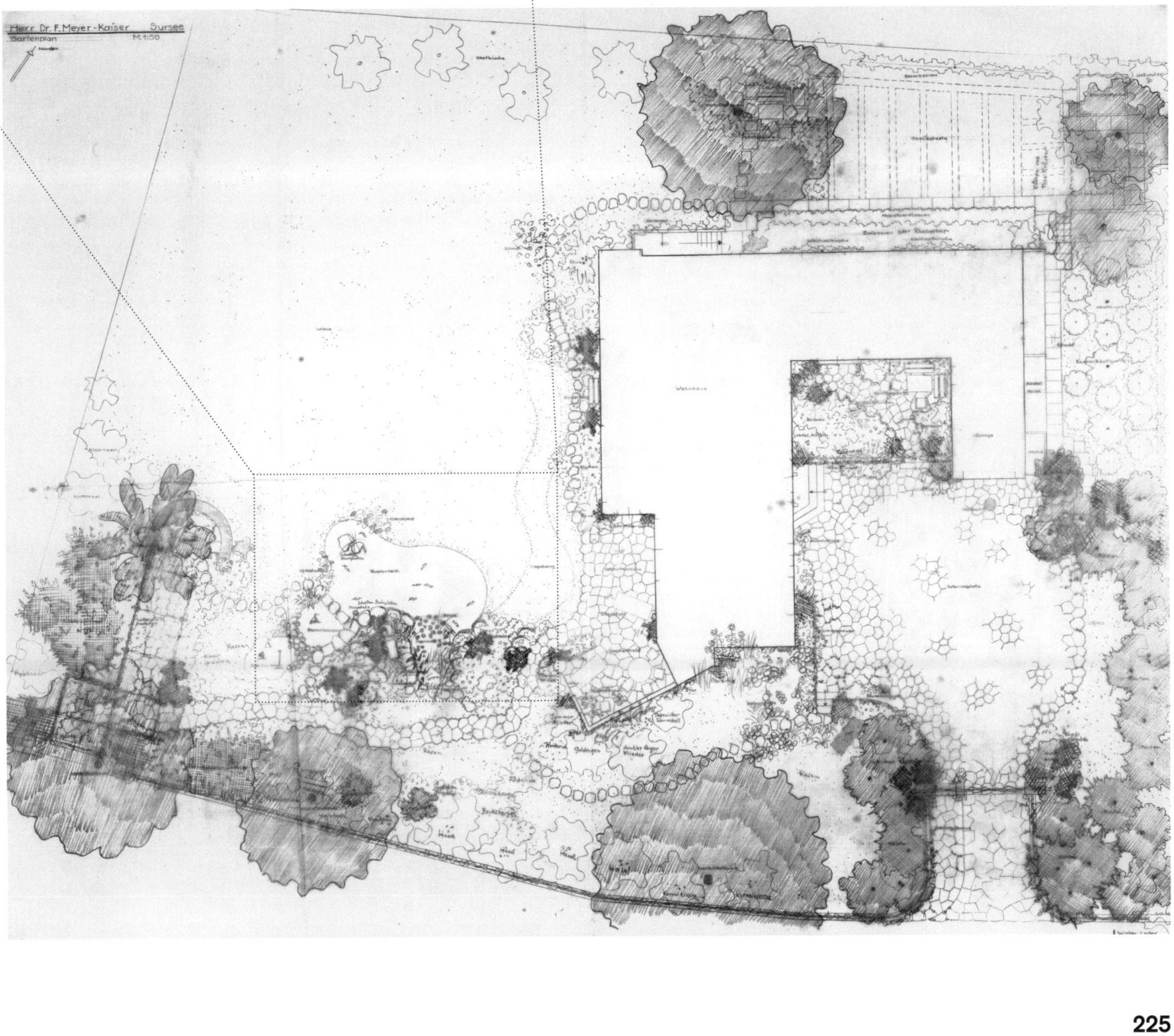

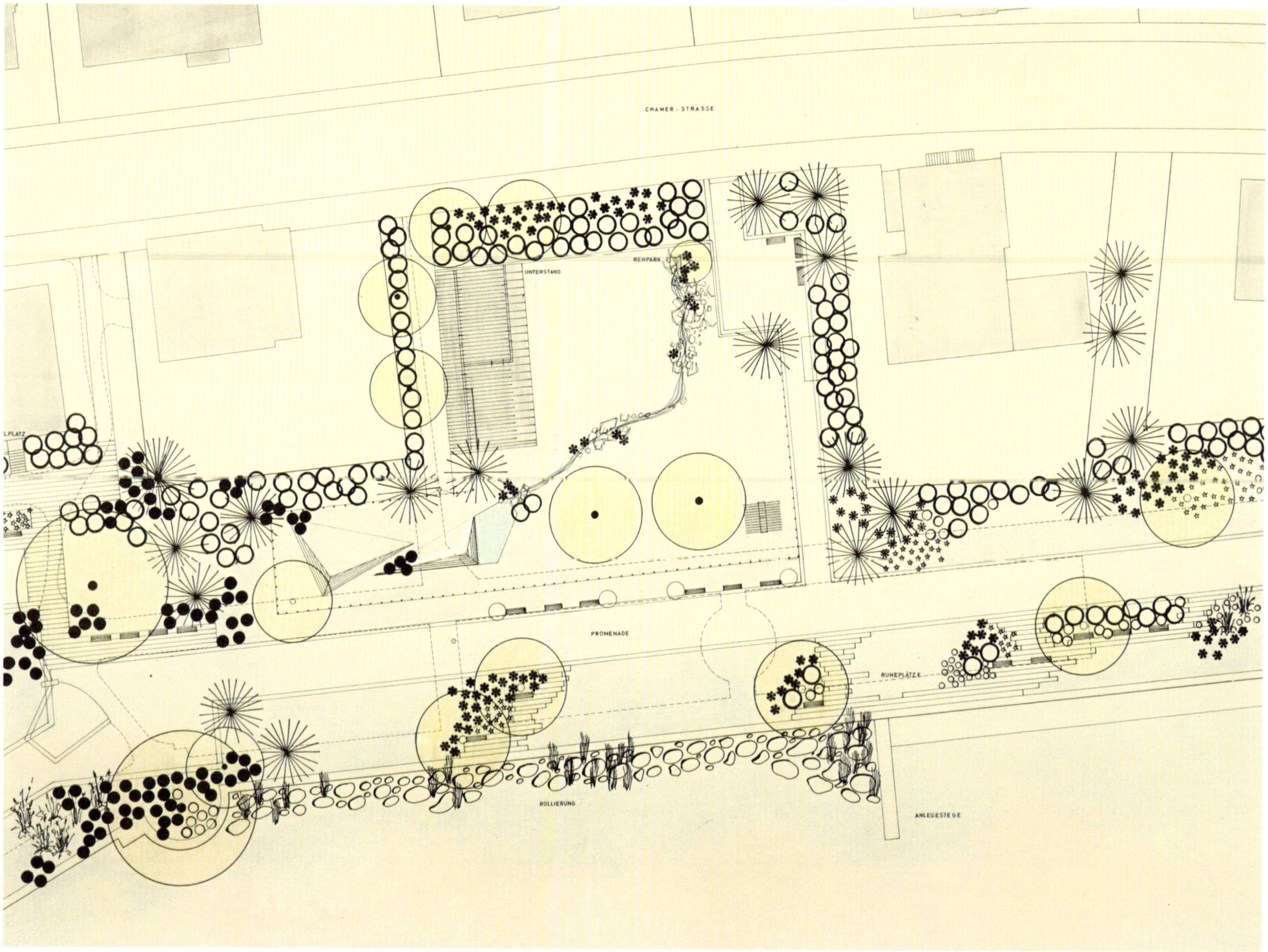

The “deer park” in the project for the lakeside design in Zug in 1965 could be a reference to Nara Park, which Leder and his colleagues visited in Japan in 1964. The stream, its course defined by stone settings in the Japanese manner, flows into a rugged rock pool.

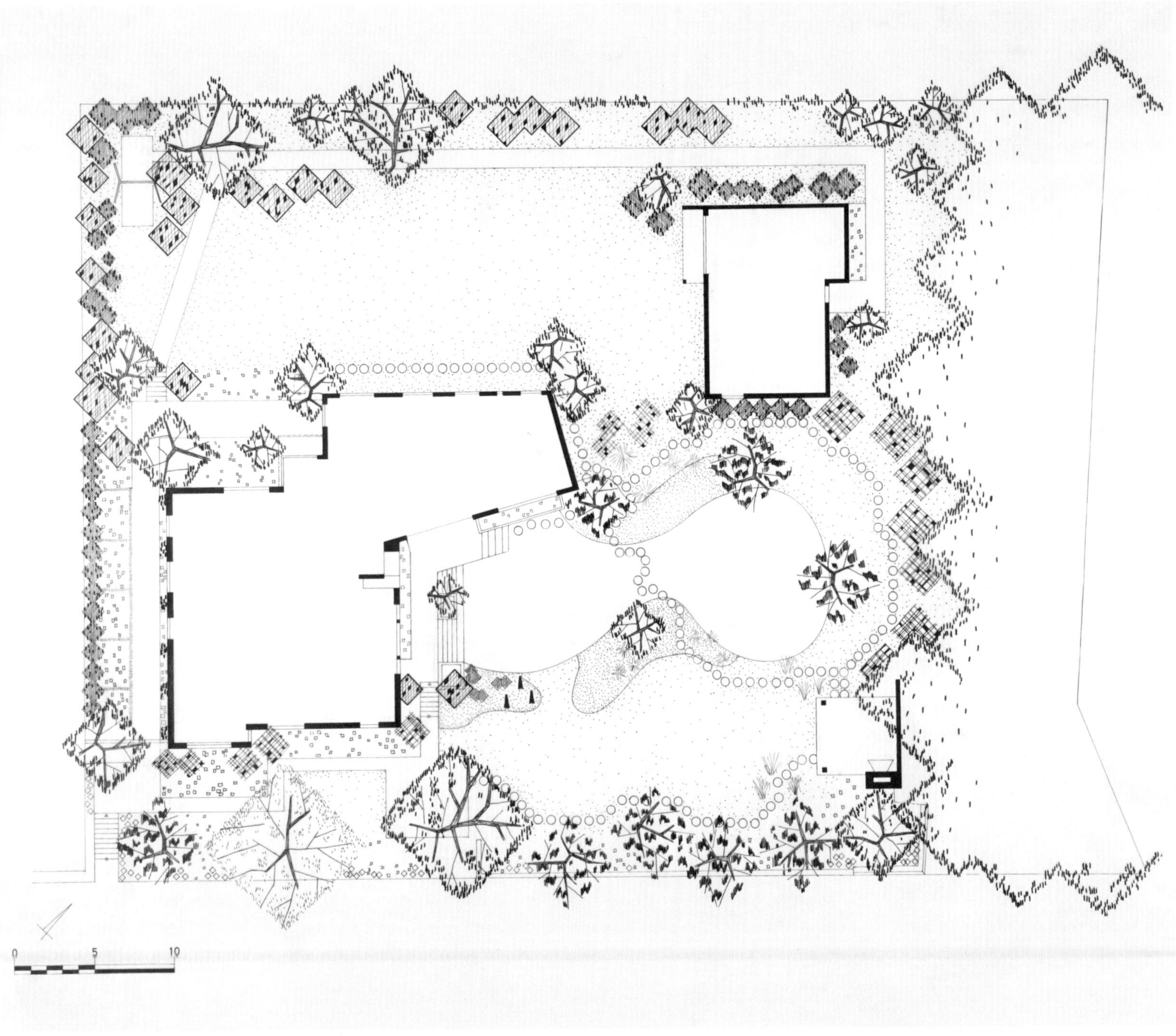

In this anonymous project in St. Gallen, abstract geometric avenues in the spirit of Mondrian are combined with organic naturalistic Japanese forms through the use of the iconic circular stepping stones.

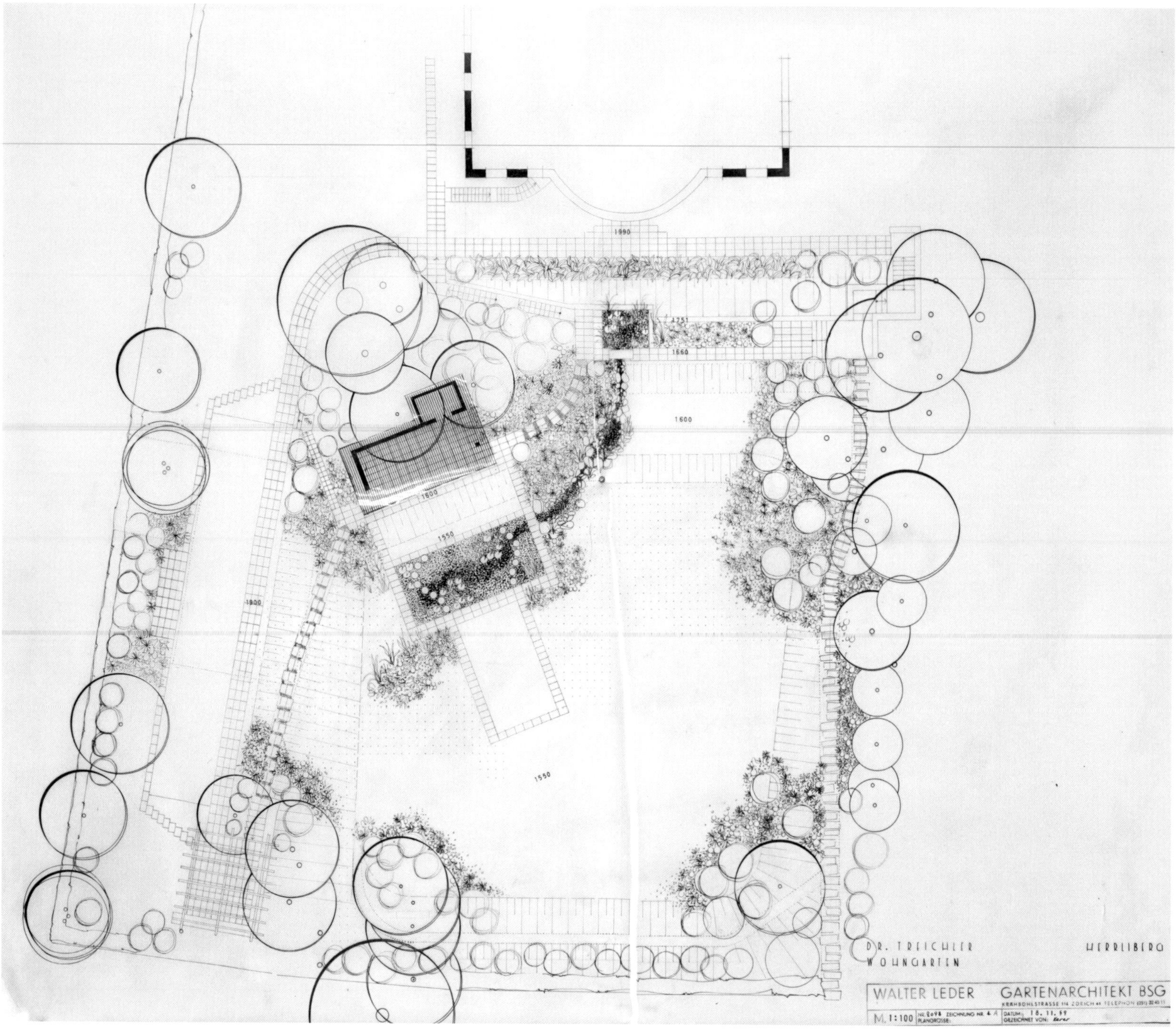

In the “Treichler Garden” in Herrliberg in 1959, Leder proposed a combination of geometric elements – spring and pool – and naturally designed elements – stream.

For the "Frey Garden" in St. Niklausen in 1945, Leder borrows from Japanese dry landscapes to create two dry streams.

Evariste Mertens 1846–1907
Walter Mertens 1885–1943
Oskar Mertens 1887–1976
Hans Nussbaumer 1913–1995

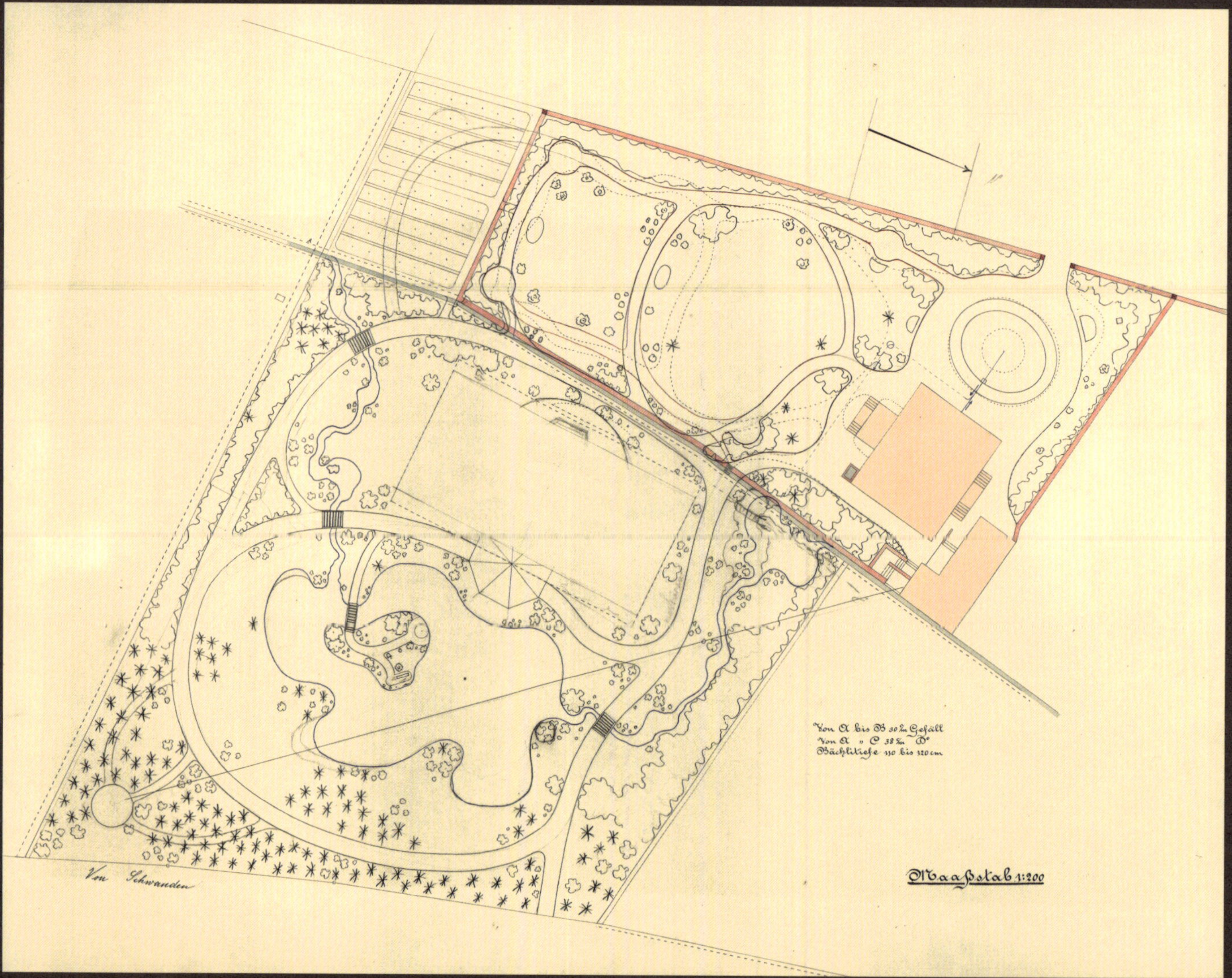

Evariste Mertens drew five designs for the garden of Jacques Trümpy-Heer in Mitlödi (Glarus) in 1895. The one shown here, with the pronounced curves of the stream and the amoeba-like outlines of the pond, is the most spectacular of the designs produced. The size of the off-centre island is also most harmoniously proportional with respect to the body of water.

Japanese masked

Create a horizon as wide as possible, which includes the most attractive motifs of the outlying landscape, mask any unattractive points and try to bring the planting into harmony with that of the shore on the other side.[1]

Evariste Mertens, who was born on 9 December 1846 in Breda, the Netherlands, and grew up in Schaerbeek, a suburb of Brussels, would probably be described today as a high achiever. He studied at the horticultural school under Louis van Houtte (1810–1876) in Ghent, Belgium, graduating with honours in 1865, and was awarded a state scholarship for further training abroad, which he completed at the English company American Nurseries in Leytonstone in East London and with Edouard André in Paris, one of the leading gardeners, garden designers, and botanists in the 19th century.

Arnold Neher (1846–1906), a fellow student from his time with van Houtte and also a friend, persuaded him to "become his associate".[2] In 1870, at the age of just 24, the two founded a gardening company on the grounds of the Sonnenburg estate[3] in Schaffhausen.[4]

After his partner withdrew from the venture and his appointment to the Swiss Federal Polytechnic as a lecturer in fruit growing and fruit science, Mertens relocated his radius of activity to Zurich in 1885, where he teamed up with Otto Froebel for three years. Together with Leopold Karl Theodor Froebel's son, he worked on the redesign of the quayside facilities at Zürichhorn before opening his own business. Through his teaching at the ETH Zurich, he imparted his knowledge to the next generation until his death in 1907, as did his two sons Walter and Oskar, born in 1885 and 1887, who followed in their father's footsteps in 1907 and 1912 respectively.[5]

His public parks and private gardens in Zurich – Villa Patumbah, Schloss Sihlberg, Platzspitz near the Landesmuseum, Belvoirpark and Zürichhorn quayside – as well as the Kurpark and Villa Boveri in Baden, Park in Bad Schinznach, Villa Sulzer-Grossmann in Winterthur (1894) can all be classified as being of the late landscape garden style.

Nevertheless, Mertens advocated "borrowing" the surrounding landscape. "Where an estate borders a lake, it is natural to want to make the most of its privileged location [...]. One creates as wide a horizon as possible, encompassing the most attractive motifs of the outlying landscape and masking any unattractive points, and then devises a planting that harmonises with that of the shore on the other side."[6] Here he combines shades of Jiejing's notion of borrowed landscapes with the concepts of English landscape gardens.

In the same issue of the *Zeitschrift des Schweizerischen Gartenbauvereins*, Georg Lasius highlights the influence of Chinese principles. "The English garden, as seen in Kent, [...] seeks to artificially bring together peculiarities of nature that may be scattered over a wide area on a small terrain, so that the garden appears almost to be a picture book of interesting regions."[7]

This approach he attributes to an awareness of the model of Chinese gardens, as propagated by William Chambers in *Designs of Chinese Buildings*[8], although it is likely he also studied the sources. In any case, he refers to the cosmic dimensions of the gardens attested to in the "Chinese annals" (see "From the plenty of paradisiacal islands to the beauty of emptiness"). At the same time, Lasius sowed the seed for the West's conflated perception of Chinese and Japanese garden art, which Mertens responded to: "Where more modest gardens were to be created, the goal became to achieve variety in the smallest possible space, to create miniature editions of larger landscapes."[9] He even touched on the art of penjing (盆景) or bonsai (盆栽). Since the beauty of a stately tree would exceed the scale of such gardens, Lasius notes, "artificially trained dwarf variants" are made in which "the more gnarled and crippled its form, the more perfect the solution".[10]

While Mertens' own writings make no reference to East Asian gardens, his recommendations for the creation of "water and rock installations" were on a par with the *Sakuteiki*. His detailed instructions for the design of "water accumulations" recommended drawing inspiration from natural lakes and islands and working with the contours of the terrain. In the case of flat terrain, "rounded forms in the largest possible curves are called for"; in the case of a depression, by contrast, an elongated outline is necessary and in the case of an undulating surface, "a jagged form with long, narrow incisions and sharp projections must be aimed for".[11]

He even specified the positioning of the islands, which were to be created "to the side of the mouth of a stream, or as a continuation of an interrupted spit of land extending into a pond"[12].

His goal in the design of Jacques Trümpy-Heer's park in Mitlödi (Glarus) in 1895 was to embody these principles. He produced several variants, of which the second and fifth projects were found in his estate. Both are dominated by an expansive water landscape formed by a pond with inflow and outflow. The stream, which starts in the south of the property and tumbles over a collection of stones, is drawn as a sinuous meandering body of water, widening into small ponds in some places, narrowing into a trickle in others, and crossed by bridges at three points – just as he had recommended. "Streams flowing through a garden" are "irregularly shaped with a few sharp turns, their beds widened or narrowed here and there, the gradient soon increased or diminished, in order to emphasise the soothing sound of the water trickling over stones in some places or to create a tranquil spot elsewhere that can become a favourite haunt for fish".[13]

At first glance, it seems strange that this water landscape is not arranged in front of the villa but surrounds three sides of a "lawn tennis court". The latter, however, can be associated with the Southern Courtyard, which in early Japanese parks laid out in the tradition of the

1 Mertens E 1881c, p. 202.
2 Anon. 1907a.
3 See anon. n. d., *Das Sonnenburggut*.
4 Anon. 1907a.
5 Ibid; Ruoff 1993; Lerch 2012.
6 Mertens E 1881c, p. 202.
7 Lasius 1881, p. 357. Georg Lasius (1835–1928), together with Julius Stadler (1828–1904), had taken over as interim successor to Gottfried Semper as head of the building school in 1871.
8 Chambers 1757. Chambers wrote that the gardens he had seen in China were very small. Ibid., p. 14.
9 Lasius 1881, p. 358.
10 Ibid.
11 Mertens E 1881c, p. 203.
12 Ibid.
13 Ibid.

Garden of Paradise was a white gravel area reserved for ceremonial festivities – religious, cultural, and sporting activities.

Mertens did offer the possibility to obtain a view of the garden "in its greatest extent" from a pavilion on the east side of the tennis court, but was nevertheless careful to avoid "that all the salient decorative elements of the grounds could be overlooked from the house". Instead, residents and guests should feel "lured into the garden by the view afforded by the window and then, wandering through it, encounter this and that unexpectedly interesting motif".[14]

Glärnisch and Klöntalersee – Shakkei and Shukkei

The park that Mertens created in 1905 for the Schuler-Ganzoni family in Glarus has elements that inevitably bring to mind Shukkei and Shakkei. A year earlier, Jakob Schuler (1865–1940) and Martha Ganzoni (1872–1944) had a neo-baroque villa built by the architects Rudolf Streiff and Gottfried Schindler next to the family's ancestral home on Gerichtshausstrasse, which dates back to the 18th century.

Evariste Mertens, whom they commissioned to plan the gardens, was undoubtedly attracted by the spectacular location on the Glärnisch mountain massif, and he used it to demonstrate the principles he had so passionately and vehemently propagated a quarter of a century earlier. The principles he had outlined for including a lake, he applied instead to the mighty north-eastern flank of the Glärnisch to utilise "a successful landscape scene" that was created "by chance"[15].

"Where the background contains beautiful aspects of very different character – e.g. […] towering masses of rock with deep dark gorges and crippled tree growth – it is advisable to treat these two elements separately by placing an artificial central backdrop that hides the space in between. The composition of this backdrop (a wooded hill or simply a group of trees) must be meticulously considered, for it should convey the transition from the rugged to the lovely in a meaningful way."[16]

Mertens accomplishes this transition with consummate skill. If one views the Glärnisch from the centre of the park, the trees in the middle ground create such a backdrop so that the mountain massif appears to rise up immediately behind the property boundary. The effect is particularly impressive in autumn, when the yellow and orange tones of the tulip and sweetgum trees merge with the autumn colours of the larches on the slope. As an object lesson in applied garden theory, it showcases Mertens' convictions in exemplary manner. His adaptation of the borrowed landscape is a veritable expression of the manner established in Japan in the Japanese Middle Ages (1185–1600) under the influence of the painting of the Chinese Sung period.

Apart from the monochrome painting style, one of the most striking features of this manner was the way the artists gave their landscapes a sense of perspective. With cloud formations or fog banks, they created a deliberate separation between a sparsely populated scene in the foreground and a majestic natural backdrop in the background. This "artificially" inserted middle ground was the fulcrum for the composition of Shakkei, except that in the realm of garden design the method was employed to, as it were, reverse the effect. Where, in landscape painting, obscuring the middle ground created the illusion of spatial depth, the effect of masking in a garden design could conceal three-dimensionality.[17] As seen here in Mertens' work, this usually meant placing woody plants in such a way as to create a frame that obscured the view, thereby visually reducing the distance to the borrowed object, and visually tying foreground and background together so that the desired backdrop was perceived as part of the garden.[18] At the same time, it suggested a continuity between man-made and natural landscape, and thus the oneness of man with nature.[19]

To construct a more perfect illusion, it could be necessary to eliminate disturbing elements, just as Mertens mentioned when he advised that one "mask any unattractive points that may be present" and strive for the "successful concealment of obstacles".[20]

Mertens may have taken up Otto Froebel's suggestion. In any case, Eeva Ruoff suspects that he chose, among other things, "17 different varieties of spruce" for the landscaping of Bühl Church in Wiedikon[21], built in 1894–1896 by the architect Paul Reber, "in order to create a convincing visual connection with the wooded Uetliberg behind it".[22]

The Uetliberg also served Froebel as a reference for the idea of using Zurich's local mountain as a kind of shukkei: "One square metre of surface, be it in the garden, or just in the courtyard, or on a flat roof, is sufficient to create a most enchanting stone arrangement […]. Let us try, for example, […] to reproduce in miniature the Uetliberg with its ribs and spurs towards the Sihl and the Reppisch. Our imagination will adapt to the available space. For a garden to be convincing, not the size is important, but the way it is constructed."[23]

Mertens also designed such a replica in miniature – that of the Klöntalersee – in the Schuler-Ganzoni Park, creating it, true to his motto that smaller ponds should be placed "in a part of the garden concealed from the house" so that one encounters them "without being prepared for them, and the charm of the unexpected" intensifies the impression.[24]

In consummate Shukkei style, one finds, among the boulders with which he fortified the pocket lake, a particular specimen that one can imagine to be a miniature Glärnisch.[25]

Mertens already had the view of the Glärnisch in mind in an earlier 1891 design for a park for the silk manufacturer Robert Schwarzenbach-Zeuner in Rüschlikon, Zurich, which in the plan is declared as the south-eastern view, with the Grossmünster visible to the north [→ p. 233].

14 Mertens E 1881b, p. 89.
15 "Who can refrain from a joyful, pleasant feeling at the sight of a successful landscape scene, whether it was created artificially or by chance?" Mertens E 1881a, p. 39.
16 Mertens E 1881c, p. 202.
17 Cf. Nute 2019, p. 21.
18 Cf. Nute 2004, p. 21.
19 Ibid.
20 Mertens E 1881, p. 202.
21 See Special Inventory 2006, p. 92.
22 Ruoff n. d.
23 Froebel 1884; Froebel 1884/85.
24 Mertens E 1881c, as note 1, pp. 202, 203.
25 Conversation between the author and Andrea Schuler, current owner and grandson of Jakob and Martha Schuler-Ganzoni, 14 Oct. 2022.

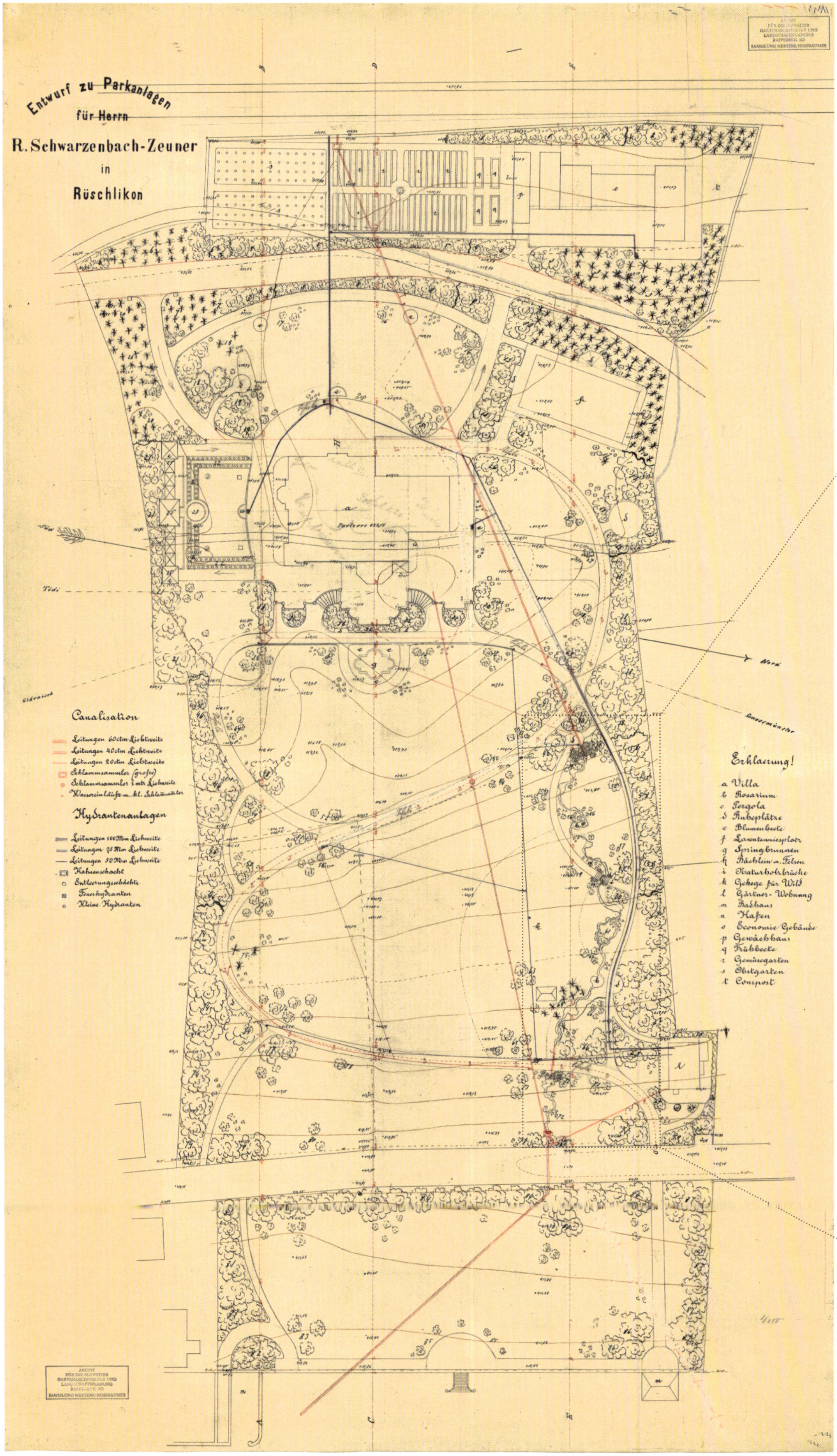

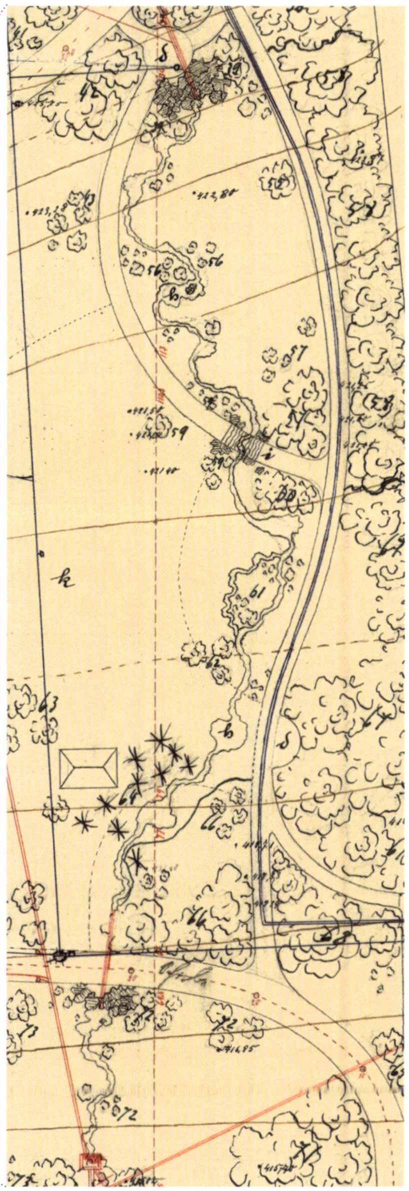

Plan of the park design for Robert Schwarzenbach-Zeuner in Rüschlikon, Zurich, 1891, on which the sight line to the Glärnisch is marked above the symbol denoting “canalisation”.

Glärnisch and Klöntalersee with tributary in a bird's-eye view photograph taken by Walter Mittelholzer in 1932. Evariste Mertens "reproduced" the scenery *en miniature* in the Schuler-Ganzoni Garden.

"The Appeal of Chinese Garden Creations"

After the death of Evariste Mertens in 1907, his oldest son, Walter, took over the horticultural firm in Zurich before Oskar also joined in 1912 and the brothers ran the business, to which they added a nursery in Meilen, under the name "Gebrüder Mertens".

Walter Mertens trained at the École cantonale d'horticulture de Châtelaine (Châtelaine School of Horticulture) near Geneva and completed an internship in England. Oskar Mertens by contrast studied at the Swiss Federal Institute of Technology (ETH) in Zurich and undertook further practical training in France, Holland, and England. Finally, he attended the Kunstgewerbeschule in Düsseldorf, where he graduated under Peter Behrens. Like his father, he passed on his knowledge to the next generation, teaching at the horticultural school in Niederlenz from 1913 until 1940.[26]

Through their work in the first decades of the 20th century, the brothers, along with Gustav Ammann and Johannes Schweizer, were among the most influential representatives of architectural gardening in Switzerland. Exhibition gardens – National Exhibition 1914, Werkbund Exhibition 1918, Züga 1933, Swiss National Exhibition 1939 – were also part of their repertoire. Like Ammann, they increasingly designed residential gardens from around 1930.

By 1933, Oskar Mertens had diagnosed a complete turnaround: "Within a few decades the main principle of all four thousand years of art, the transformation of models from nature into the artistic work of man, was overthrown". In his eyes, this had been sparked by English poets and painters, Jean-Jacques Rousseau, and "the appeal of Chinese garden creations".[27]

This still comparatively vague allusion to the East Asian cultural sphere by Oskar Mertens in 1916 became a more concrete one in the context of the Züga in 1933, when he wrote "Japan and China are sending their influence and even their plants".[28]

With regard to their own design approaches, while they did not openly cite Japanese garden design as a model, they did take note of it[29] and lived by its principles. In 1937, they created a garden expressly designated as a "Japanese garden" for the arms manufacturer Emil Georg Bührle in Zurich, after he had taken over the property from Herbert Klotz [→ pp. 238, 239].[30]

"Creating from the spirit of the place"

The Mertens brothers' design of the special garden at the Zurich Horticultural Exhibition (Züga) in 1933 did not just have Japanese undertones. Walter and Oskar Mertens, along with their colleague Josef Träger, visited the site frequently to familiarise themselves with the situation to the point that its elements began to "come alive" and "speak for themselves". "Rather than imposing creative intentions born of the mind, the process shifted

26 Roth G 1977.
27 Mertens O 1916, p. 67, 72.
28 Mertens O 1933.
29 As the author of an article about a garden at the Wroclaw Horticultural Exhibition for the 1913 Centenary, where a Japanese garden had become a small sensation, Walter Mertens had likely already gained a good impression of such gardens. See Mertens W 1913.
30 The Wehrmacht Colonel Herbert Klotz had entrusted the architect William Dunkel and the garden designer Gustav Ammann with the layout in 1936, but it was not carried out.

Garden plan drawn around 1905 for Jacques Schuler-Ganzoni, Obere Gerichtshausstrasse 58, Glarus. The miniature scenery can be seen on the lower left of the plan (see vignette), and the legend reads: "G: Pond with spring, stream and rocks, alpine garden".

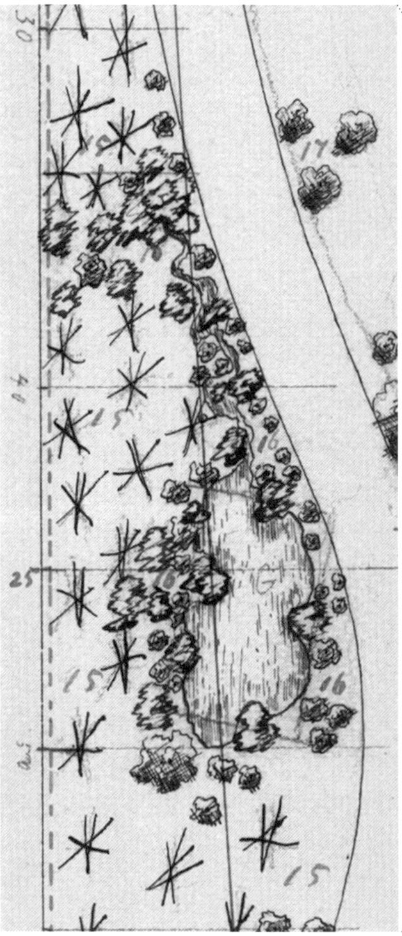

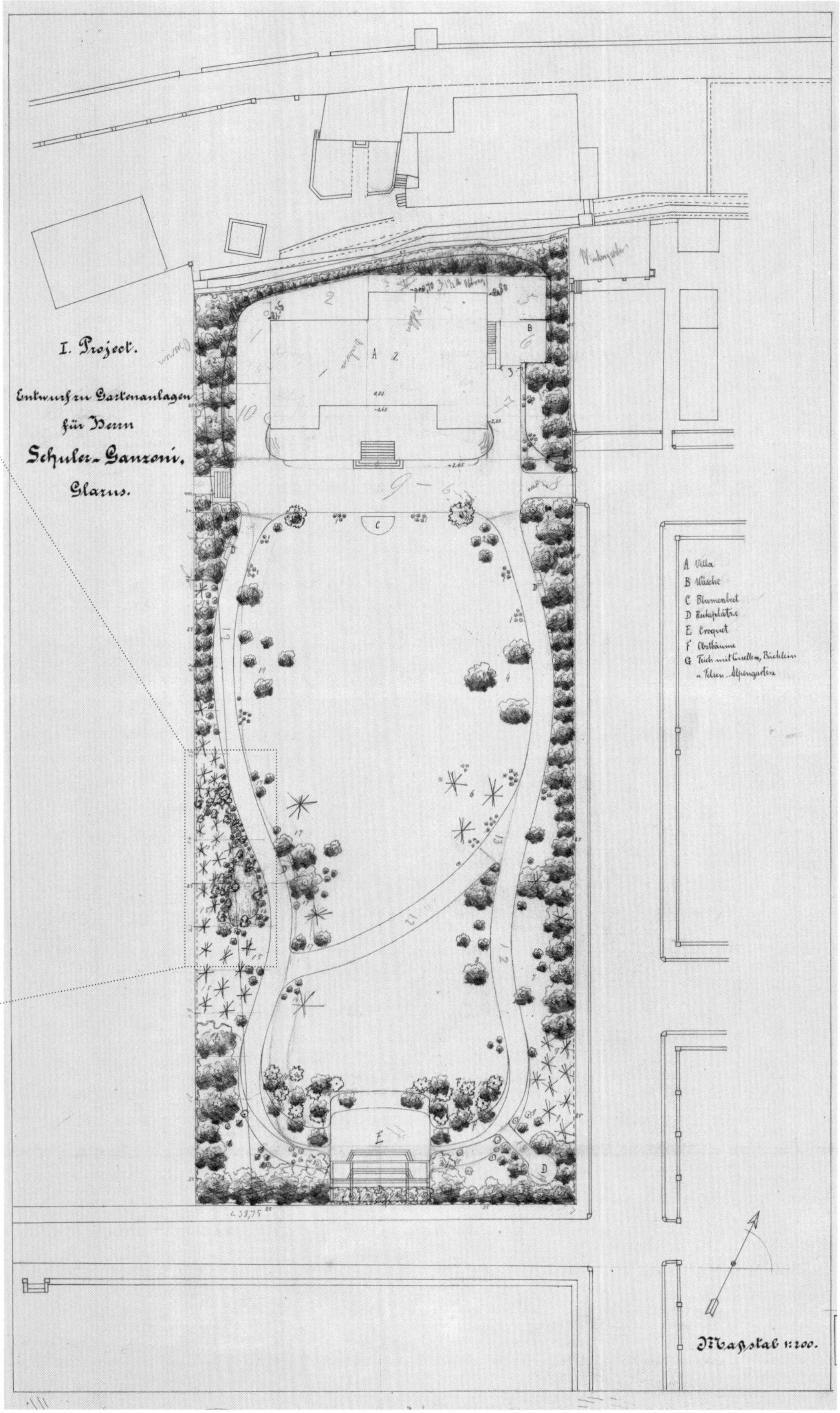

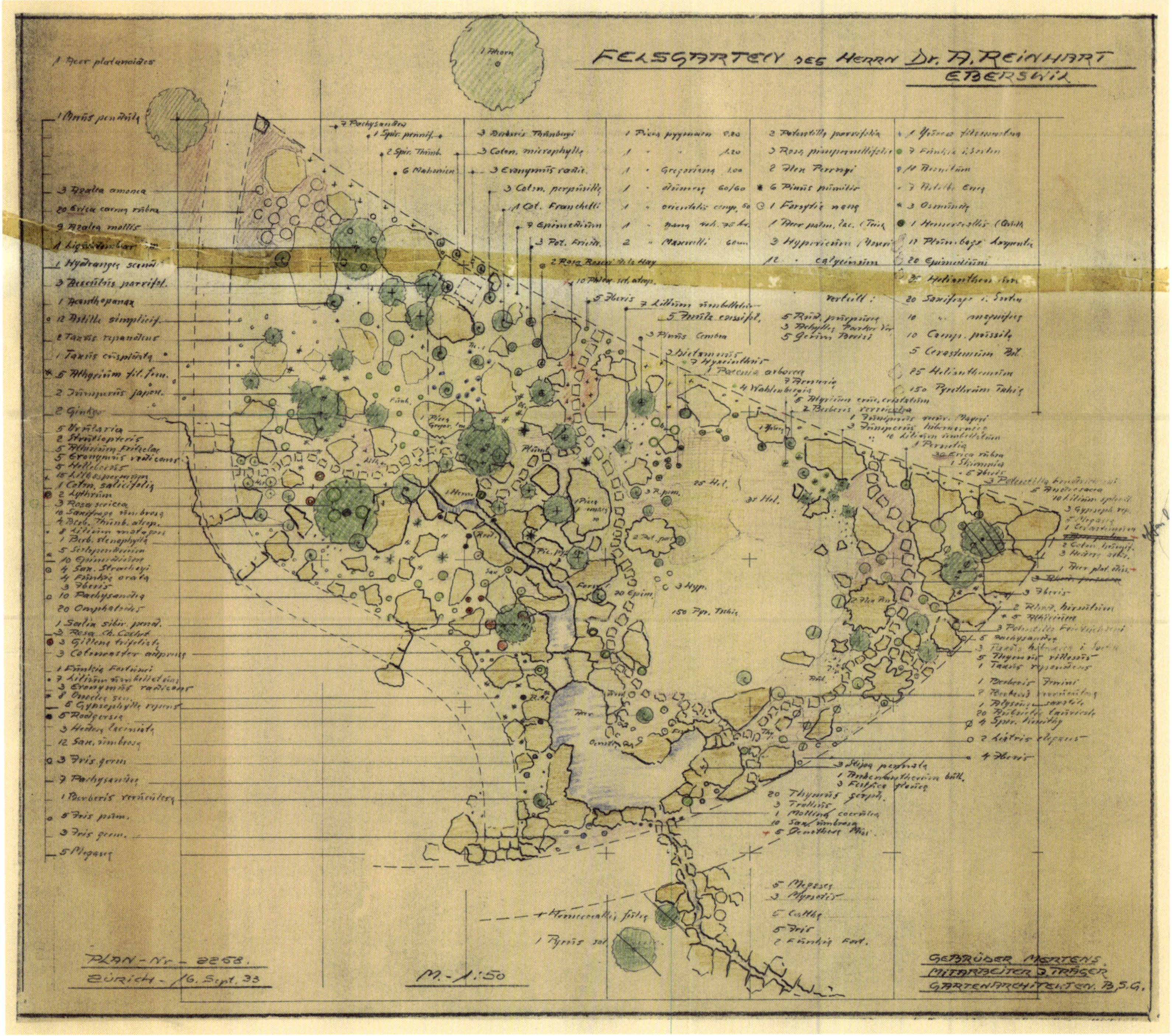

Planting plan of the rock garden designed by the Mertens brothers in Hausen am Albis, 16 September 1933: The organically shaped miniature is placed alongside a bend in the path and measures 35 × 20 metres from end to end.

to listening inwards and allowing creative impulses to arise from the spirit of the place".

Walter Mertens' account makes it seem like they had adopted Japanese principles. It is the "natural constitution" and the "innate disposition",[31] that the brothers were interested in revealing. They saw how the most distinctive trees were becoming increasingly prominent and recognised the "years of silent struggle" endured by long-present shrubs "in their attempt to withstand the constraints placed on them". "Their weather-beaten silhouettes now became the most valuable structural elements of the garden space."[32]

"The swaying forms of the old park trees became an animate foreground against a backdrop of clouds drifting by in blue skies. The weedy ground cleansed and intensified [...], the drooping branches of an olive willow reached over a pool of water, and on the other side a ditch transitioned into a small canal, that reflected the modest shade vegetation of the forest floor in its still waters."[33]

A "World Garden"

An exceptional testimony to the art of garden design is the Villa Tannhof in Hausen am Albis, Zurich. After Theodor Froebel in the years before 1870 and Ernst Klingelfuss in 1913–1914, the Mertens brothers were the third firm to be entrusted with its design.[34] The crowning glory of the intervention carried out in 1932–1933 is the rock garden located in the west of the estate, whose Japanese atmosphere has remained palpable to this day, even more so since its restoration in 2018–2020. It is a miniature "mountain-water" landscape formed of a constellation of rocks and boulders. The brothers adhered to their father's advice: "Wherever possible, the rock section should consist exclusively of externally weathered blocks and, where these are not available, the fragments should be joined together in such a way that the work offers a true-to-nature imitation of the formation in question.[35] The slightly sloping terrain provided sufficient gradient for a stream to flow, its course and velocity directed by red sandstone blocks, splashing over mossy stones, and washing out a furrow through which it trickles as a rivulet before feeding a shallow lake and finally breaking through the path that marks the border of the garden where it pours into the park and thus into the world. The rock garden is a delightful microcosm that corresponds to the idea of capturing the universe in miniature that so fascinated the early Japanese garden designers. It is a landscape of desire in the spirit of the Shukkei, a concentrated replication of the world in the smallest space à la japonaise.

As a "world garden"[36] it also embodies what Karl Förster enthused about in 1917 when he senses "cosmic-artistic transfiguration" in the "natural flora of the Carpathians, the Himalayas, Taurus, Japan, Dalmatia, Transylvania, Greece, Croatia, North America, Mexico, Turkestan, Siberia and Kamchatka"[37]. In a comparatively modest space, the Mertens brothers succeeded in gathering together a wealth of plants from almost all parts of the world. With the exception of Oceania and Antarctica, the garden included specimens from all continents. Their original habitats ranged from Argentina to Canada, from Morocco to Scandinavia, from the Iberian Peninsula to Anatolia, from Mesopotamia via Pakistan to Mongolia and to Siberia as well as China and Japan. 124 different plants – 20 trees, 37 shrubs, and 67 perennials – are listed on the plan of 16 September 1933, whereby the brothers listed between one and 25 specimens of each variety – with the exception of the chrysanthemums: for the Japanese national flower, they calculated they would need 150 seedlings and concentrated them in the largest contiguous planting area. Moreover, the number of 27 plants originating from the East Asian region was almost as much as that of the 30 species native to Europe.

Closing the circle

After Walter Mertens' death in 1943, his brother hired the 30-year-old Hans Nussbaumer as his partner. The latter had graduated from Oeschberg Canotonal Horticultural School and studied garden architecture for three years in Berlin-Dahlem.[38] The two then continued to trade as "Mertens and Nussbaumer" until Oskar Mertens retired in 1951 for health reasons and sold the company to his partner, who in turn, years later in 1978, sold it to Spross Gartenbau.[39] Coincidentally, this brings this story full circle, as Spross had also once created a Japanese-inspired miniature garden at the Züga in 1933.[40]

31 According to Jirō Takei and Marc P. Keane, the term for "nature" used in *Sakuteiki* is not just senzui ("mountain-water"), as is usually abbreviated, but Shotoku no senzui, where Takei/Keane translate Shotoku as "natural constitution" or "innate disposition". See Takei/Keane 2004, p. 164, note 3.
32 Mertens W 1934, p. 120.
33 Ibid.
34 Cf. Nyffenegger/Gonzalez/Sanchez 2016, p. 5.
35 Mertens E 1881c, p. 202.
36 Kühn 2012, p. 25.
37 Foerster 1917, p. 39.
38 See Beglinger 1995, p. 52.
39 See sda 1978.
40 See anon. 1933.

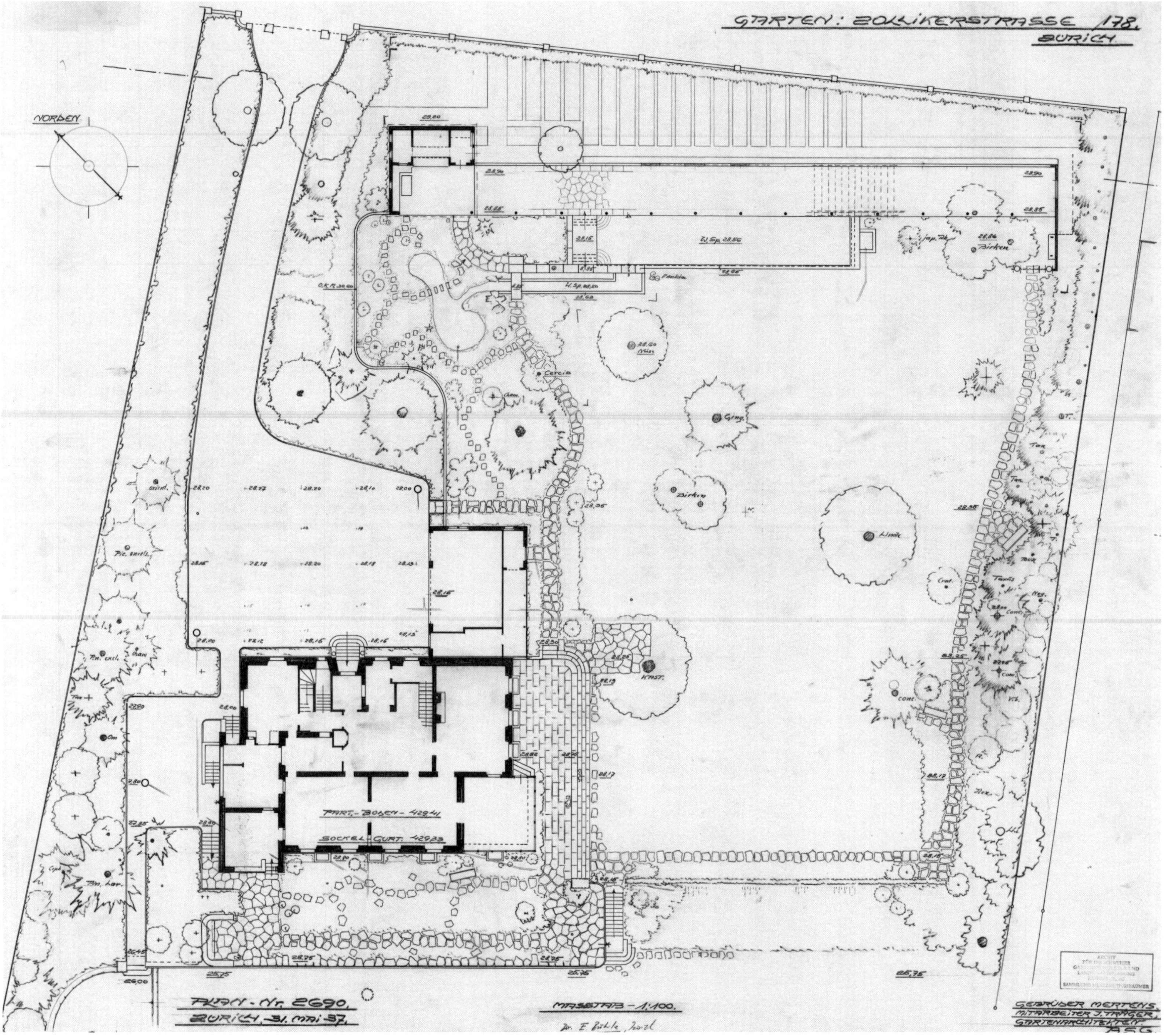

Garden plan for Emil Georg Bührle, 31 May 1937: In an earlier preliminary study drawn up on 3 April (archive signature: asla_gm_1133_6-0002), the Mertens brothers labelled the area of the basin, which describes a figure of eight, as a "Japan garden".

The "Japan garden" was realised almost unchanged
and still exists today.

Willi Neukom 1917–1983

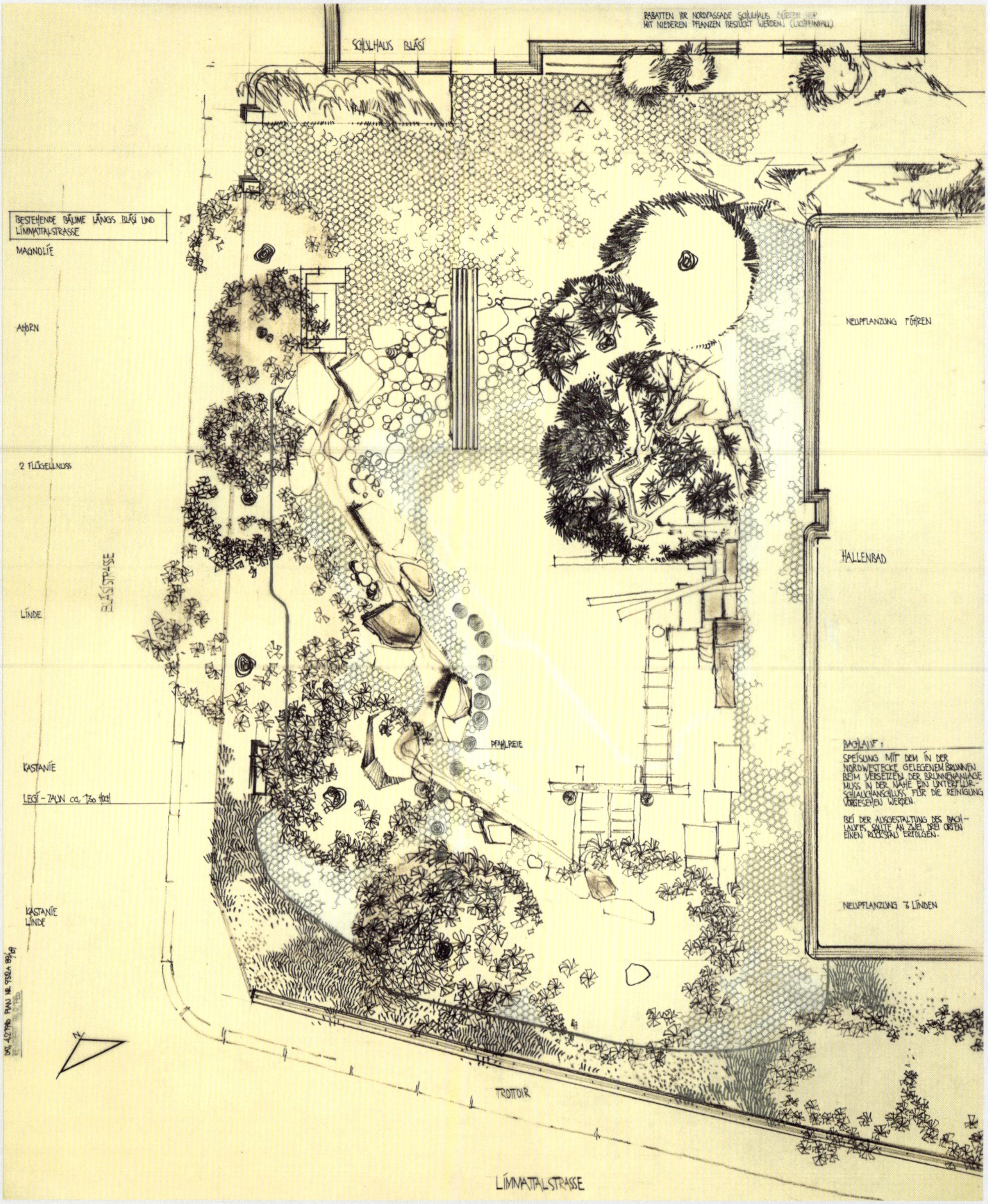

For Bläsi indoor swimming pool in Zurich, Neukom designed a landscape in 1980 that featured a stream contained by boulders and stones and bridged by oak planks. Its planting with Japanese kerria, which Sei Shōnagon mentioned as Yamabuki in *The Pillow Book*, along with Japanese maple, magnolia, wing nut, Japanese wild apple, and thorny olive willow underlined its Far Eastern atmosphere.

Japan as a basso continuo

Lots of water in a small space – a stroke of design genius as poetic as a Japanese garden.[1]

Willi Neukom first completed an apprenticeship as a gardener before becoming a self-taught landscape architect. In 1938[2] or 1939[3], he joined the office of Ernst Cramer, where he was able to develop his talent for drawing, his aptitude for project management, and his interest in architecture, fine arts, and graphics. In 1951, he founded his own "Studio for Garden Architecture and Landscape Design" in Zurich.

In the field of garden architecture, he was inspired by Roberto Burle Marx, Eywin Langkilde[4], and Ingwer Ingwersen, and in the field of art, by figures from the circle of the Zurich School of Concretists. He undertook study trips to Japan (1964), Scandinavia (1967), England (around 1970), and Turkey (late 1970s).[5]

Three years before Neukom set out on his own in 1951, Ernst Cramer had designed the most Japanese of all his works: the "Hegner Garden". But Neukom was by no means dependent on this experience to be able to adapt Japanese elements. Awareness of traditional Japanese garden design had been widespread since the 1930s. In his first years of independent work, Neukom designed gardens in which he experimented with Far Eastern motifs: curved pools planted with water lilies or with islands, the banks of which were lined with rocks, irregularly laid stepping stones, and bamboo. In these first projects, one can already see his enthusiasm for woody plants with bizarre patterns of growth, which would become a hallmark of his work at the ETH Hönggerberg in the 1970s.

Musically speaking, Japanese garden design figured in Neukom's creations as a *basso continuo*. This he would suffuse sometimes with Scandinavian influences, combine with arrangements of Burle Marx's both organically exuberant and geometrically defined approach, or – artistically speaking – overlay with schemes from the concrete art of a Max Bill or the kinetic art of an Alexander Calder.[6]

Geometric-organic, positive-negative

He oscillated between geometric and organic concepts, opting for the former for the headquarters of the International Business Machines Corporation (IBM) in Rüschlikon, Zurich, in 1962 [→ p. 134–136, 254, 256, bottom], for the Fondli School grounds in Dietikon in 1966, [→ p. 244, top] and for the courtyard of the mechanical-technical department of the Gewerbeschule in Zurich in the same year [→ p. 299], and for the latter for the grounds of the housing estate for the Swiss Life Insurance Company Pax in Wohlen, Aargau, in 1983 [→ p. 241], and for the Ungricht Garden around a single-family house in 1982 [→ p. 245]. In some cases, he would work with both approaches in one and the same project – most evidently in the design of the grounds of the Swiss Federal Institute of Technology (ETH) on the Hönggerberg in Zurich [→ pp. 252, 253]. Or he would combine near-natural with purist creations, as in the gardens of the nurses' accommodation at Bülach district hospital in 1962 [→ p. 244, bottom].

He likewise had an affinity for positive-negative compositions. The Japanese custom of expressing water through sand fascinated him, but he drew on it to develop his own interpretation, turning paths into dry streams – for example, in the design for Bruno Scheiwiler in Zurich in 1977 [→ p. 243] – or inverting its principle, by expressing islands through water – for example, in the courtyard of the research building at the ETH in 1976 [→ p. 253]. Another trick in his repertoire was to concentrate the three elements of water, stone, and plants in separate spaces – as seen in the atriums of the Swiss Accident Insurance Institution (SUVA) in Bellikon in 1973, which has a "Green courtyard", "Water courtyard", and "Stone courtyard" [→ p. 300].

One of the most prestigious projects, which remained oft-cited right up to the recent past, was for a villa in Feldmeilen (ZH) designed by Marcel Breuer (1902–1981) for the lawyer and art collector Willy Staehelin-Peyer (1917–1996) in 1956–1957 and built in 1957–1958.[7] The two-storey house is divided into three offset cubes, not unlike the structure of the main building of the Katsura Villa, which are each accentuated by three courtyards inscribed within them – a children's courtyard, living courtyard, and studio courtyard [→ p. 248].

A pond crossed by stones à la japonaise is the centrepiece of the 1983 housing estate for the Swiss Life Insurance Company Pax in the town of Wohlen in Aargau.

1 o. A. 1977a, p. 22.
2 See Stoffler 2016b, p. 22.
3 See Weilacher 2001, p. 39.
4 Langkilde (1919–1997) was responsible for the Danish contribution to the 1963 Hamburg Horticultural Exhibition. See Arioli 1963a.
5 See anon. 1983, p. 35; Nyffenegger 2004, p. 27.
6 The first indication of any involvement with Max Bill is a card inviting Neukom to his exhibition at the Galerie des Eaux Vives at Seefeldstrasse 48 in Zurich from 20 Oct. to 15 Nov. 1946: ASLA, NL WNE, no rec. An article on Alexander Calder, which Neukom kept, dates from the same year: Näf 1946; ASLA, NL WNE, no rec.
7 Füeg 1959; Mathys 1960, p. 28; Anon. 2002. Remmele/Vegesack 2003; Maurer 2013.

Heini Mathys remarks on the studio courtyard's "unmistakable Japanese influence", an impression further highlighted by the Akari lamp inside the studio created by the designer Isamu Noguchi. The "strict concentration and intellectual rigour" that he identified in the courtyard caused him to posit it as "a new variant of the 'Philosopher's Garden'"[8] that Neukom had created for the G59 – and where he most publicly demonstrated his affinity with Japanese garden design and its transformation in a way not seen before.

Zen miniature

An inner courtyard was also the pivotal point of the "Landhaus auf der Stotzweid" project at Bergstrasse 77 in Horgen, whose garden Neukom designed for Adolf Feller AG immediately after returning from his trip to Japan for the IFLA Congress in 1964 [→ pp. 246, 247]. The plans for it are among the most enchanting in Neukom's estate, and the Japanese influence is beguiling.

The building, constructed on an approximately square ground plan, rises gently out of the slightly sloping terrain. In the centre, it encloses a courtyard that is also roughly square. Neukom designed two different versions. They differed in the way they used water as an intrinsic element and as a component defining the outdoor space. One variant proposed a water courtyard with islands made of stones. Two wall planes extend into the outdoor space, creating a transition between architecture and nature while also delineating two further courtyards. The north-western of these was intended as a "fountain and kitchen garden", the south-eastern courtyard as a charming outdoor living area accentuated by Japanese stone settings.

Neukom worked up the second variant in more detail, which he sets out in three coloured plans and one uncoloured pencil-drawing dated 16 July 1964.

The inner courtyard now figures as an open-air green room, where a bird bath carved out of a stone recalls the motif of the traditional cleansing basin of a tea garden. The living area is replaced by a pond landscape in which Neukom draws on the entire repertoire of Japanese elements. He borrowed the Mühlebach stream as the source for an organically meandering water landscape, bounded on the southeast side by a crescent-shaped mound. In the centre of the water, he placed a triangular island with a plant and three stones. At its narrowest point, stepping stones cross as a bridge. On the shore he sketched what appear to be marsh plants and bamboo.[9]

In the coloured drawings, he varied the shapes of the bridge and the island, in one case adding a slender bridge to the stones. The island now consisted of a boomerang-like section and a Z-shaped section.

The plan that most closely resembles the Japanese model shows variations of both elements. The island now takes the form of an elongated triangle and runs parallel to a raised hill-like section. It serves as the connection between the banks, with a footbridge leading from the edge of the hill to the island and from there to the house.

Hans Koella was the architect responsible for the design of the house at Bergstrasse 27 in Küsnacht for Albert Lins. The inventor and industrialist Lins was a member of the board of directors of Armaturenfabrik Wallisellen AG, founded in 1955, and held several patents for electrically operated valves, mixer taps, and radiator valves. Alongside his technical expertise, Lins was also a "pronounced lover of plants and gardens"[10], and in response, Neukom designed "a rich plant world with many exquisite and precious species"[11] in 1965, which was realised in 1966 [→ p. 249].

The main attraction was a water basin that although orthogonal in construction appeared organic in shape due to the use of stones and plantings. The pond was fed by a "small waterfall" splashing over "Guntliweider hard sandstone"[12] – an article in *Anthos* even described it as "rushing"[13]. The sound dissipated in small rippling waves that played around islands planted with water lilies. Its intimate seclusion only added to its Japanese charm.

In addition to marsh and water plants, Neukom included several plants from the East Asian region: Japanese wax bell, Chinese gylcinia, privet honeysuckle, alternate-leaved buddleia, fortune's spindle, trumpet-shaped campsis along with unspecified azalea variants and bamboo, although here it seems he wavered between three species, of which only arrow bamboo and ruscus-leaf bamboo can be deciphered.

Neukom also experimented with the contrasting juxtaposition of geometric and organic for his design for the Swiss Garden at the Wiener Internationale Gartenbauausstellung 1964 (Vienna International Horticultural Exhibition, WIG64), which was held in the Donaupark from 16 April to 11 October 1964. Although by the end of 1963, he had already drawn up detailed plans that included sharp-edged hewn stone blocks in cyclopean dimensions[14], in late February 1964, he summarily rejected his geometric approach and developed an organic design for a water landscape lined with bamboo and replete with stone islands.[15]

Natural garden and dry landscape at ETH

Johannes Stoffler saw in the design of the grounds of the ETH Zurich on the Hönggerberg, which Albert Heinrich Steiner (1905–1996) commissioned Willi Neukom to plan in 1967, an adaptation of Far Eastern models, noting especially "the clustered planting of 'Japanese'-looking Scots and black pines, for which Neukom deliberately selected crooked and crippled specimens to heighten the picturesque effect."[16]. Steiner, former city architect of Zurich (1943–1957) and professor at the ETH since 1957, was responsible for the overall planning. In a joint venture with Hans Epprecht (1917–1980), and from 1971 assisted by his son Tobias (* 1952)[17], Neukom worked on the project from 1967–1977[18], with the second phase from 1972 onwards under the direction of the architects Max Ziegler and Erik Lanter.

8 Mathys 1960, p. 28.
9 wne.426.1.2-6
10 Anon. 1977a, p. 20.
11 Ibid., p. 21.
12 Guntliweid hard sandstone is popular in gardening and landscaping, and is a homogeneous, fine-grained sandstone of blue-grey colouring, which takes its name from the Guntliweid quarry in Wangen-Nuolen on the Upper Lake of Zurich, where it is quarried. https://material-archiv.ch/ma:material_1949 (retrieved 14 Dec. 2022)
13 Anon. 1977a, p. 21.
14 Wien, Gärten für Nationen, Schweiz, Gesamtsituation, Vermassung (Vienna, Gardens for Nations, Switzerland, overall situation, survey), 4 Nov 1963, ASLA, NL WNE, wne.426.1.2-7
15 Wien, Gärten für Nationen, Schweiz, Grundrissplan, Variante (Vienna, Gardens for Nations, Switzerland, floor plan, variant), 24 Feb 1964, ASLA, NL WNE, wne.426.1.2-10.
16 Stoffler 2016b, p. 25.
17 Cf. Stoffler 2016b, p. 23.
18 Construction was completed in 1976 and Neukom's involvement ended with the maintenance plan of 15 Aug 1977; wne.78.1.2-32.

A dry stream between the house and the arbour at the end of the garden that flows into a "pond" beneath the canopy of a weeping birch, Chinese wisteria (*Wisteria sinensis*), golden kiwi vine (*Actinidia chinensis*), Persian ironwood (*Parrotia persica*) as well as Japanese fan and threadleaf maples (*Acer palmatum*, *Acer palmatum 'Dissectum'*) give the Schweiwiler Garden in Zurich a distinctly Far Eastern atmosphere.

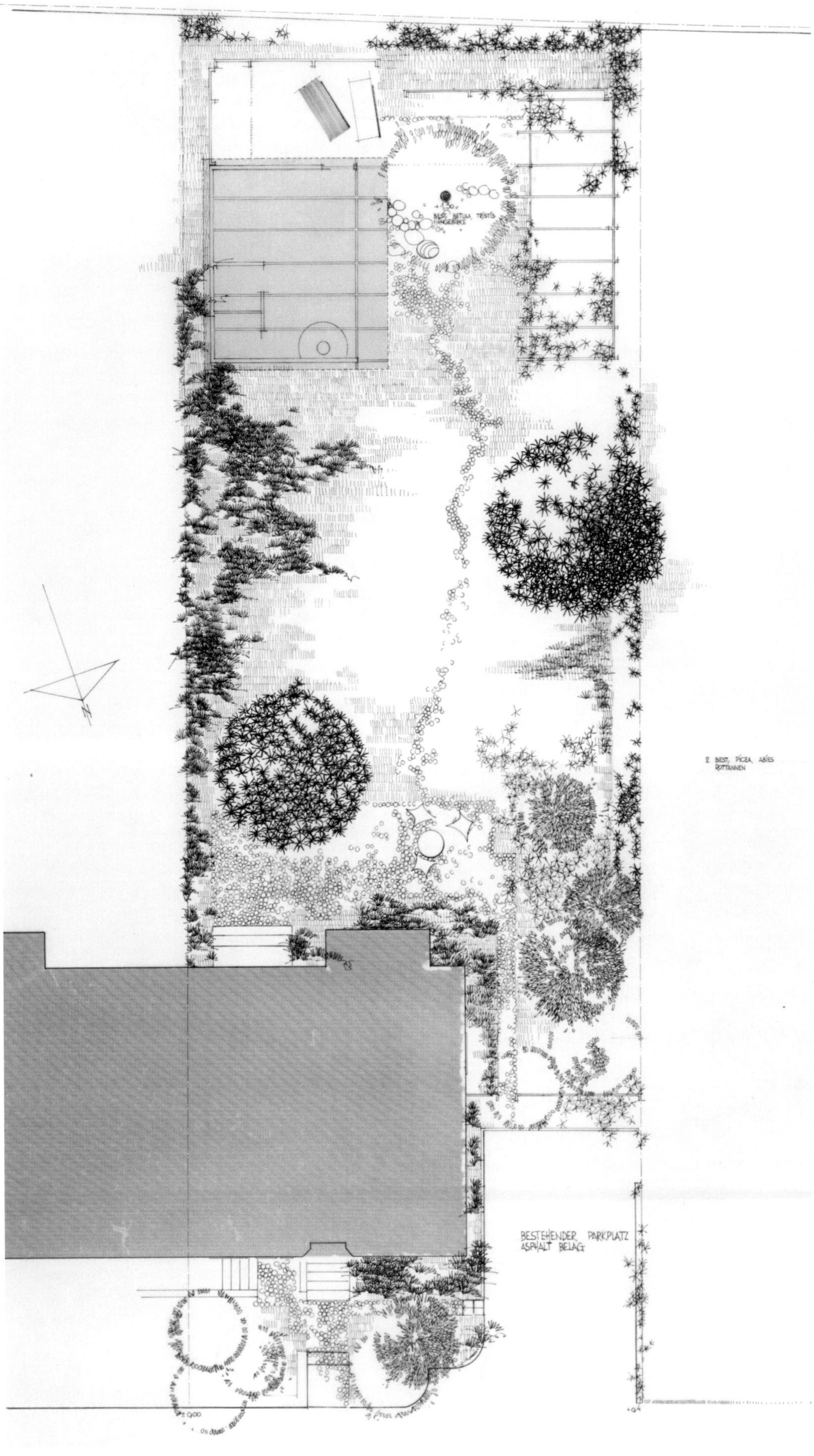

→ Fondli School building, Dietikon, 1966, site plan: An organically shaped tongue of land made of river pebbles spills over into the square pond; while the stone setting picks up the geometry of the basin.

↓ In the garden of the nurses' accommodation at Bülach district hospital, the waterfall cascades over square basins into a naturally shaped pond filled with islands made of plants and stones.

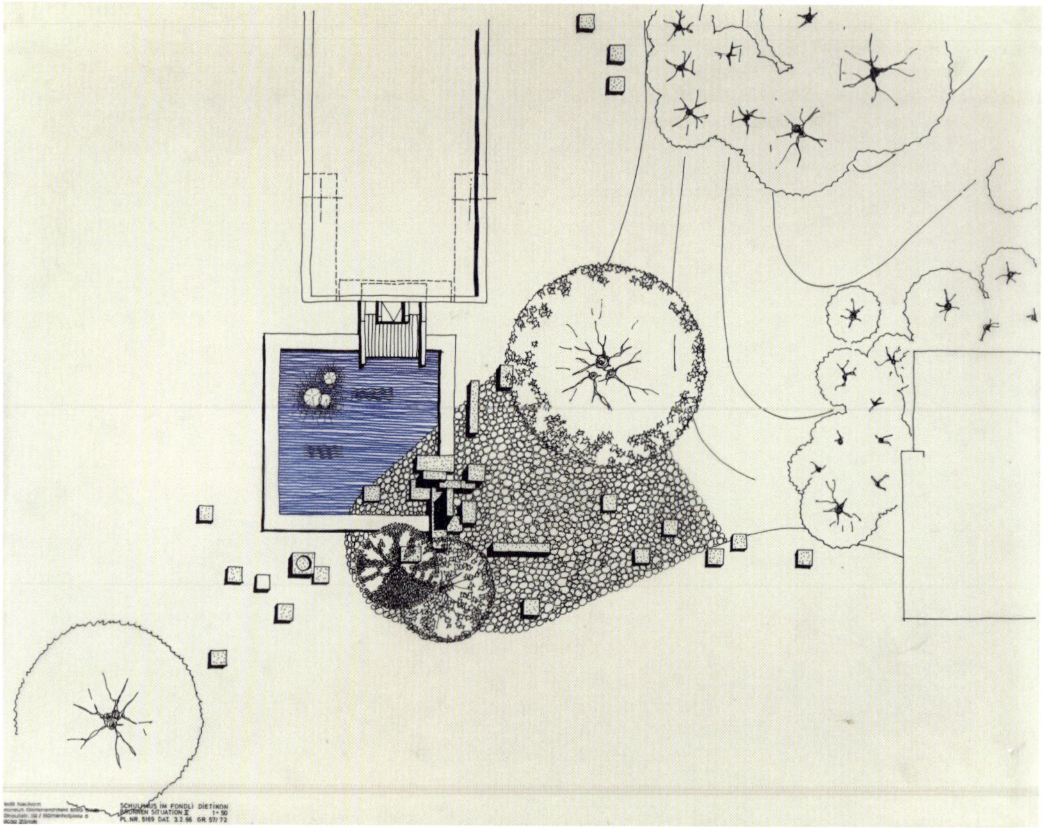

Japanese ingredients – pond, island, stone settings, and a bridge made of staggered boards – inform the design of the Ungricht Garden in Dietikon from 1982.

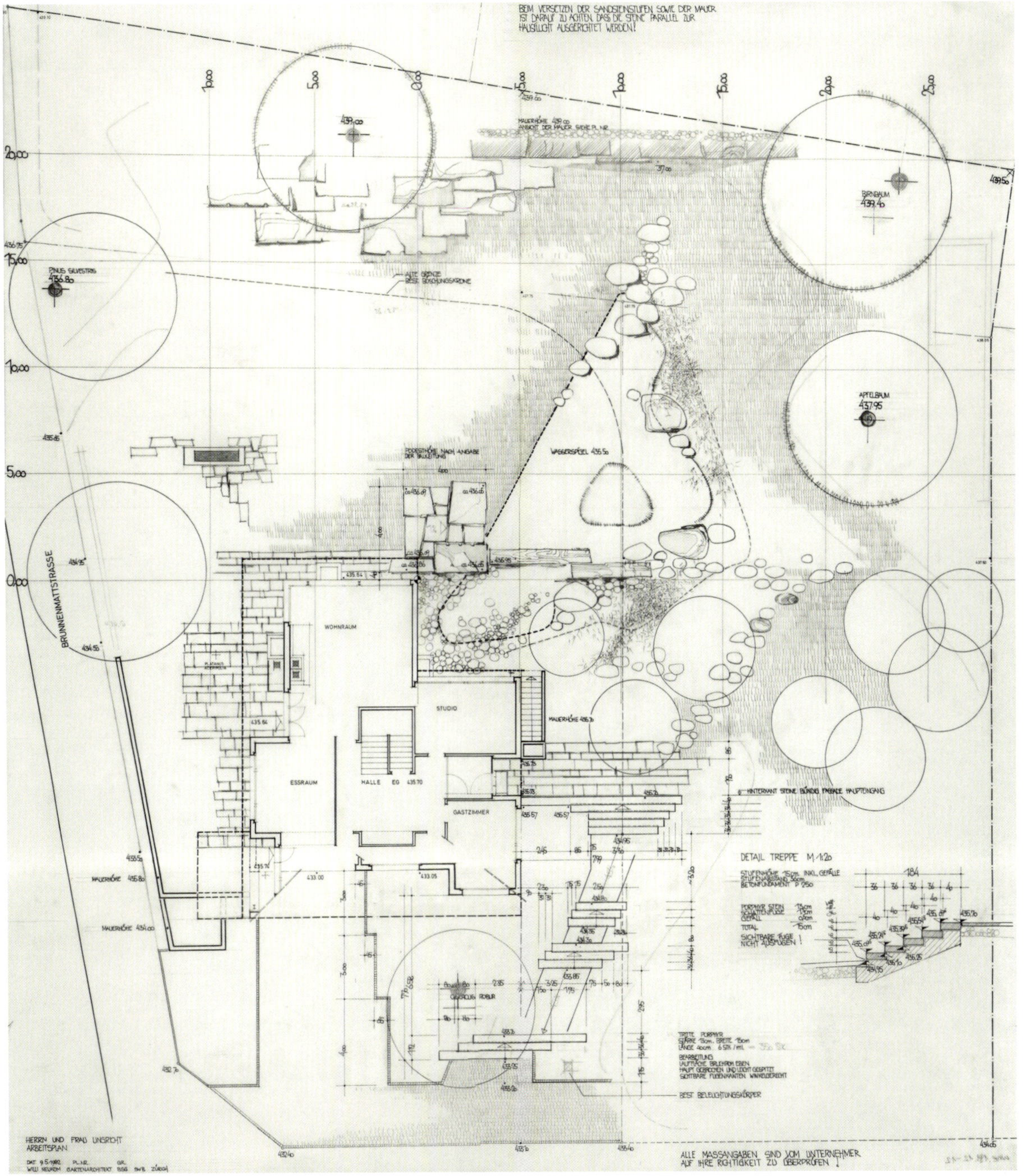

Feller Garden, Horgen

← While the water courtyard appears as an introspective Zen miniature, individual clusters of trees inside and outside the enclosure tie together the designed and natural landscape.

↓ The contours of the islands are connected, but the space in between is shown in blue. Depending on the water level, the island would have been washed over at its lowest point leaving two islands showing or would have risen out of the water as a single island.

→ The bridge walkways are made of staggered boards or stone slabs, as is typical for bridges in Japanese gardens.

19 See Steiner A-H 1974.
20 See B. O. 1976.
21 Stoffler 2016b, p. 23.
22 Stoffler 2016b, p. 27.
23 Neukom marked the immediate surroundings of the pond as the "planting area of Prof. Landolt"; Elias Landolt (1926–2013) was a geobotanist who taught at the ETH, ASLA, NL WNE, wne.78.1.2-180.
24 ASLA, NL WNE, wne.209.1.2.
25 "Platzgestaltung bei Lehrgebäude Abt. II + VIII", 1:200 (Design of forecourt for Teaching Buildings Depts. II + VIII), Arbeitsgemeinschaft Willi Neukom Hans Epprecht, ASLA NL WNE, wne.78.1.2-205.

Some 240 sketches and plans document the development of the project through the first phase around the physics buildings from 1965 to 1973[19] to the second phase for the grounds of the building sciences and architecture departments completed in 1976.[20] Neukom's designs for the first period can be described as "abstract naturalism"[21], visible, for example, in the geometry of the enclosure of the fire water pond – today the Albert Heinrich Steiner Garden – which was liberally planted along its banks to soften its sharp edges. In the second phase, the influence of the "natural garden movement"[22] is more evident, and ecological aspects inform the design of the pond as a biotope and of the outdoor "surroundings between the buildings for teaching and research".[23] This green space today bears the name of the first woman professor of architecture at the ETH Flora Ruchat-Roncati (1937–2012).

Its organic contours blend both with the Japanese borrowed landscape as well as the promenade-like pathway, a few accumulations of stones and the planting of the modelled terrain and its different vegetation areas. These were home to numerous ferns, grasses, and rushes – such as Japanese sedge – as well as water, marsh, and reed plants, depending on their location. Species of Japanese origin – Japanese larch, Japanese snowball, and hydrangea – are also named among the shrubs and woody plants in Neukom's plan.

For the roof of the underground car park, he conceived a dry landscape as a terrain that paired its Japanese pedigree with an ecological impetus, combining a dry stream meandering through a gravel bed lined with grasses, mullein, and silver thistle with a flower meadow that was allowed to run wild. The rock garden remained a lifelong fascination for Neukom, and even in the last year of his life, he was planning an alpine garden for the Internationale Gartenausstellung (International Horticultural Exhibition, IGA) in Munich.[24]

Assimilated nature and designed garden

The work at the ETH provided the first real opportunity for Neukom to translate the impressions of Japanese gardens he had gained first-hand during his trip to Japan at a large scale. His design drawings, in particular, reveal a deeply felt empathy with the art of Japanese garden design that comes across more strongly than in the design that was eventually realised.

This is especially true of his design for the garden courtyard of the teaching building, which Neukom explored in meticulous detail. Sixty plans testify to this phase, 24 of which exhibit strong Japanese tendencies. That is twice as many as he produced when working up the definitive design. The first three sketches from 1972 are revealing.[25] A "Central Square" in the form of an extensive water landscape spills into the garden courtyard. It comprises several orthogonal pools, offset to one another so that they meander, bridged in places by

← The reductive design of the studio courtyard in the Villa Staehelin is a fitting counterpart to the Japanese dry garden. On a bed of round pebbles from which hints of green sprout here and there, Neukom placed rocks of different shapes, sizes, and colours, along with a small conifer.

↓ While Neukom signals the Japanese atmosphere of the children's courtyard with two low maples, two boulders, and a ground-covering juniper, the courtyard adjoining the living area was transformed into a Japanese-inspired miniature landscape planted with low conifers, iris, lily, and hosta.

› Lins Garden, Küsnacht: The waterfall splashes over hollowed-out stone blocks. As a contrast to the landscape-like backdrop, Neukom laid square slabs as stepping stones and created islands from plants growing out of the top of concrete pipes.

narrow footbridges. Two of the pools widen out into ponds – one outside on the square, the other within the garden courtyard, which is bounded on three sides. Square stepping slabs lead across both ponds, set either in a loosely scattered diagonal pattern in the outlying section or accurately lined up parallel to the edge of the pool in the courtyard – an allusion to the differing characters of assimilated nature and designed garden [→ p. 300].

The sketch recalls his design for the nymph pond of the G59 and foretells his first drafts for the Grün 80 (see "Inspiration, imitation, integration"). The impression of the G59 design becomes more evident in nine further plans created between April and December 1974. Here the design now concentrates on the sharply defined courtyard, while the stepping stones have become "roundels" that run parallel to each boundary of the courtyard, meeting on a rocky island of marsh and water plants [→ p. 252]. Some plans also include a smaller, inaccessible island and stone settings clustered into groups of two, three, or five to balance the composition.

Like a Japanese folding screen

The way Neukom played with the elements of stone, water, and plants in the inner courtyard of the research building was remarkable. Adapting the Japanese custom of representing water through sand or gravel, he used "rounded cobblestones" to represent water, which he then bridged using natural, "flat stepping stones as a walkway". This inversion is applied consistently throughout the design, where circular water basins with water features serve as islands. While the courtyard did featured planting, he radically reduced its colour palette.[26] The plan also reveals Neukom's aptitude for creating balanced compositions – a skill that testifies equally to his familiarity with the mobiles of Alexander Calder and with the Ryōan-ji garden [→ p. 253].

Neukom's design also shows his affinity with art. Perhaps more so even than the notion of the window as a picture – a familiar artistic motif since antiquity – where the viewer has a horizontal perspective, a courtyard acts as a frame for a view from above. Here Neukom chose to make the landscape seem artificially cut off, as if it extends beyond the edges of the courtyard, a device he also employed for a design produced in the same year for the Strickhof site at the University of Irchel [→ pp. 258, 259].

By contrast, he conceived of the courtyard between the buildings of the Technical Physics and the Industrial Research Institute (AFIF) as an "emptied" counterpart. Here the range of plants has been reduced to only Persian ironwood tree and East Asian snowball – either *Davidii* or *Mariesii* – and the courtyard features a rectangular water basin equipped with three stones – two overlapping and one slightly set apart – just as Tetsurō Yoshida had depicted the design principle of "main stone, auxiliary stone and supplementary stone".[27]

His love of stone, as anchored in East Asian cultures where it serves as the nucleus of all garden designs, is – to strike a musical note once more – the *cantus firmus* in Neukom's oeuvre. He developed and perfected it over the course of his professional life, as can be seen in the comparison of two drawings created between 1942 and 1950 [→ pp. 250, 251].[28] The stones he sketches there bear resemblance to depictions on woodcuts that Jan Tschichold published as facsimiles in 1940 under the title *Der frühe chinesische Farbendruck* (Early Chinese Colour Prints).[29] Neukom wrote approvingly: "To gaze upon at a stone like this, is to think of eternity."

26 Of the total of 2000 seedlings, the blues and reds were modestly represented by 80 campanula (bellflowers), 30 globularia (globe flowers), 150 daphne seedlings, and three red-flowering daylilies.
27 Yoshida 1957, p. 11.
28 They were found in a folder entitled "Cramer Surbeck". Since Surbeck worked with Cramer from 1942 to 1950, Neukom's drawings must date from this period.
29 Tschichold 1940, plates 4, 15. Tschichold's publications were the talk of the town at the time. "By contrast, the reprints of the illustrations for the 'Ten Bamboo Hall' and the 'Mustard Seed Garden' seem almost familiar, as a result of new reproductions," wrote the reviewer of an exhibition at the Basel Gewerbemuseum. Ma. 1942, p. XXII.

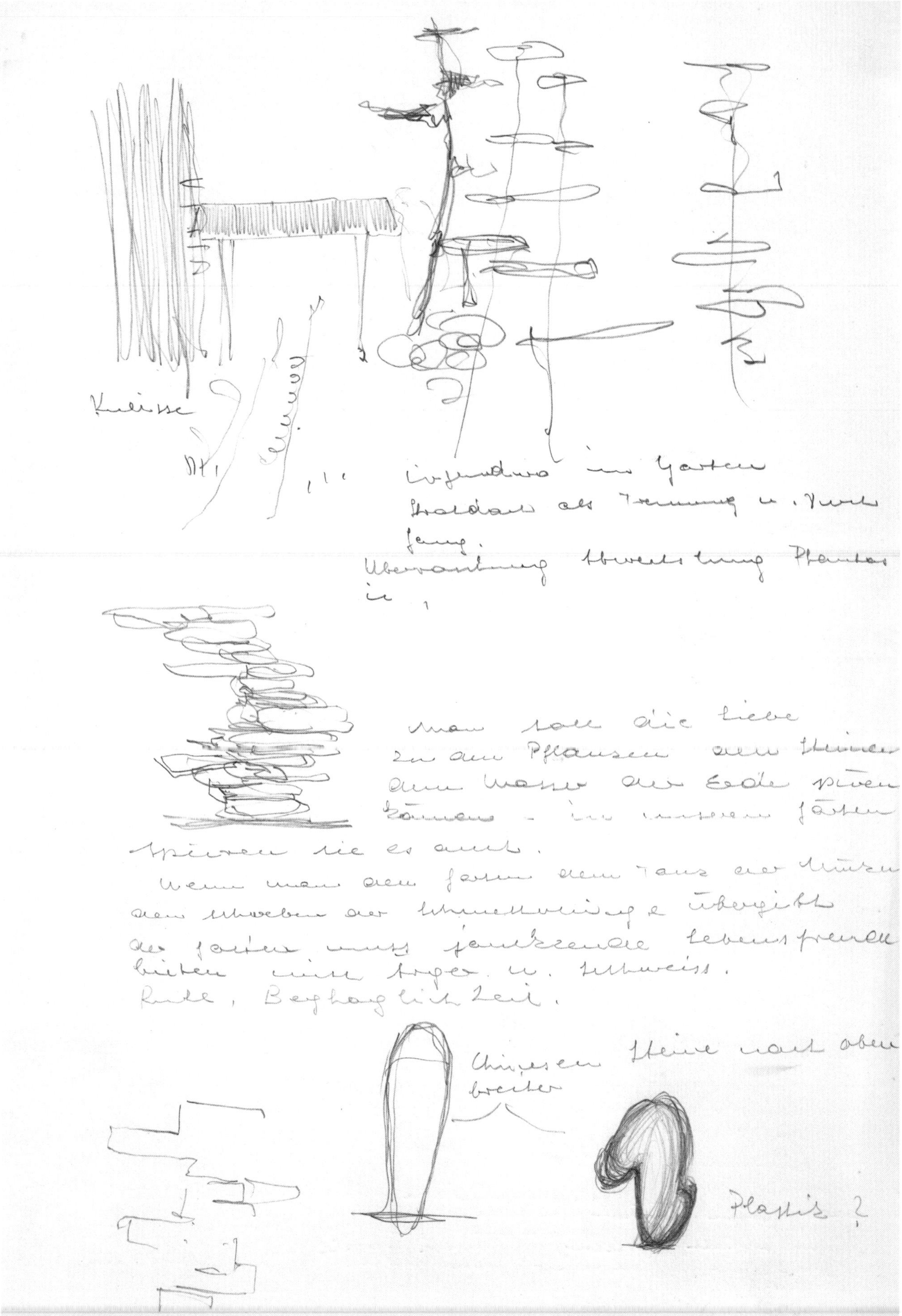

The folds of stone resemble those immortalised by the artist of the woodcut on plate 4 of Tschichold's *Early Chinese Colour Prints*.

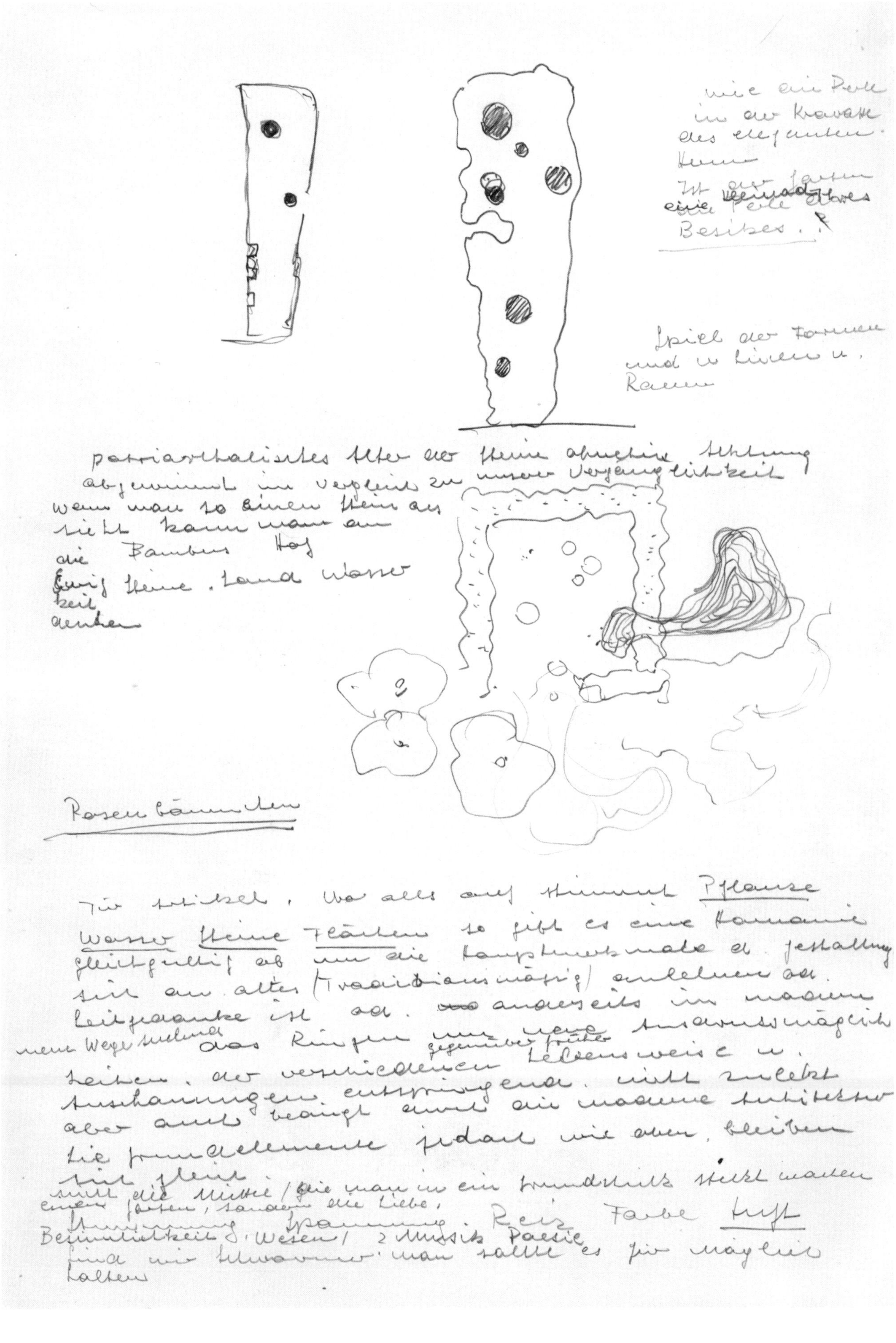

The stone erosions bear resemblance to the cavities depicted in plate 15 of Tschichold's book.

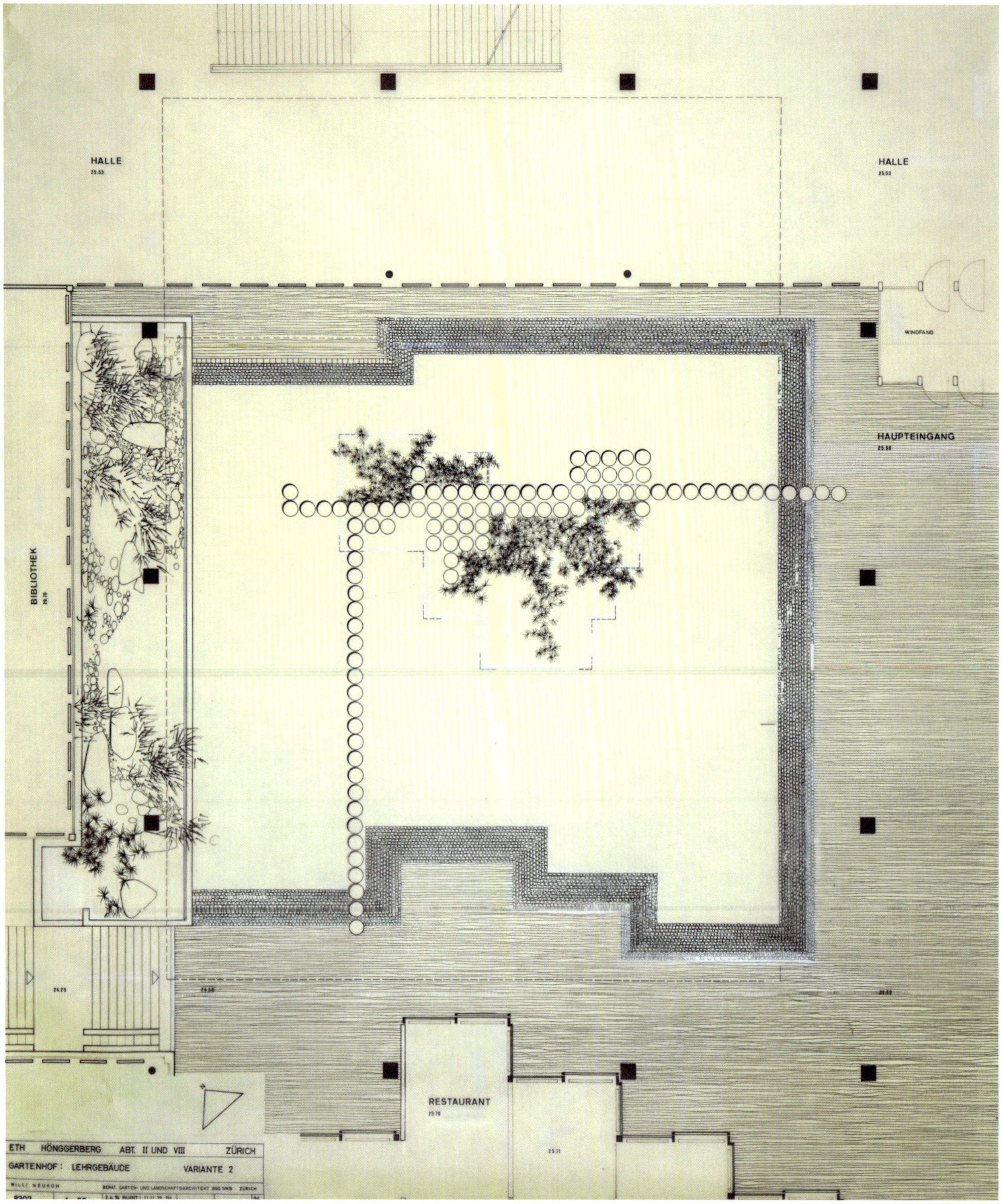

↑ Study of a variant for the forecourt of the teaching building undertaken between 2 April and 12 December 1974: The narrow section at the edge adjoining the library (left) with pebbles, groups of rocks, and what looks like bamboo serves as a dry counterpart to the water courtyard.

→ Proposal 3 for the inner courtyard of the research department at the ETH from 19 March 1976: The bank sections were to be lined with grasses, ground cover, perennials, and woody plants, flowering in shades of white and yellow, following the lead of the Japanese golden maple with its creamy white flowers and bright foliage.

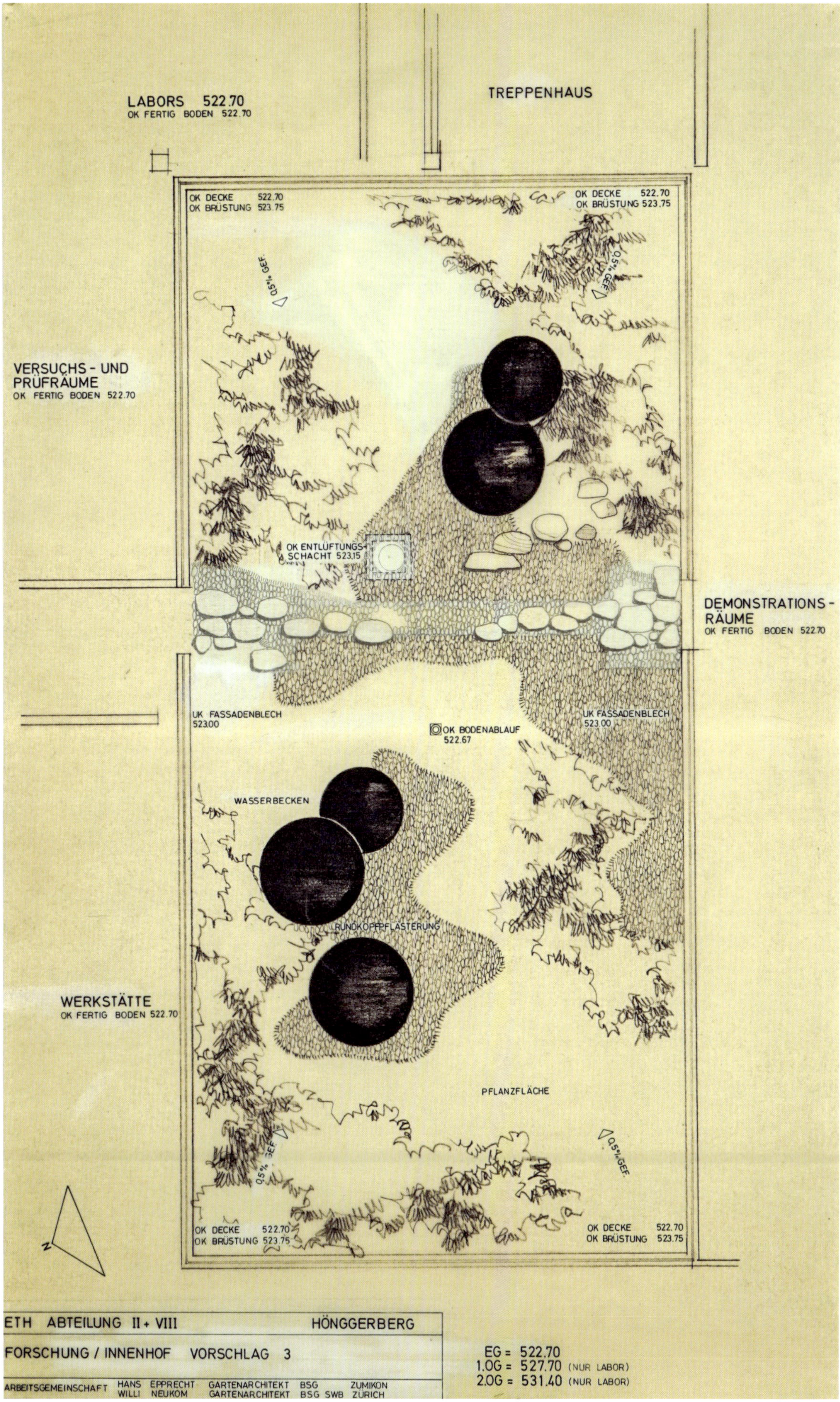

LABORS 522.70
OK FERTIG BODEN 522.70
TREPPENHAUS
OK DECKE 522.70
OK BRÜSTUNG 523.75
OK DECKE 522.70
OK BRÜSTUNG 523.75
0.5% GEF.
0.5% GEF.
VERSUCHS - UND
PRÜFRÄUME
OK FERTIG BODEN 522.70
OK ENTLÜFTUNGS-
SCHACHT 523.15
DEMONSTRATIONS-
RÄUME
OK FERTIG BODEN 522.70
UK FASSADENBLECH
523.00
OK BODENABLAUF
522.67
UK FASSADENBLECH
523.00
WASSERBECKEN
RUNDKOPFPFLÄSTERUNG
WERKSTÄTTE
OK FERTIG BODEN 522.70
PFLANZFLÄCHE
0.5% GEF.
0.5% GEF.
N
OK DECKE 522.70
OK BRÜSTUNG 523.75
OK DECKE 522.70
OK BRÜSTUNG 523.75
ETH ABTEILUNG II + VIII
HÖNGGERBERG
FORSCHUNG / INNENHOF VORSCHLAG 3
ARBEITSGEMEINSCHAFT HANS EPPRECHT GARTENARCHITEKT BSG ZUMIKON
WILLI NEUKOM GARTENARCHITEKT BSG SWB ZÜRICH
EG = 522.70
1.OG = 527.70 (NUR LABOR)
2.OG = 531.40 (NUR LABOR)

← Sketch for the garden courtyard of Wolfsberg Training Centre in Ermatingen from 23 May 1974: The original concept drawing has a stronger Japanese atmosphere than the final design as built. The basin was given organic contours and featured an island and stone settings.

↓ The sketch for the IBM headquarters in Rüschlikon, Zurich, in 1962, shows water “washing beneath” the walkway, stone settings, a waterfall and dry bed, and has an intensely Japanese feel to it.

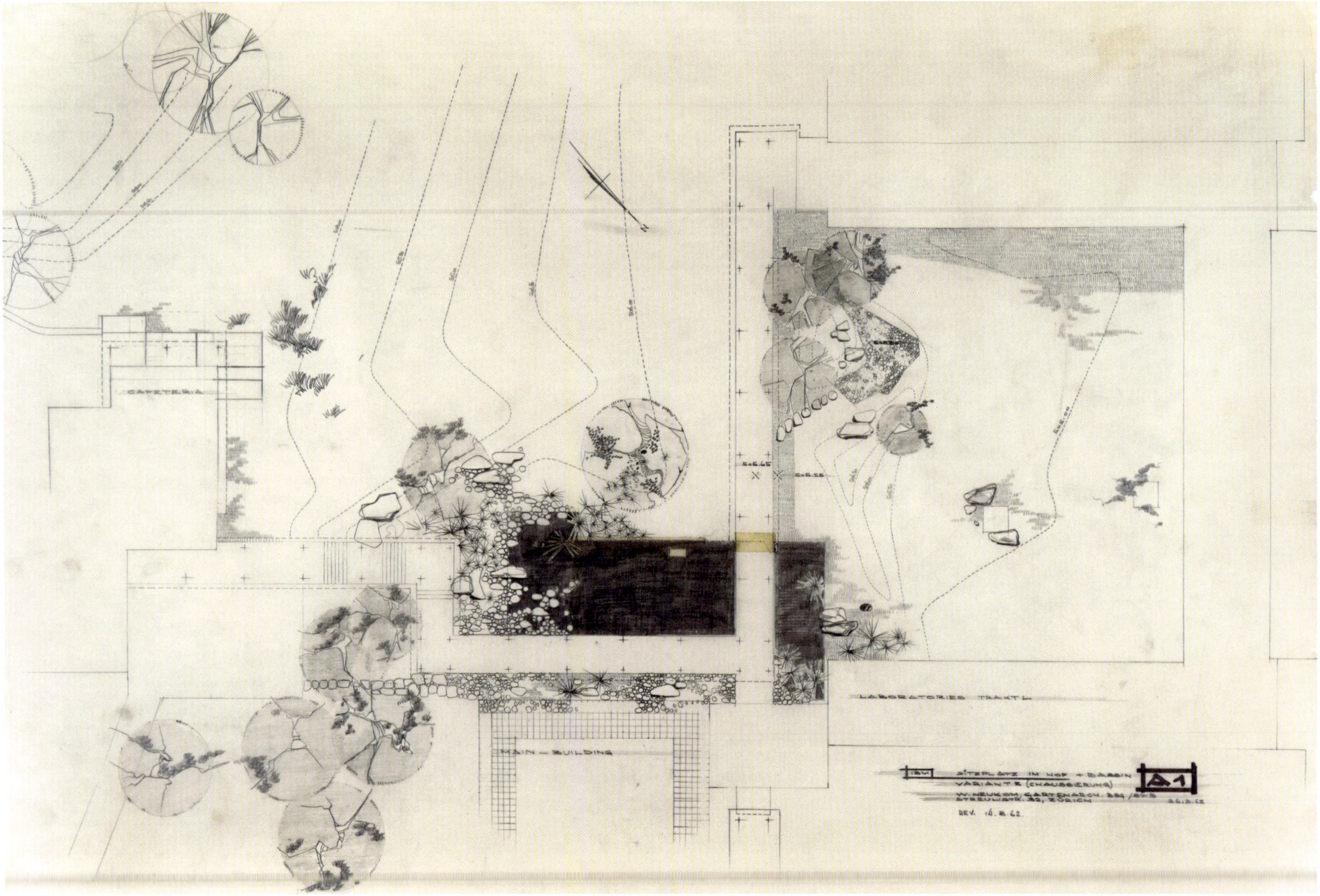

→ Neukom created a delightful hidden Japanese garden alongside the existing "Kindlistein" at the north-eastern edge of the site at the ETH. It features a crescent-shaped dry pond fed by an equally dry tributary.

↓ The bust of Buddha on the house wall in the entrance area of the atrium house built for Michael Meyer in Kilchberg in 1960 signals its Far Eastern inspiration.

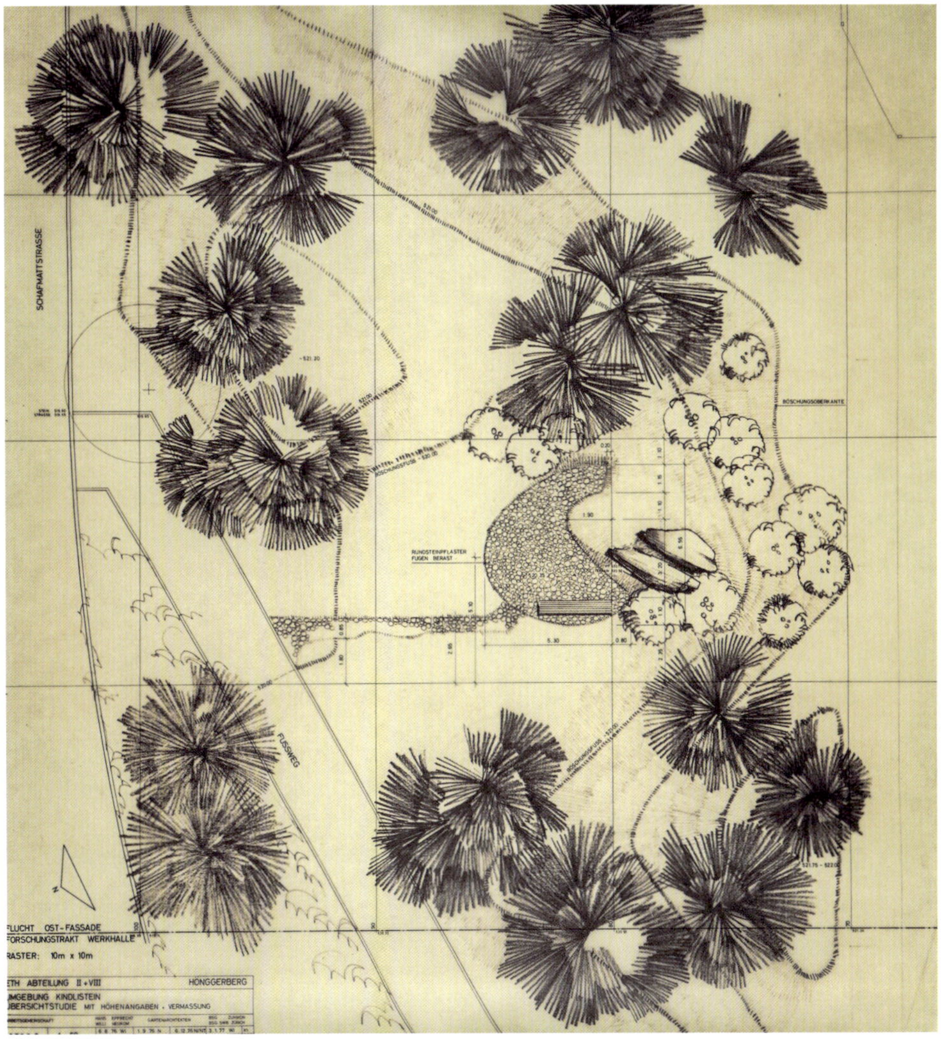

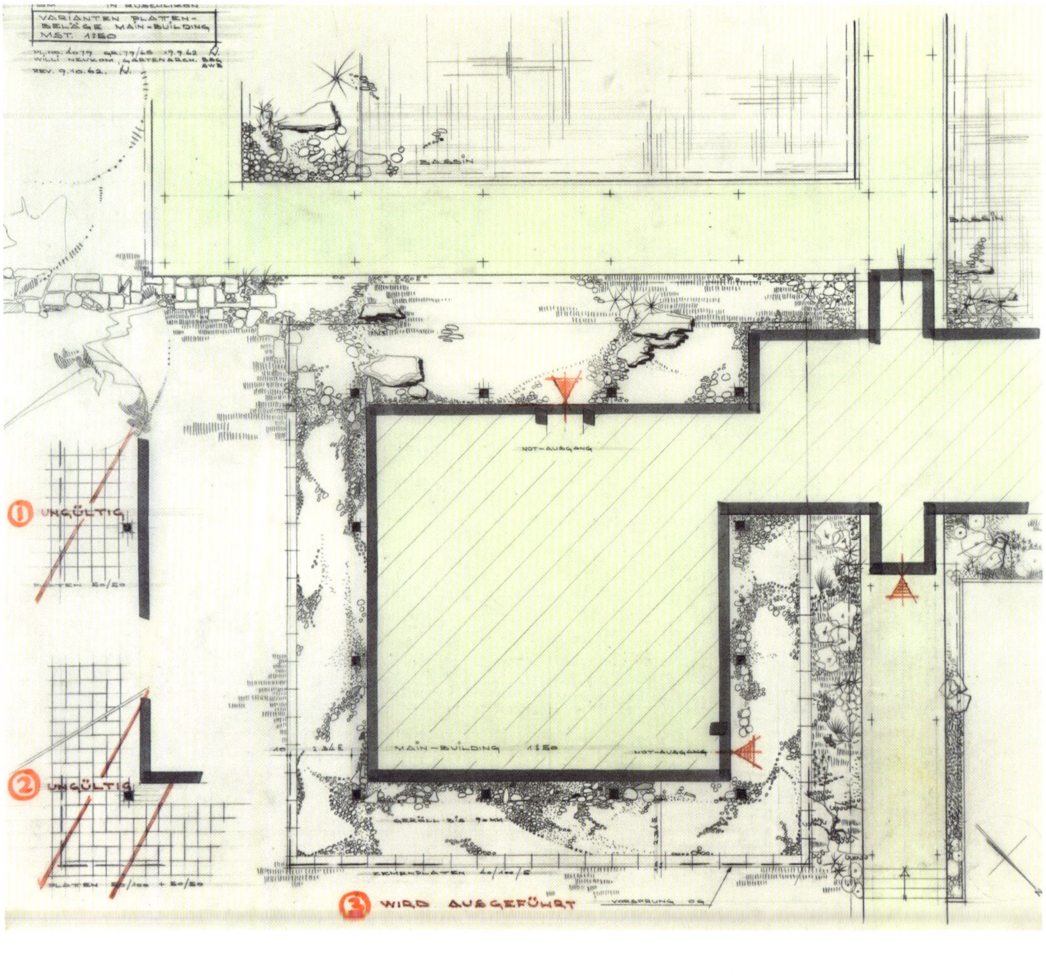

↑ The garden for Esther and Rudolf Guyer in the Seldwyla estate in Zumikon is a delightful Japanese-inspired dwelling garden.

→ Neukom's 1962 design draped a dry landscape around the IBM headquarters building in Rüschlikon, Zurich.

Design for an intimate garden courtyard for René Wettstein in Zumikon in 1964 that, much like an expanse of water, is crossed by broken stone fragments.

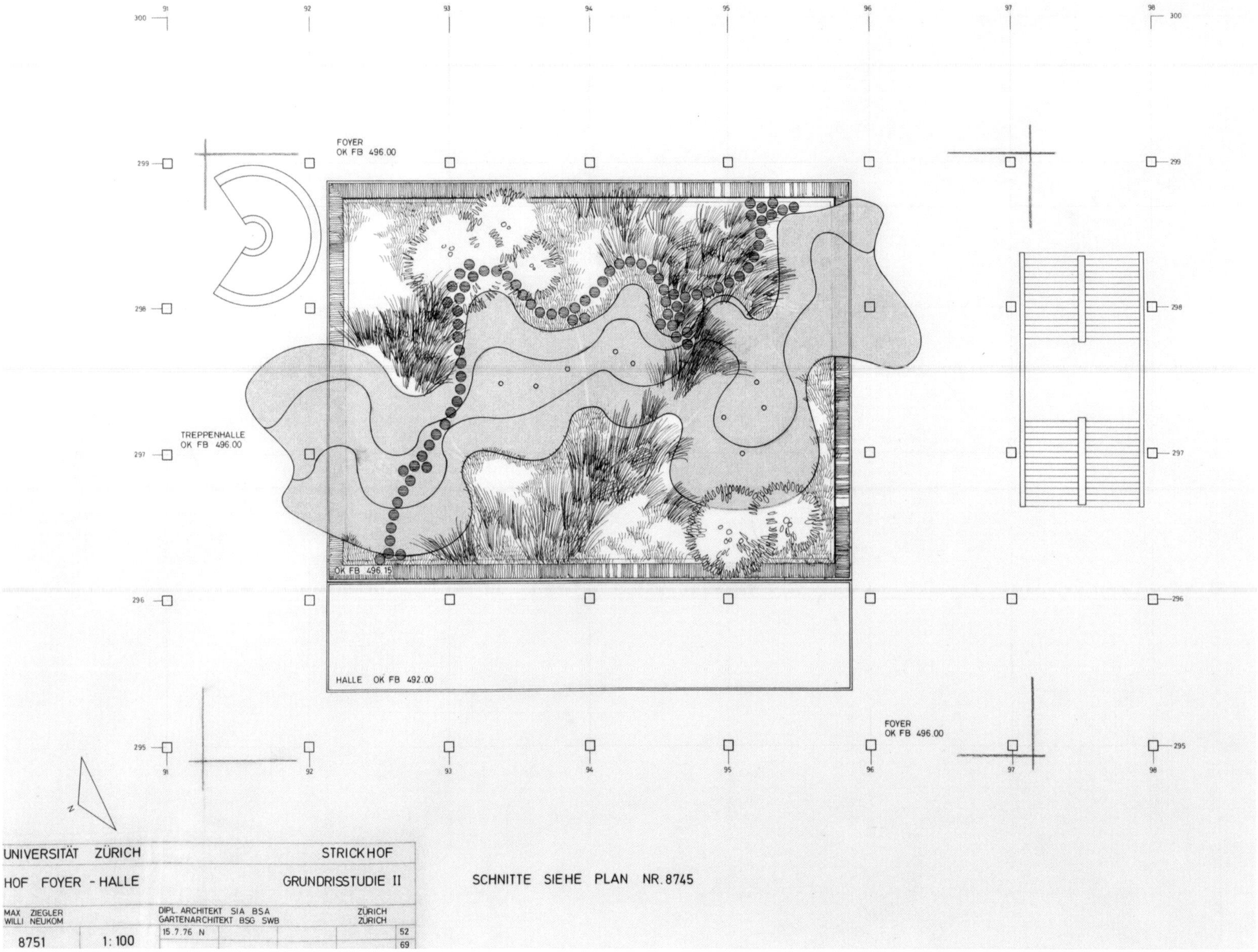

University of Irchel, Strickhof site, Zurich – Courtyard of the foyer hall, ground plan study II, 15 July 1976: The body of water in the water courtyard appears to flow beyond its architectural container, as if painted on a Japanese folding screen or a monogatari scroll.

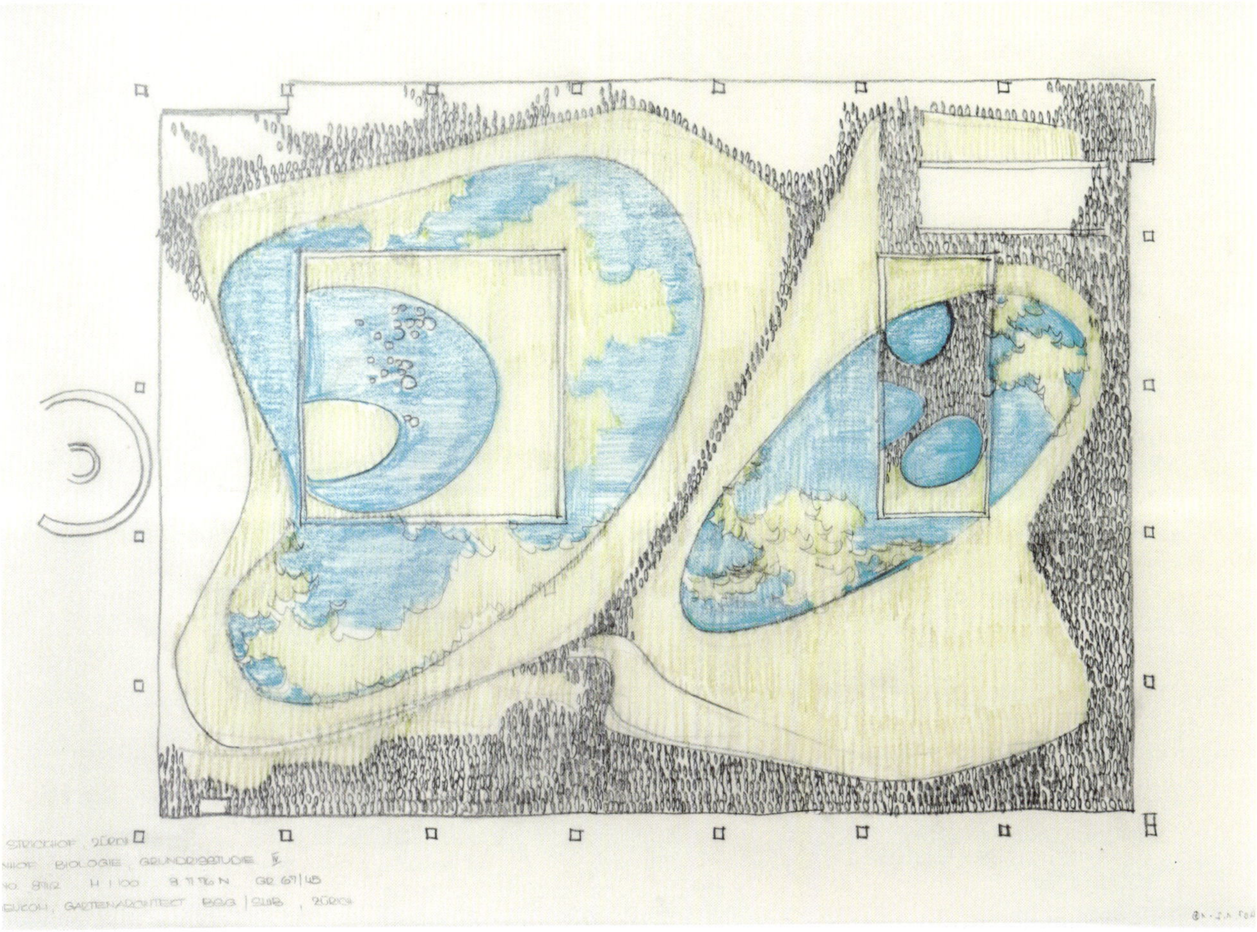

University of Irchel, Strickhof site, Zurich – sketch of green courtyard of the biology building, floor plan study IV, 8 July 1976: The quadrangles inscribed in the courtyard function as interior spaces whose boundaries Neukom in turn broke down and intertwined with the overall design of the surrounding space.

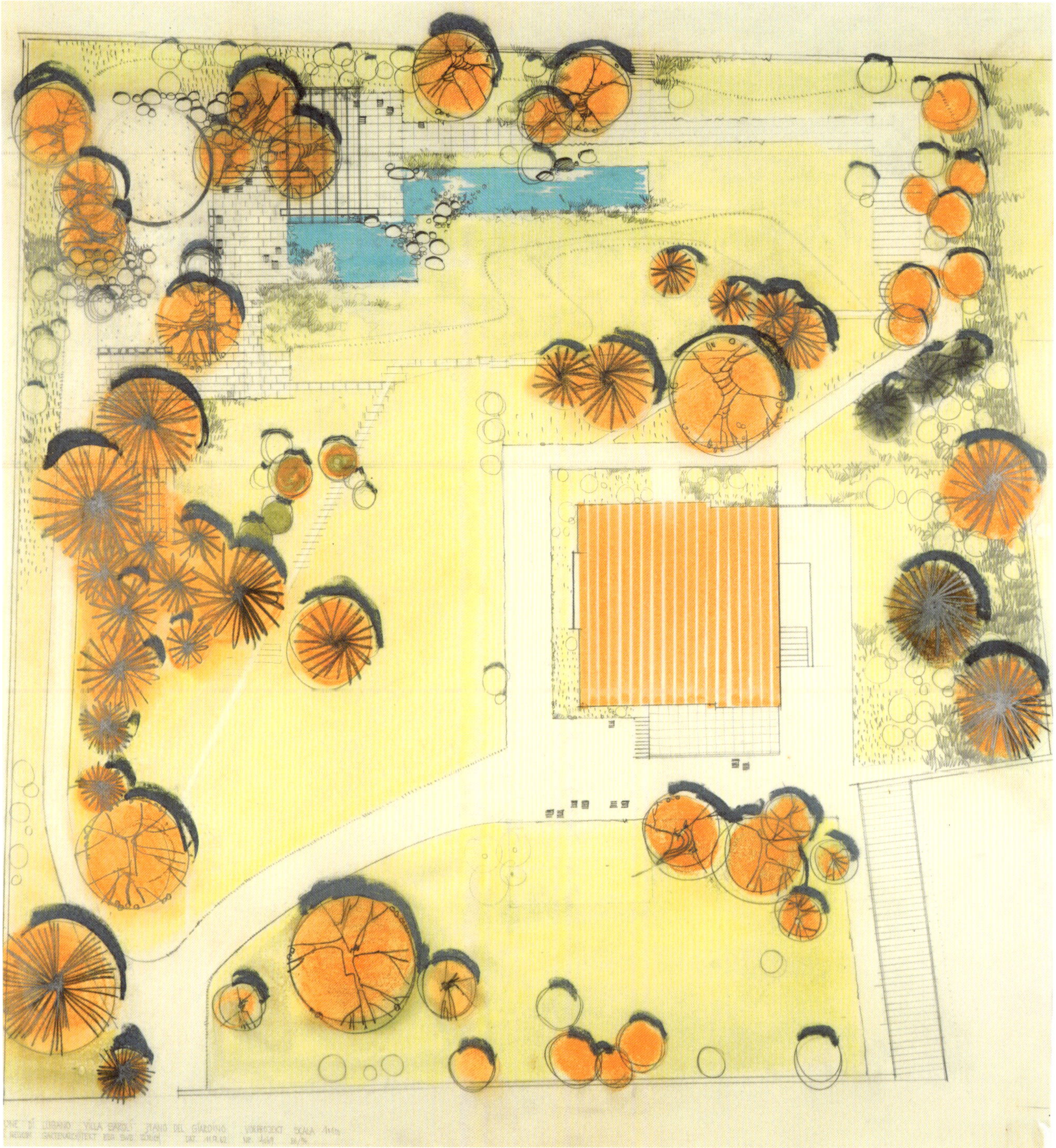

The colour study of the park of the Villa Saroli in Lugano is dominated by its water landscape. Stone settings break up the geometry of the pools and set up a dialogue with the paved sections that accentuate the trees.

→ Rolf Wirz Garden, Münchenstein, 1962: A stretch of irregular terrain, reminiscent of that in the Katsura garden, spills into the rectangular pool. On the opposite side of the house, a dry stream pours into a square "pond".

↓ The sketch Neukom made in 1960–1961 for E. Halpern in Zurich-Witikon shows a dry zone adjacent to a geometrically delineated water landscape with a Japanese feel. The bounding wall has an opening that affords a view of a sliver of the garden that recalls a photograph of the dry garden of the Daisen-in published in K. Takashima's *Die Gartenschönheit* and labelled "Japanese Garden Ideals II" in 1937.

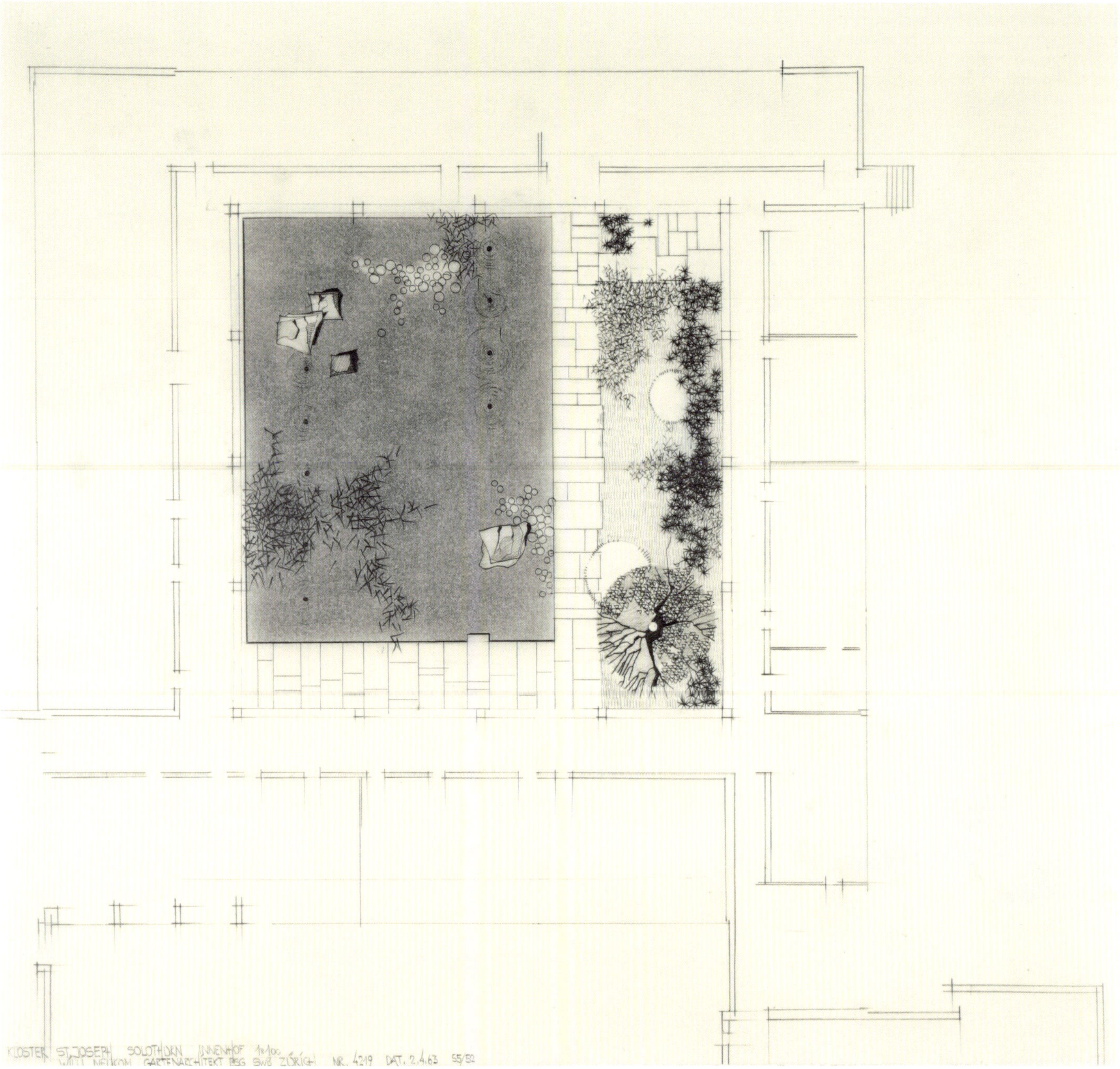

In his definitive design for the cloister of Solothurn Monastery in 1963, Neukom combined a rectangular water basin, bordered with what appear to be small willows, bamboo, and an ornamental shrub, with a courtyard comprising only stone settings and bamboo.

The drawing of the inner courtyard of the Nursing Training School at Zurich's Triemli Hospital is a study in Zen Buddhist concentration.

Adolf Zürcher 1934–2000

Sketches and a photograph of the roof terrace of the Alpenblick project in Cham show the building comprised of geometric, modular components, as well as the waterfall garden, stream, boulders, and water plants.

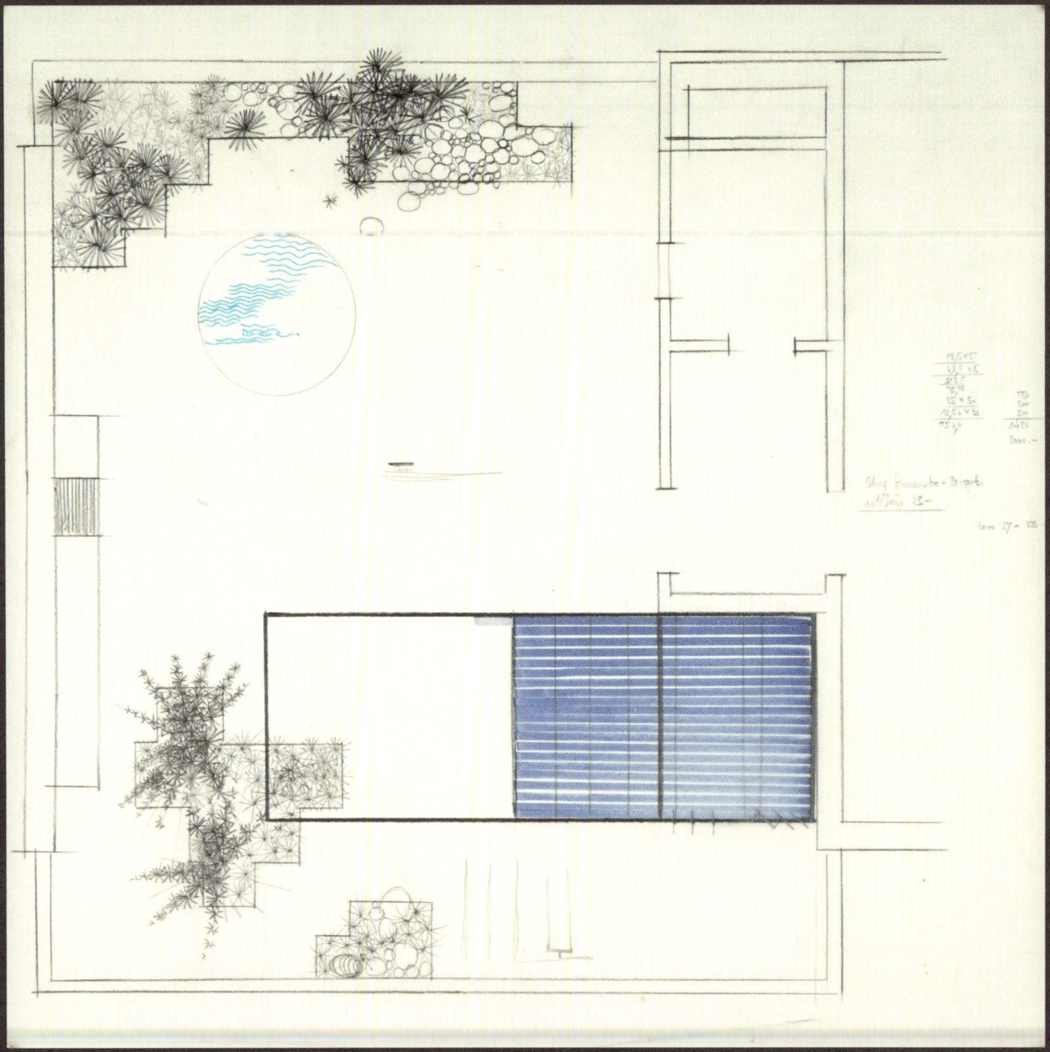

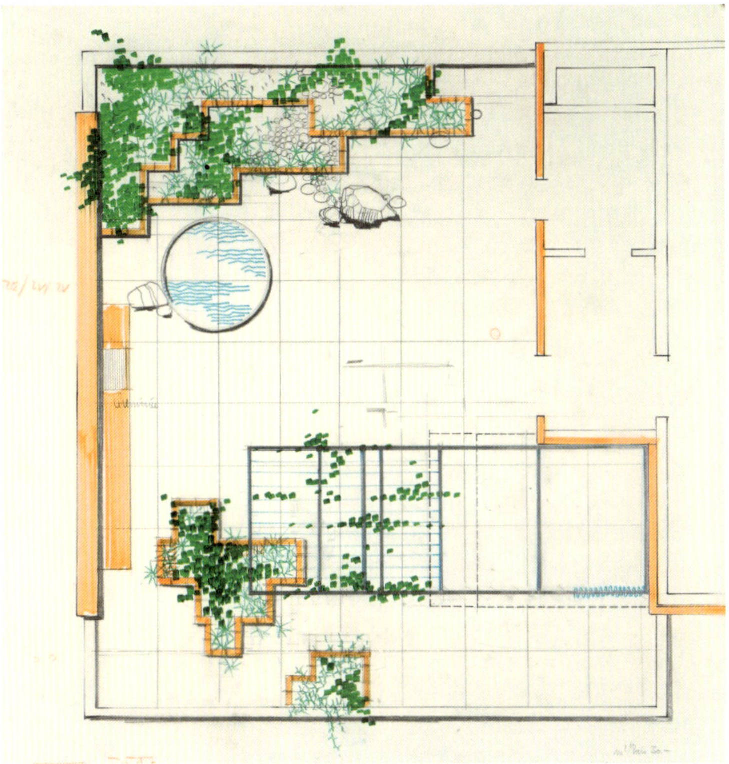

> With increasing land scarcity, garden space around urban settlements is becoming ever smaller, and the task for garden designers ever more difficult. Our task is to make even the smallest garden habitable, to provide an outdoor 'green room' as the Chinese have mastered for millennia.[1]

At first glance, it is surprising that Adolf Zürcher, in a conversation with the veteran Swiss journalist Walter Schnieper, attributed a Japanese phenomenon to Chinese tradition – thus propagating a mistaken impression from the early days of the reception of Japan in the West that was thought to have since been rectified. As the youngest of the garden designers portrayed in this book – almost a generation later than most of the others – Zürcher entered professional life at a time when the perception of Japan was already well established. In 1954, *Das Japanische Wohnhaus* by Tetsurō Yoshida was reprinted, and three years later his work *Der japanische Garten* was published.

The reason for Zürcher's hazy attribution probably lies with his teacher at Kassel, where Zürcher, who came from Sempach near Lucerne, went to study after completing his training as a landscape gardener in Frauenfeld in 1959. At the Werkakademie, he earned the title of garden architect under Professor Hermann Mattern (1902–1971). Mattern's relationship to the Asian continent stems from a period towards the end of its early reception. He was fascinated by Marie Luise Gothein's book on Indian gardens published in 1926[2] and – like Hugo Häring and Hans Scharoun, with whom he felt intellectually connected – shared her interest in Chinese architecture, which she in turn had gleaned from Ernst Boerschmann.[3] It was only later that Mattern began to examine Japan, first discovering Yoshida's work when he followed it up in 1952 with *Japanische Architektur*[4], about which Mattern wrote a review.[5] It was not until 1970 that he would return to the subject.[6]

Naturalistic and abstract

As a pupil of Willi Neukom, with whom he worked before opening his own office in Oberwil, Zug, in 1962, Zürcher experienced a generous interpretation of Japanese garden design, which Neukom had demonstrated with a garden of hitherto unattained richness at the G59 (see "Inspiration, imitation, integration"). Zürcher learned to master both approaches, the naturalistic and the abstract, the organic-biomorphic and geometric-reductionist sceneries. Two examples, created around the same time in 1965–1966 in the canton of Zug, illustrate this virtuosity: a private garden for the Biland-Wyttenbachs in Holzhäusern-Risch and a concept for the grounds of St. Franziskus hospital in Menzingen.[7]

A stone's throw away, at the Bernarda teacher training seminary, Ernst Cramer had designed two courtyards in 1955–1958 that Zürcher was undoubtedly familiar with and appreciated (see "Ernst Cramer – Japanese tempered"). Zürcher's two quadrangles, which like Cramer's propose modern interpretations of cloister gardens, can be read as interrelated but compositionally independent counterparts. Zürcher sketched a pond landscape and a dry landscape for the courtyards, which are proportioned with respect to each other in the ratio of the golden section [→ p. 269]. One comprises a square water basin, visible on the coloured ground plan, which is bridged by a series of stepping stones that progress into a surface of pebbles. These in turn continue their journey, now as a dry stream, into the smaller segment between two headlands. Bamboo and pine are indicated as plants in both courtyards.

The final execution lacked the atmospheric intensity of the design. The water was now "symbolised" using the interlocking pavers typical of the time – possibly under Cramer's influence.[8] The relationship between these two designers is worth more than a footnote. A year after the Menzing "tête-à-tête", the two met as competitors for the design of Julius Hodel's garden in Zug – Zürcher proposing a circular basin and stone settings, Cramer a stone "trinity" in a rectangular frame. Three years after Cramer's death, Zürcher would develop a Japanese-style redesign of the pool at Andreas C. Brunner-Gyr's property in Oberwil, which Cramer had originally created in 1957–1958.[9] Zürcher's homage to the creator of "visionary gardens" can be seen in a much later plan from 1989 for the "Metro Park" in

1 Schnieper 1980.
2 Mattern 1927 NHH, see Hopstock 2015, p. 180; Gothein 1926.
3 See Blundell-Jones 2008; Boerschmann 1923. In a letter to the landscape architect Herta Hammerbacher, his later wife, in the mid-1920s, Mattern drew a correspondingly vague distinction between China and Japan: "Not dissolution and merging of nature and manmade [...]. In the 'plant garden' in contrast to this a laisser-faire, a huddling of plants against each other, left to randomness and to themselves, while the selection is calculated for a specific area, the way the Japanese treat it in their gardens in contrast with the Chinese. Here, too, no imitated nature but man's hand perceivable in the selection and combination." Mattern 1926 NHH, see Hopstock 2015, p. 136.
4 See Hopstock 2015, p. 180.
5 Mattern 1953.
6 Mattern/Mattern 1970a; Mattern/Mattern 1970b. Mattern's rather vague distinction between China and Japan was probably also a consequence of his work with his daughter Merete in the 1960s, where they collaborated on projects with the Chinese architect Chen Kuen Lee (1915–2003). "During the same period" – in 1965 – he travelled to Japan. See Hopstock, p. 181.
7 See anon. 1970a.
8 Cramer's episodic fascination with modular building blocks was one of the themes in the autumn 2022 exhibition "The Organisation of the Void – with Jürg Altherr through the Archive of Swiss Landscape Architecture (ASLA)", "Alte Fabrik", Rapperswil, 3 September – 16 October 2022.

Rochester (NY), which bears more than just a passing resemblance – right down to the drawing style – to Cramer's "Poet's Garden" at the G59 [→ p. 268].[10]

For the realisation of the "Biland-Wyttenbach Garden", Zürcher – in a manner worthy of his Japanese predecessors – was not above lending a hand in "giving shape to the brook trickling through the garden" and to placing "each stone and each plant".[11] He constructed a light pergola, reduced to a skeleton of filigree posts and beams, through which the garden flowed – comparable to the depictions on the Japanese scrolls of the Monogatari Emaki showing pavilions hovering over stones, streams, ponds, copses, and lilies [→ pp. 32–37]. A stream gushed from a spring, made its way between stones and over fast-flowing sections, before finally pouring into three successive circular basins [→ p. 272]. These acted like mirrors, reflecting a "constant playful change [between] grasses, blossom shrubs, bamboo and bushes".[12]

Like his "patron and friend"[13] Willi Neukom, Zürcher was equally at home with the grand gesture – such as the lakeside area at the G59 – and the small-scale, intimate garden room, as seen in the Jardin d'amour, the "Philosopher's Garden", the "Water Garden" and "Crystal Garden" in the same show. Sometimes Zürcher combined both. For the Alpenblick development in Cham, Zug, Zürcher on the one hand incorporated the existing marshland stream to the east of the development, continuing its planting down to a lake. On the other, he created several courtyards as Japanese-style dry landscapes, along with roof gardens with water basins, stone settings, pines, bamboo, grasses, and swamp irises [→ pp. 264, 265].[14]

In the Bruggenmatt garden settlement, he created an overall "site-appropriate planting concept"[15] and incorporated the Schachenbach, which flows into a pond, as a borrowed landscape [→ p. 284]. He took a similar approach for the grounds of the Hübeli School in Emmenbrücke in 1966, "doubling" the existing stream and creating pond-like water areas [→ pp. 270, bottom, 271].[16]

For his 1983 competition project for the Birkenwäldli in Unterägeri, he literally borrowed a body of water: tapping into the Nübächli, which flows into Lake Ägeri, he used it to feed a stream that was to run around the entire perimeter and flow into a pond planted with marsh and aquatic flora. The stream, which divides in places into several rivulets, was filled with agglomerations of boulders reminiscent of the delta-like gravel accumulations of Japanese gardens, such as in the Katsura residence.

In the competition project for the 1969 Federal Garden Show in Dortmund, Zürcher sketched a playground with an extensive dry landscape of pebbles interspersed with boulders [→ p. 279]. It flowed around a sand pit that mimicked a pond. "The landscape begins at the children's playground," Zürcher said, and accordingly he adapted the Japanese elements to this playground.[17]

Zürcher's bravura in lending green spaces a Japanese habitus is documented in numerous plans and photographs. One of Zürcher's photographs of the "Müller-Caviezel Garden" in the Schönau estate in Zug shows a curved pond strewn with cherry blossom petals [→ p. 283] and evokes the apocryphal anecdote about the Japanese tea master Sen no Rikyū (千利休, 1522–1591), that came to epitomise the wabi-sabi ideal of imperfection. Irritated by the accuracy of a meticulously swept tea garden, he vigorously shook a tree so that its leaves scattered across the ground.[18] The fact that the clients Müller-Caviezel have since had the garden redesigned according to the principles of Feng Shui, originally a Taoist harmony theory, brings us full circle, so to speak, to Zürcher's association with China.[19]

By contrast, the courtyard he redesigned for the architect Mittler in Beckenried, which was shielded from the outside world, caused Zürcher to "think of Japanese living culture" [→ p. 267].[20] With its platform of wooden decking and walls of various types of bamboo, it may have served as the model for the bamboo courtyard at Grün 80 in Münchenstein – which Zürcher together with Neukom developed for the conclusion of a cascade of *horti conclusi* from which the so-called "urban structure" was assembled[21] [→ p. 117].

The design of both these projects came at a time when the Japanese Garden created by Akira Satō at the Federal Garden Show in Bonn in 1979 was attracting attention. Zürcher was impressed by the pond landscape with waterfall, stone settings, islands, and a delta accentuated by a stone lantern – a blatant quotation of the corresponding scene in the "Katsura Garden" [→ p. 113].

9 Cramer's design for the "Hodel Garden" survives in a single, undated plan, asla_cre_6191. For the "Brunner-Gyr Garden", Landhaus Gimmenen, see Stucky 1959; Weilacher 2001, pp. 100–101.
10 Adolf Zürcher, "Metro Park Rochester NY USA Conceptual Design Outdoor Space Var. A 1:200", 27 February 1989.
11 Anon. 1966a, p. 10
12 Ibid.
13 Andermatt 2001, p. 75.
14 See anon. 1965; Anon. 1968; Anon. 1967b.
15 Anon. 1983b, p. 19.
16 Since the existing stream was to be partially covered as part of the motorway construction, Zürcher had to create a new stream bed two metres above the old bed. Anon. 1969a, p. 18.
17 Anon. 1970b, p. 262.
18 中村昌生 / Nakamura M 1971; cited in Goto/Naka 2016, p. 30.
19 Interview with Renata Müller-Caviezel, 8 July 2022.
20 Zürcher 1978, p. 16.
21 Zürcher 1980b, p. 45.

Zürcher designed an intimate courtyard with a moon-viewing deck for the architect Mittler in Beckenried.

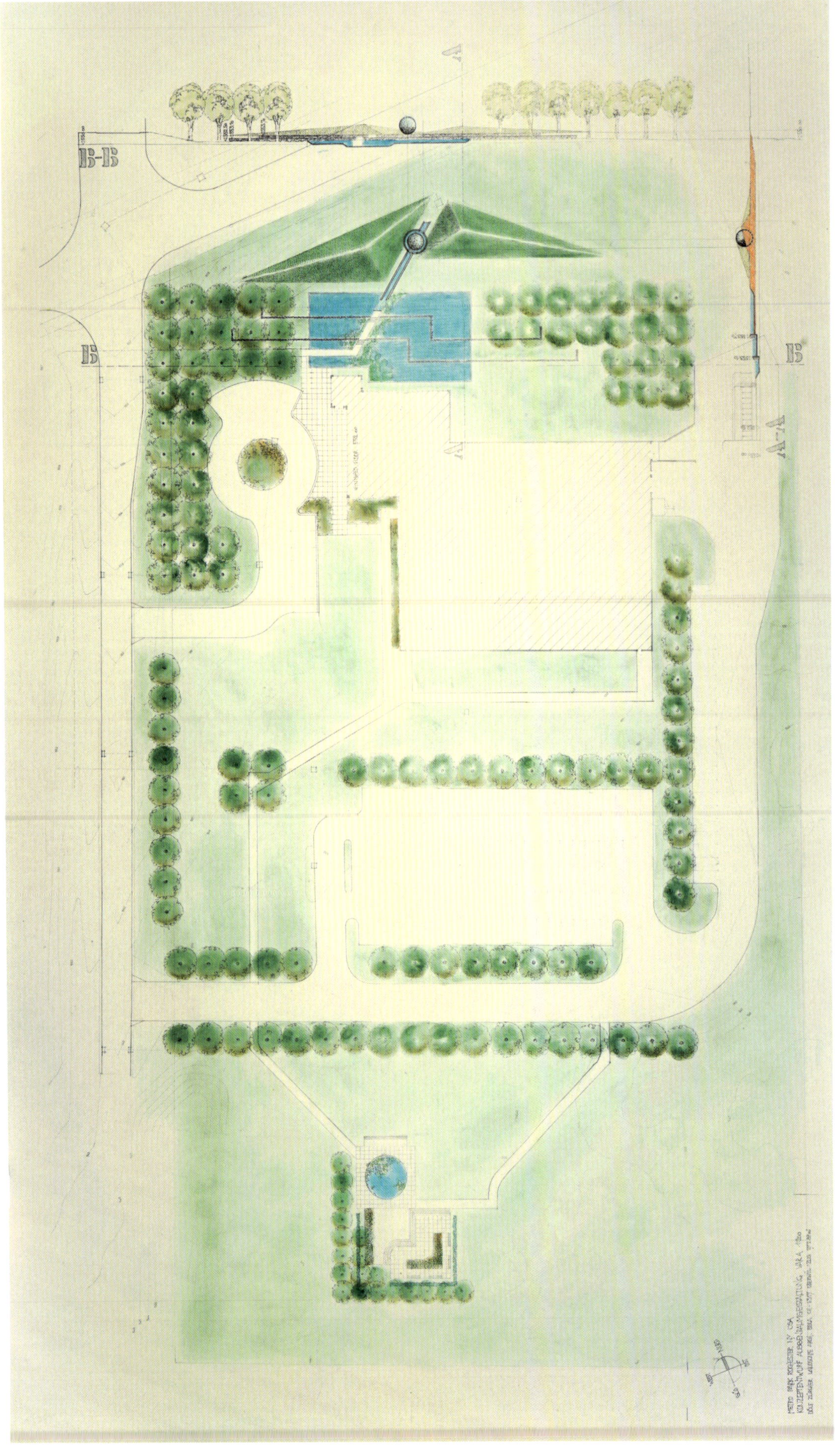

← Zürcher's plan for the Metro Park in Rochester in the US state of New York, 1989: A green courtyard and water courtyard on circular ground plans act as reciprocal counter-figures of emptiness and fullness, of yin and yang. The two geometrically contoured hills and a water landscape in front of them recall Ernst Cramer's "Poet's Garden" created for the G59.

→ The floor plan of the courtyards of the St. Franziskus Hospital complex in Menzingen, designed in 1966, corresponds to the perspective from which the two segments are viewed – from the terrace above them. They are in essence three-dimensional paintings, or framed landscapes, and evoke associations with looking at the panels of a painted screen or successive frames of a monogatari emaki (narrative hand scroll).

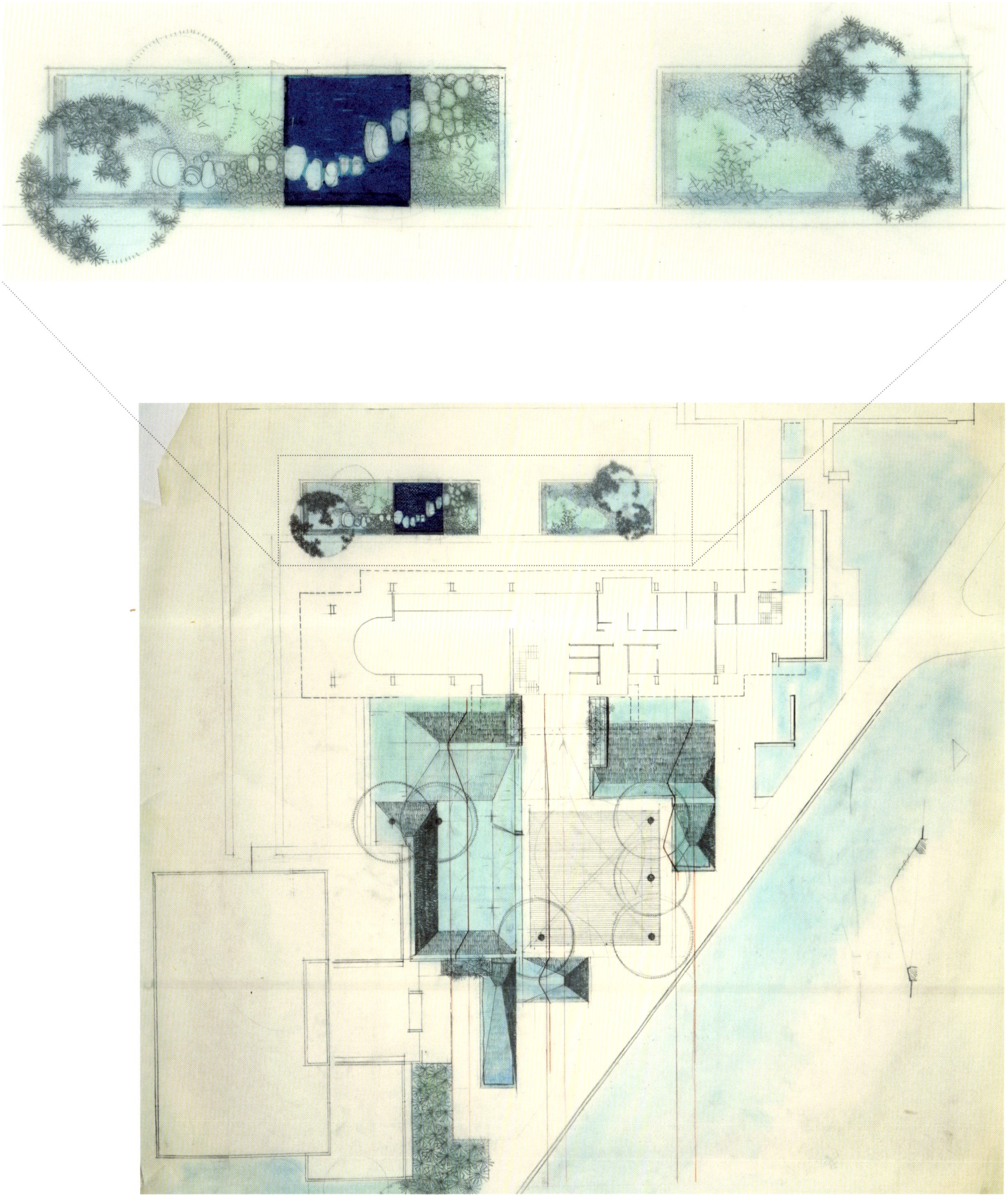

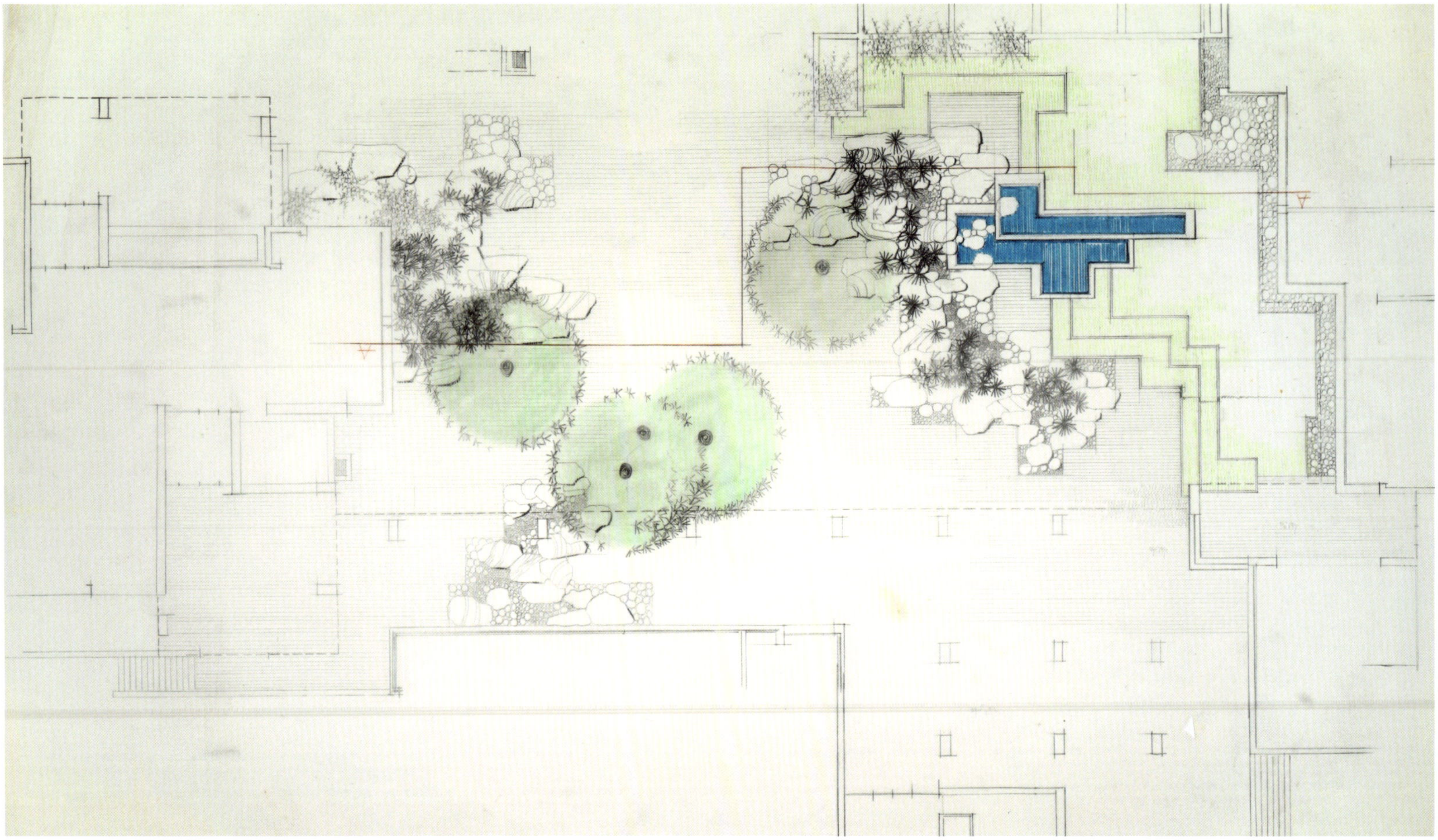

↑ For the Loreto School in Zug, Zürcher combined a geometrically delineated water basin with a biomorphically composed dry stream bed.

↓ Sketched ground plan and section of the stream and pond landscape designed for Hübeli School in Emmenbrücke in 1966.

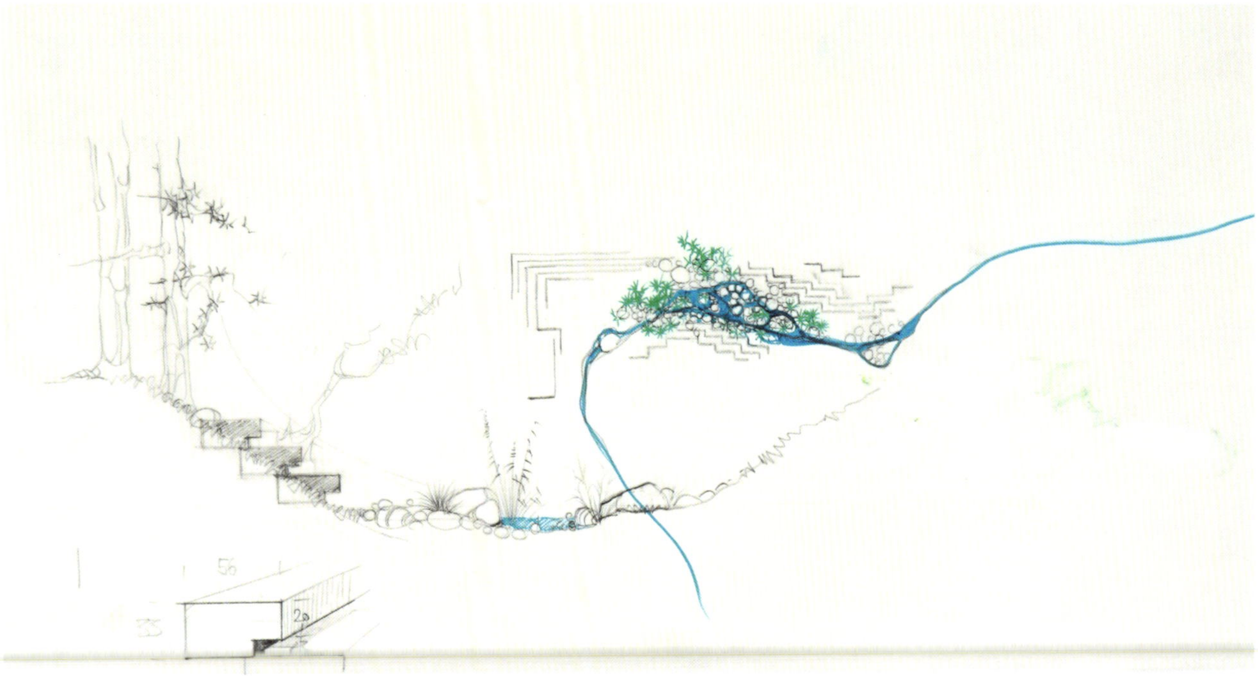

Zürcher added a Japanese touch to the Hübeli School in Emmenbrücke in 1966 through the use of swamp irises, bamboo, and water lilies, as well as stone settings and islands.

It is possible that, like Cramer, Zürcher was inspired by Richard Neutra in his design for the Biland-Wyttenbach perennials garden. In any case, the circular basins appear to strongly resemble Neutra's garden design for the Nesbitt House.

A kidney-shaped basin, "boulders with bamboo", and iconic circular stepping panels feature in the design for the "Troxler Garden" in Bellikon in 1964.

A meandering stream characterizes the green space in "Vasella Garden" in Risch, Zug.

The stream feeds an expansive basin with water lilies, swamp irises, and natural stone blocks in the "Vasella Garden" in Risch, Zug.

← EAWAG Kastanienbaum: In 1979, three years after it was designed, the waterfall tumbles over several steps before flowing into a lily pond.

↓ For Philipp Schnadt in Zug, Zürcher created a geometric reductionist version of a Japanese-inspired design.

"Swiss pine tussling near Grünsee, above Zermatt", 1981: Zürcher's freehand drawing has a distinctively Japanese quality.

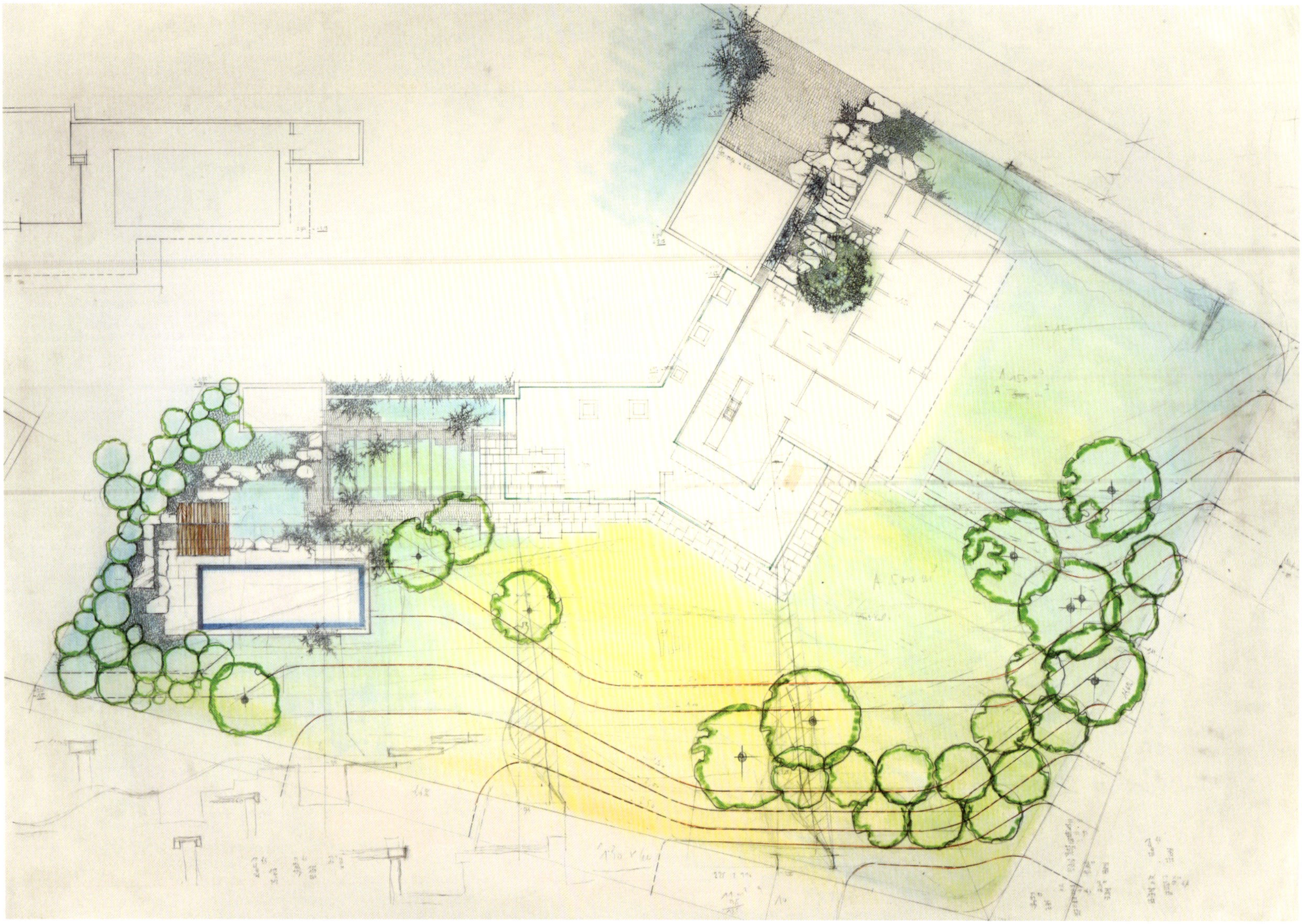

The sketch for the “Bezzola Garden” exhibits Japanese-style stone settings and sections of boulders, pebble, and bamboo.

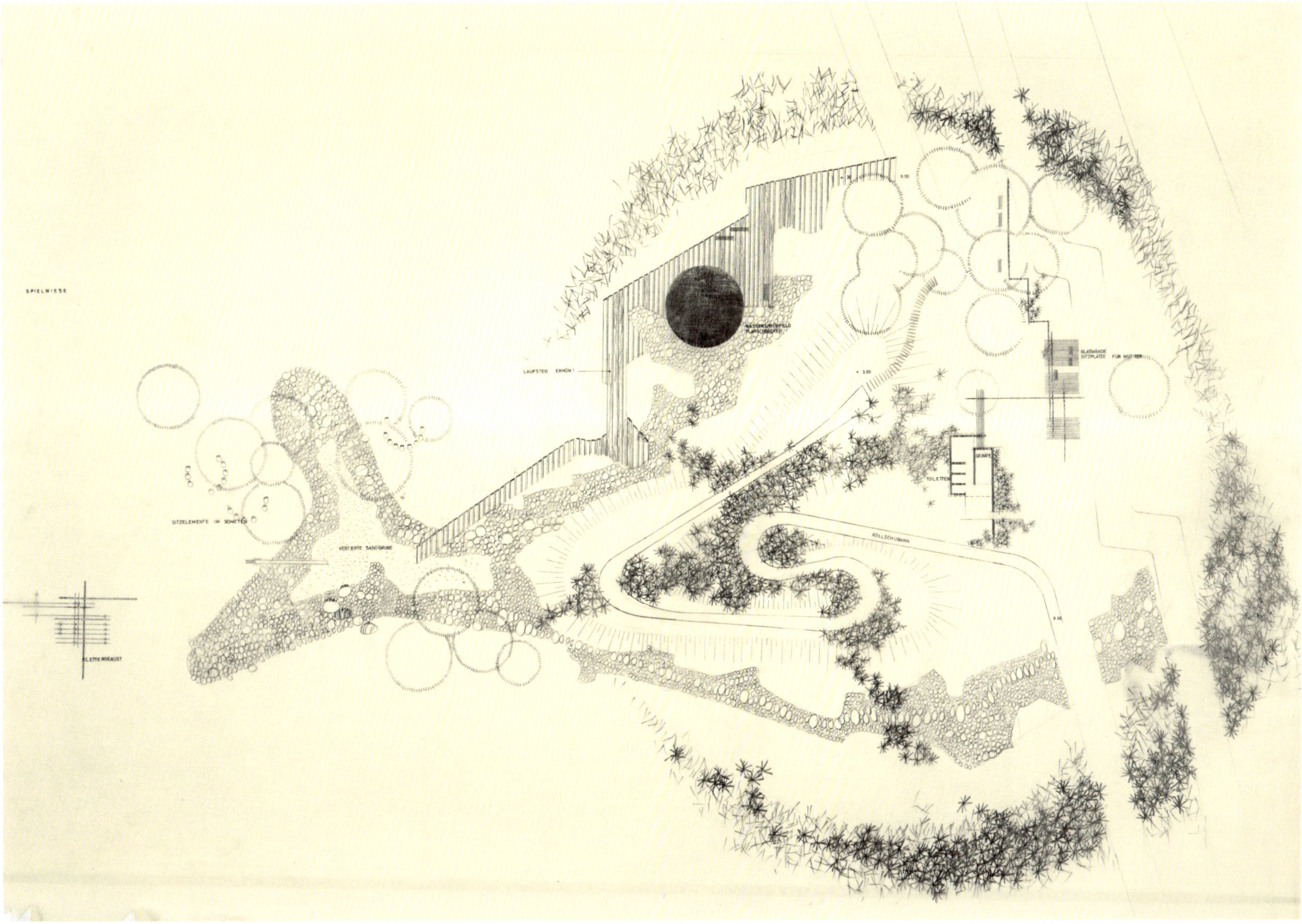

In the competition project for the 1969 Federal Garden Show in Dortmund, Zürcher sketched a dry landscape of pebbles and a sand pit that imitates a basin of water.

↑ Photograph taken in 1989 of a water landscape in the garden of the Hermans' terrace house in Oberägeri. Zürcher turned to Chinese models to create a garden within a very small space.

→ A stepping stone lurches out of line in the "Girschweiler Garden". Zürcher used such techniques to celebrate the beauty of imperfection.

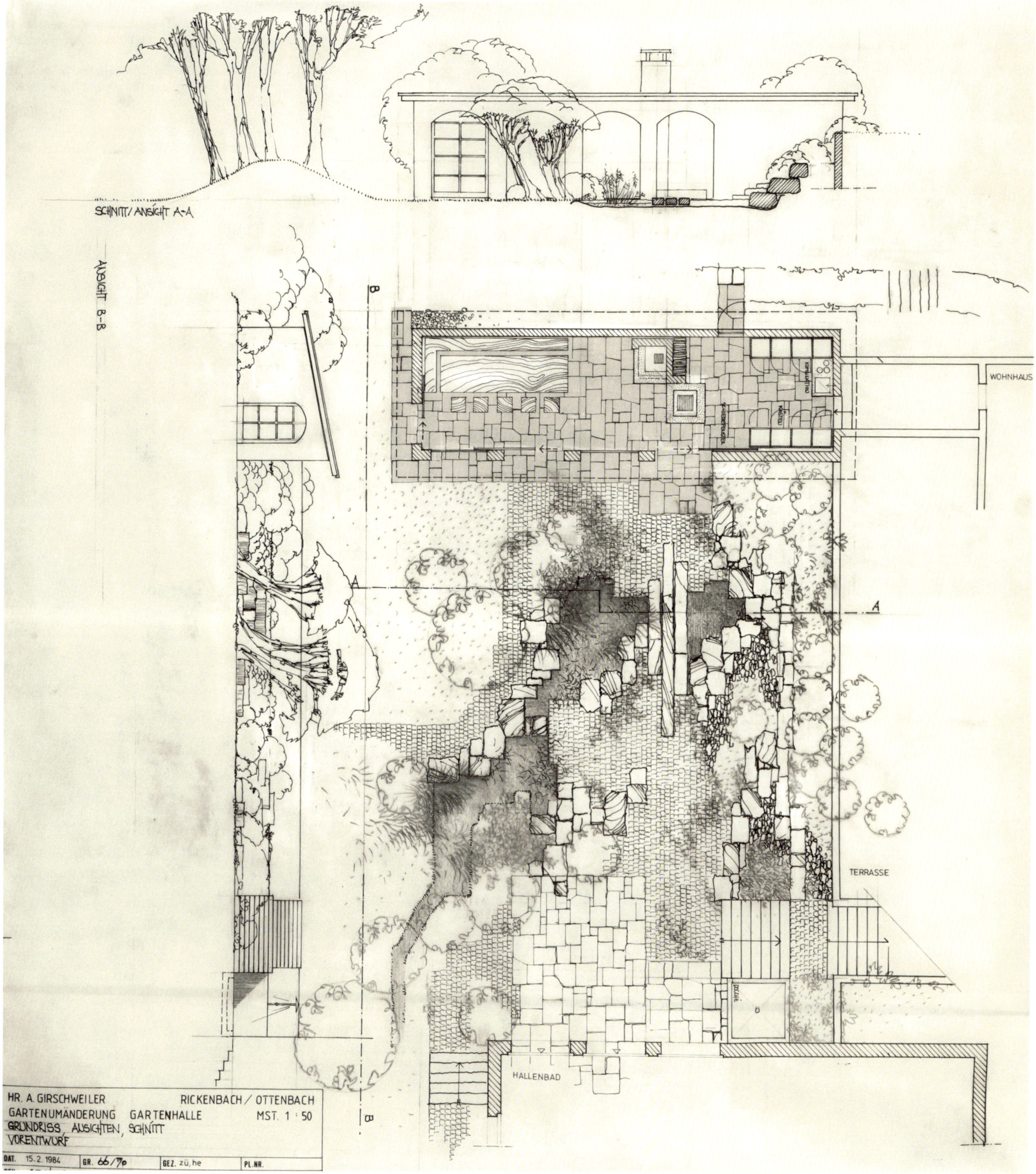

The pond in the "Girschweiler Garden" was probably fed by the nearby Lindenbach stream.

← The "Müller-Caviezel Garden" in Zug is dominated by a Japanese-inspired miniature landscape, with a water lily pond whose banks are articulated as a kind of delta.

↓ The design of the bridge over the pond in the "Wiede Garden" could almost have been drawn directly from one of Albert Baumann's tabular classifications.

→ In the "Müller-Caviezel Garden", the aesthetics of imperfection coincide with a heightened awareness of impermanence.

In the Bruggenmatt garden settlement Zürcher incorporated the Schachenbach, which flows into a pond, as a borrowed landscape.

Lists of works

Selection of the most impressive
Japanese-influenced projects

Legend

▸ Garden preserved completely or in its basic features

▹ Garden not localizable because address not specified (n. s.), not executed, heavily redesigned or destroyed

Gustav Ammann

1923–1943
Ammann, Gustav → p. 146
▸ Hirslanderstrasse 37, Zurich

1928–1947
Eugster, Josef → pp. 41, 42, 124
▸ Bahnhofstrasse 3, Altstätten

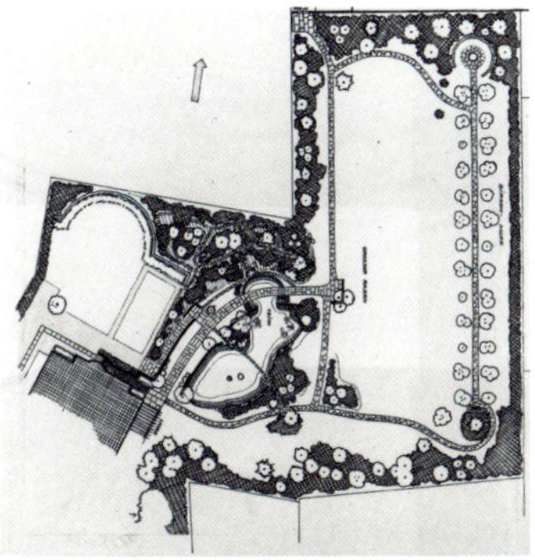

Ten years after the villa was built in 1918 by the St. Gallen architect Johann Scheier (1886–1945) for the embroidery factory owner Josef Eugster, Ammann tackled the first stage of the garden design.

1933–1938
Bosshard, H. U.
▹ Scheideggstrasse 80, Zurich
The rectangular basin, which is divided into areas for water lilies, marsh plants, and for bathing, is bridged by square stepping stones in a geometric interpretation of Japanese models.

1934–1935
Debrunner, Alfred
▹ Bergstrasse 41, Zollikon

A miniature garden that combines Japanese ingredients in a tight space – embedded in an almost wild pattern of paving slabs.

1934–1935
Schmid, Hans
▸ n. s., Embrach
"A freeform fish and water lily basin", is the caption Ammann chose for his "typified" water basin.
Lit. Ammann 1955, pp. 138–139.

1934–1936
Egger, Hermann
▹ Zelgweg [6], Worb

The geometric version of the typified water lily basin and bathing pool adjoin at right angles and are separated by square stepping stones.

1934–1936
Stadler-Gujer, E.
▹ Brunnenstrasse, Uster
The organic-geometric basin is crossed by stepping stones and planted with water lilies.

1934–1937
Schleiniger, Max → p. 140
▹ n. s., Wohlen

1934–1938
Endemann-Suidter, Anna
▹ Sternhalde 1, Lucerne
The layout features a geometric cascade in the style of the terraced rising water lily ponds created by Paul Schädlich at the Züga in 1933.
Lit. Ammann 1955, p. 150.

1935/1940–1941
Spinner-Debord, W.
▹ Zollikerstrasse 97, Zollikon

1936
Borsari, E.
▹ Binderstrasse 40, Zollikon

1936–1945
Hauser-Studer, Rudolf
▹ Schreberweg 8, Zurich

The garden around the water basin crossed by square stepping stones is planted with numerous plants of East Asian origin, including Japanese maple, olive willow, spiny-leaved scented flower, Chinese reed, cedar, azalea, rhododendron, winter jasmine, Japanese skimmia, Japanese aukube, Japanese flowering cherry, wisteria, yellow willow as well as Japanese and Korean snowball.
Lit. Stoffler 2006b.

1937–1952
Cemetery
▸ Obere Wiltisgasse 40, Küsnacht
A small waterfall feeds an organically curved basin with an island of reeds.

1939
Blattmann, W.
▹ n. s., Wädenswil
A series of iconic circular stepping stones lead to an island of water plants within the virtually oval pond.

1939
Landesausstellung (National Exhibition) 1939 → p. 89
▹ Zurich
Lit. Ammann 1939, p. 210.

1940–1941
Walder-Hasler, F.
▹ Seeblick, Männedorf

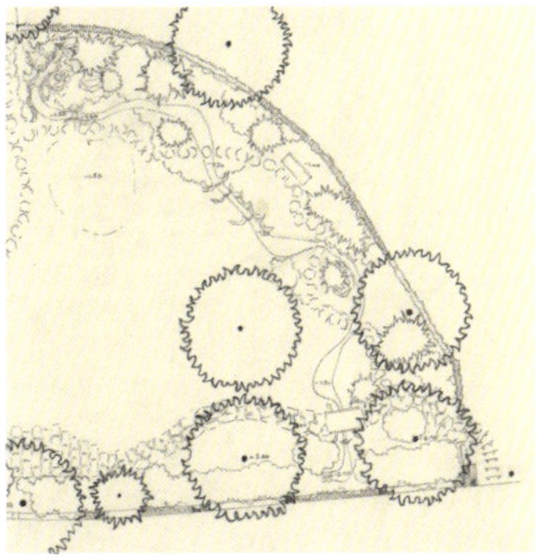

Faithfully interpreting the Japanese model, the stream gushes out from between stones, meanders through the meadow, its bed widening or narrowing here and there.

1941
Müller
▹ n. s., Embrach

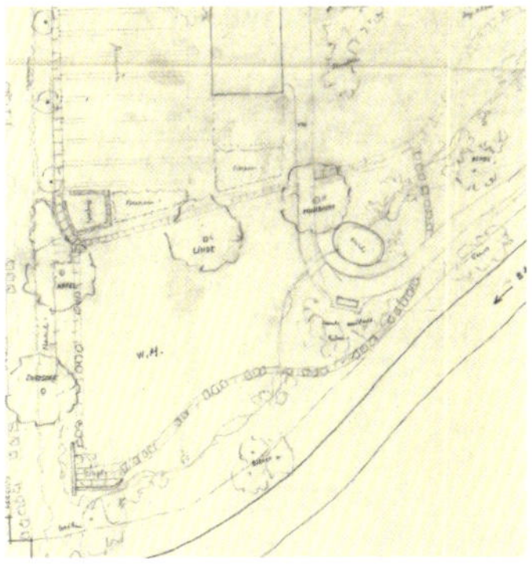

Similar to the Walder Garden, Ammann created a stream bed that weaves its way excitedly between boulders, bridges, and ponds.

1941–1943
Wohlfahrtshaus Werkzeugmaschinenfabrik Oerlikon, Bührle & Co. → pp. 122, 123, 140
▸ Langwiesstrasse 30, Oerlikon
Lit. Ammann 1955, pp. 188–189.

1942–1951
Letzigraben Lido → p. 146
▸ Edelweissstrasse 5, Zurich
Lit. Ammann 1955, pp. 158–159.

1943/1949–1953/1955–1958
Kantonsspital → p. 142
▹ Brauerstrasse 15, Winterthur
After the ideas competition in 1943 Gustav and Peter Ammann sketched designs for the cantonal hospital in 1949–1953 and 1955–1958 respectively with a Japanese influence but neither design was selected. The complex was realised in stages in 1962 by Ernst Meili and in 1968 by Christian Stern.

1944
Müller-Keyser, Robert → p. 144
▹ Zollikerstrasse 44, Zurich

1945
Deller, W. → p. 139
▹ n. s., Winterthur-Wülflingen

1945–1946
Bischofberger, K.
▹ In der Hub 12, Zurich
A channelled stream passes three small fountains and two miniature bridges on its way to a circular basin, revealing the hand of its creator. The garden, however, was realised to a design by Walter Leder.

1946–1948
Küderli, O.
▸ Waldgartenstrasse 14, Zollikerberg
The two-part basin, composed of a circle and rectangle, is surrounded by three Japanese threadleaf maples, with bamboo and an empress tree adding to its Asian character.

1946–1950
König-Moll, Werner
▸ Äussere Ringstrasse 36, Thun

1947–1948
Bertschy-Ringier, Herbert
▸ n. s., Zofingen

1947–1953
Jordan, C → p. 145
▹ Zürcherstrasse, Uitikon

1947–1956
Waid city hospital → p. 143
▸ Tièchestrasse 99, Zurich

1948–1950
Bergmann, C.
▸ Öscherstrasse 4, Zollikon

1949–1953
District hospital
(today: hospital and nursing care centre)
▸ Mühletalstrasse 27, Zofingen

The park's character is determined by a stream flowing in natural curves, stone settings, and a pond. In addition to lotuses, swamp irises, and bulrushes, the pond and stream are populated by Japanese brocaded carp.

1952–1953
Sennhauser, A.
▹ n. s., Egg

1954
Baumgartner, H.
▸ Zumikerstrasse 38, Zollikon
The curved lily pond is crossed by a Japanese-inspired bridge and augmented by constellations of rocks and boulders.

Albert Baumann

1931
Director's residence, Witzwil prison
▹ Lindenhof 10, Gampelen (BE)
Baumann designed a rock garden to appear as if it were a stretch of water crossed by stepping stones.

1931
Oeschberg Cantonal Horticultural School, Japanese Garden → p. 148
▹ Bern-Zürich-Strasse 14, Koppigen

1938
Bichsel
▹ n. s. [Emmentalstrasse 49], Lützelflüh-Goldbach

Baumann used a meandering stream to evoke Japanese associations.

1938
Gerber
▹ n. s., Lyss
Ground plan and perspective of a section of the stream.

undated
Nyffenegger, Hans
▹ n. s., Kirchberg

undated
Japanese garden with lantern and fountain → p. 148
Sketch

undated
Japanese garden with two lanterns and stepping stones planted with Japanese swamp irises, rhododendrons, and Japanese maiden pine → p. 149
▹ n. s., location unknown

undated
Liestal prison
▹ Rheinstrasse [27], Liestal

Ernst Baumann

1947
Schmid-Jenny, N.
▹ Gütstrasse 10, Küsnacht

1956/1966
Dieterle, Rudolf → p. 160
▹ Langägertenstrasse, Zollikerberg

1958
Braendlin-Blumer, R.
▹ n. s., Jona
A courtyard laid out as a miniature garden with a water basin, Japanese ornamental cherry, and stone settings contributing to its Japanese flair.

1958
Rotz von, Otto
▹ n. s., Cham

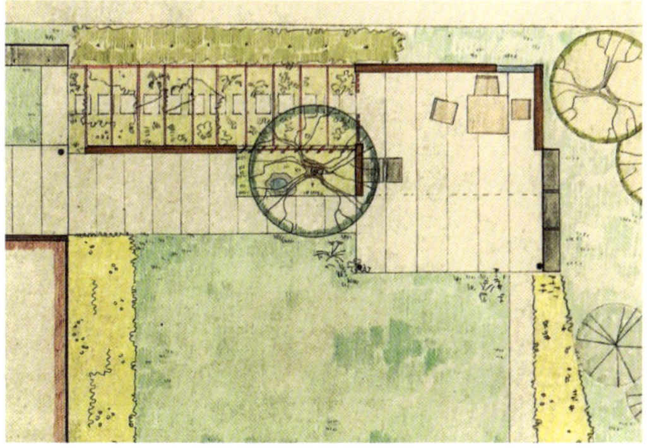

Baumann's design for Otto von Rotz completely blurs the boundaries between inside and outside. The intimate courtyard, reduced to a pond and individual solitary trees, is both an interior and exterior space: as a cut-out of the veranda, it is an outdoor space; at the same time, the courtyard is a space within the outdoor space of the veranda.

1959
G59 – 1. Schweizerische Gartenbau-Ausstellung
(1st Swiss Horticultural Exhibition) → pp. 98–101
Bellerivestrasse / Tiefenbrunnen Lido / Höschgasse, Zurich
with Willi Neukom

1960–1961
Leutert, Jakob and Marlies → p. 161
▸ Aurorastrasse 95, Zurich
Architect: Ernst Gisel

1961
Staub, Th.
▹ n. s., Richterswil

1963
Puent Primary School
▹ Püntstrasse 8–12, Oberrieden

1963
Speiser
▹ n. s., Oberrieden

1964
Siegenthaler, W.
▹ n. s., Kilchberg

The garden detail could have been photographed in a Japanese garden. It resembles a photograph taken by Tetsurō Yoshida in 1957.
Japanese paintings like to depict houses hidden behind trees. Baumann evidently had an eye for similar situations, as other photos of the house illustrate.
Lit. Yoshida 1957, p. 58.

1964
Retirement home in Ruggacher
▹ Bremgartenstrasse 39, Dietikon
A pond with a stream, bridges, and stone settings draws on and adapts Japanese inspirations.

1964
Stähli, Adelbert → p. 159
▸ n. s. [Zeughausstrasse 12], Lachen
The flowing transition between garden and house can still be discerned, but the details of its design have been lost.

1965
Hegibach building development
▹ Hegibachstrasse 54–58, Zurich
Architect: Ernst Gisel
A stone bed simulated a stream that is accentuated by stone settings.

1967
School complex
▸ Reckenstrasse 7, Thayngen
Baumann designed the six inner courtyards of the complex as miniaturised typological landscapes with the themes "water", "stone", "sand", "moss", "rocks", and "hedges" each with a specific thematic planting: water plants, tropical plants, bamboo, forest flowers, cushion plants, garden flowers, and lilies.

1968
Schulhaus Altweg
▹ Letzigraben 208, Zurich

1969
Schlieren School
▸ Schürrainweg 4, Schlieren

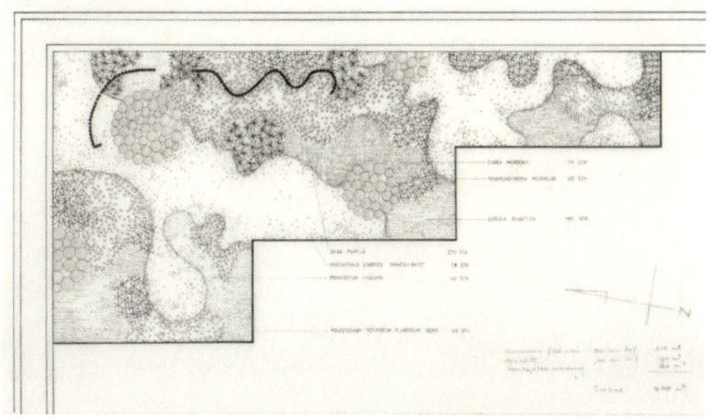

Baumann created two miniature gardens for the school, an "activities courtyard" and an "inner courtyard". The "activities courtyard" has boulder-like stone settings and a comparatively colourful planting scheme. In the "inner courtyard", a bed of gravel and pebbles flowing in curved contours imitates a body of water. It is lined with mossy soft shield fern (*Polystichum setiferum plumosum densum*), Chinese maidenhair grass (*Miscanthus sinensis Gracillimus*), feathertop grass (*Pennisetum villosum*), evergreen dwarf bamboo (*Sasa pumila*), Japanese sedge (*Carex morrowii*), umbrella bamboo (*Sinarundinaria murielae*), and greater woodrush (*Luzula sylvatica*).

1969
Fuchs, R. and Moos, F.
▹ Soorhaldenstrasse, Illnau

1969
Staub, M.
▹ Alpenstrasse 35, Thalwil
Baumann designed the inner courtyard for the atrium house, which still exists today. He divided it into two areas with accurately laid granite slabs, two sections with flowering perennials, an area planted with white-flowering spurium (*Sedum spurium album superbum*), and three junipers (*Juniperus*), as well as a still discernable Japanese-inspired section with water basins, reeds, and stone settings.

1970
School
▹ Breitistrasse 13, Greifensee

1971
Haling Secondary School
▹ [Thundorferstrasse 72], Matzingen

1971
Weibel, H.
▸ Staldenbachstrasse 17, Pfäffikon (SZ)
Baumann borrowed the stream running down the valley as a source for the pond, which flowed into a stream bordered by stones and tiered with occasional steps.

1973–1975
Hospital
(today: 'Serata retirement home')
▹ Tischenloostrasse 55, Thalwil
with Christoph Baumann

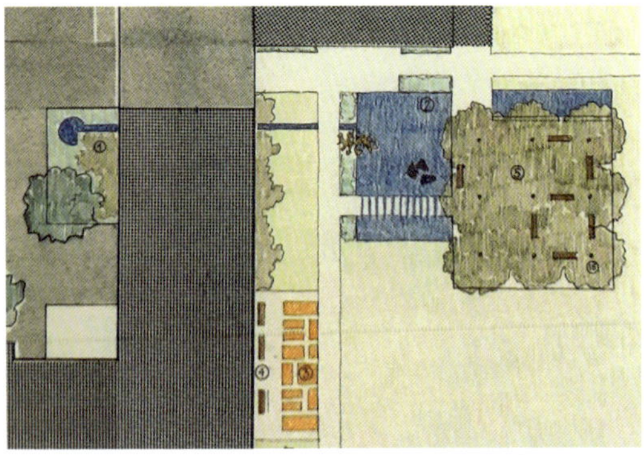

Baumann senior and his son incorporated the stream that ran along the edge of the property, widening partially into a pond, into the concept for the park design. To this end, they sketched a geometrically articulated counterpart in the courtyard situation between the buildings: an orthogonally delineated waterscape "with water plants and goldfish" bridged by narrow stone slabs, in which three stone blocks are placed that echo the Buddhist trinity of stone setting. A straight rivulet leads into the inner courtyard of the foyer, where it pours into a circular basin.
The archives also contain a parallel design by Ernst Cramer from 1973. The only aspect common to both projects was the inner courtyard, for which Cramer also proposed a round water basin. The present situation shows only a long rectangular basin with a few water plants in the outdoor area.

1974
Sihlweid School → p. 162
▹ Sihlweidstrasse 5, Zurich Leimbach

1975
Retirement home
▹ Heimstrasse 15, Amriswil

1977
Vocational training centre
▸ Schützenstrasse 9, Weinfelden
Baumann designed the inner courtyard as a dry landscape in which pebbles simulate a stream bed that meanders between ivy and various ferns. The majority of the plants are of East Asian origin: sweet olive (*Osmanthus burkwoodii*), Japanese autumn anemone 'Honorine Jobert' (*Viburnum davidii*), blue-flowering plumbago (*Ceratostigma plumbaginoides*), and Christmas rose (*Helleborus niger*). He added Canadian hemlock, holly, boxwood, and creeping yew. Yellow and red barrenwort and small periwinkle gave extra splashes of colour.

1977
Irchelpark, Competition
▹ Hirschwiesenstrasse 10, Zurich

undated [1982]
SUVA Swiss Accident Insurance Institution → p. 163
▸ Brambergstrasse 4, Lucerne
Lit. Mathys 1982, p. 31.

undated
Rossel → p. 156, 157
▹ n. s., Baar

undated.
n. N. → p. 155
▹ Small dwelling garden

undated
n. N.

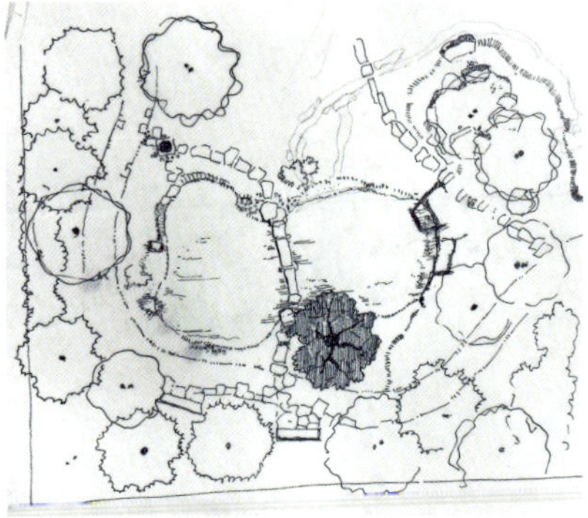

Sketch of a pond crossed in Japanese fashion by stepping stones that lead to a ritual washing basin.

Ernst Cramer

1937
Siegrist, Rudolf → p. 168
▹ n. s., Aarau
Lit. Weilacher, n. pag. [p. 267, List of works]

1947
Bechtler, W.
Weekend house
▹ n. s., Freienbach

A dry landscape on the lakeshore with distinctive stone settings, including a composition in a style that the *Sakuteiki* describes as the "Buddhist Trinity".

1947
Zürcher Kantonale Landwirtschafts- und Gewerbe-Ausstellung
(Zurich Cantonal Trade and Agricultural Exhibition, Züka) → p. 92
▹ Zurich

1947
Aargauische Gartenbau-Ausstellung
(Aargau Horticultural Exhibition, AGA) → p. 93
▹ Aarau

1948
Hegner, Wolfgang E. → pp. 129–131, 170, 171
▸ Seestrasse 334a & 342, Feldmeilen
with Willi Neukom

1948
School
▹ Spitalstrasse 20, Rüti

1949
Gut School
▸ Gutstrasse 103, 105, 107, 109, Zurich
Architect: Christian Trippel
The hexagonal pavilion still exists today. The kidney-shaped pond is likewise probably original. However, in contrast to Cramer's plans, where it is positioned directly next to the pavilion on the east side, the existing pond is situated on the opposite, western edge of the site.
Lit. Spezialinventar 2008.

1949
Zendralli, Ugo
▹ n. s., Roveredo

1950
Härdy, Jakob
▹ n. s., Oberentfelden

1950
Immobilien AG
▹ n. s., Olten

1950
Schibler, Urban
▹ n. s., Aarau

1953
Frey, H.
▹ n. s., Wangen b. Olten

1954
Amsler-Merz, E.
▸ Im Häsi, [Rebhaldenweg 30], Aarau

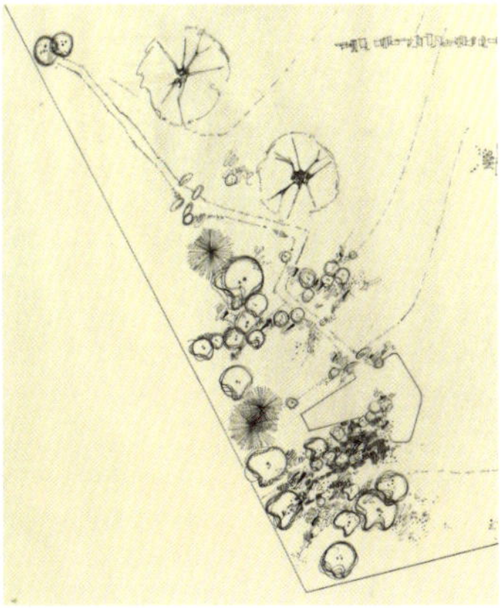

As in the *Sakuteiki* described, the stream's zigzag course was a product of the stones set in its path. It flowed into a pond.

1954–1955
Apfelbaum School
▸ Apfelbaumstrasse 31, Zurich
Lit. Wasserfallen 1957, ill., p. 409.

1955–1958
Bernarda teacher training seminary
(today: cantonal school) → pp. 76–79, 164
▸ Seminarstrasse 14, Menzingen
Architecture: Hanns Anton Brütsch, Alois Stadler, Leo Hafner, Alfons Wiederkehr
Lit. Gadient 2009.

1960–1964
Church parish centre
▸ Hauriweg 17–23, Wollishofen
The courtyard's Japanese character is most evident around the water basin, which, although geometrically delineated, has a distinctly Japanese flair due to its islands of piled stones and plants, as well as a pine tree at the edge of the courtyard and scattered stone settings throughout.
Lit. Inventarergänzung 2013.

1961
Schmidlin, Josef & Agnes → pp. 174, 175
▸ Wasserfluhweg 5, Aarau

1961
Ilgen
▹ Roethel, [Im Rötel], Zug

1958–1964
Spirig, Hugo → p. 176
▹ n. s. [Sonnmattweg 1], Wil nr. Olten [Starrkirch-Wil]

1962
Wartmann & Cie. AG → p. 172
▹ n. s. [Stahlrain 8], Brugg

1963
Open-air dance floor
▹ n. s., Thun

1963
Migros → p. 185
▹ n. s., Urdorf

1963
Frawa AG → p. 184
▹ Weltistrasse [12], Zollikon

1964
Bertschinger, Max
▹ n. s., Lenzburg

1964–1965
Cantonal school
▹ Bühlstrasse 36, Wetzikon

1965
District school
▸ n. s. [Tramstrasse 20], Suhr
Two inner courtyards as miniature gardens.

1965
Prina, Aldo → pp. 74, 75, 133, 177
▸ Raffelen 2, Dulliken

1966–1967
Meister, W. → p. 167
▹ n. s., Zollikon

1967
Retirement home
Bifangstrasse 8, Wohlen

1967
n. N.
Humboldtstrasse 13, Bern
Studio terrace

1969
Mann, Hugo → pp. 182, 183
▹ n. s. [Kaiser-Wilhelmstrasse 26], Baden-Baden (D)

1960–1977
Cemetery → p. 178, bottom
▸ Schützenweg 2, Gränichen
The Japanese aspect lies less in the layout and more in how the garden plan is drawn.

undated
Ammann, S.
▹ n. s., Zurich

undated
Angst → p. 178, top, 179
▹ n. s., Herrliberg

undated
Baumann, Rob[ert]
▹ n. s., Lenzburg

undated
Brändli
▹ Freudenbergstrasse 88, Zurich

undated
Cattaneo, Fausto
▹ n. s., Giubiasco

undated
Disteli, O.
▸ Holiday home, n. s. [Seeweg 26], Schenkon

undated
Klameth, H.
▹ n. s., Küsnacht

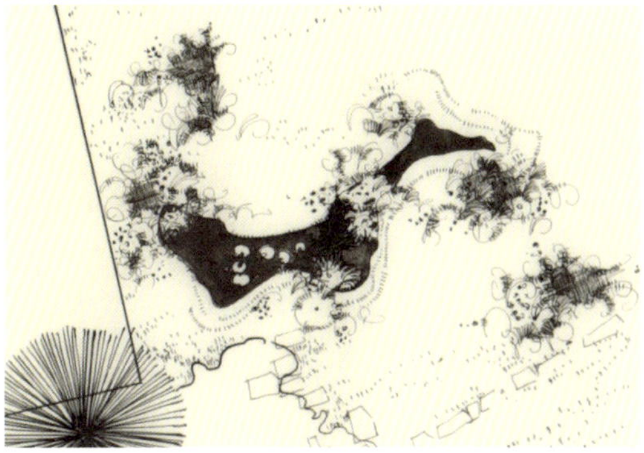

The design features a dramatically naturalistic pond form with carefully placed stone settings.

undated
Spälti, A.
▹ n.s.

undated
Building development
▹ Parkweg [Aarauerstrasse 12, 14, 16, 18], Buchs

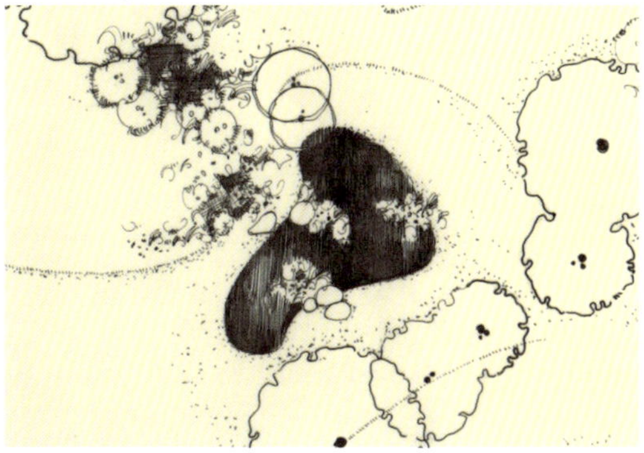

In the heart of the project, Cramer drew a secluded pond with stone settings and islands.

undated
n. N. → p. 181
▹ n. s., location unknown

Hans Graf

1952
Exhibition of Bern Master Gardeners
▹ Bern
with Hans Jakob Barth
The areas featuring Japanese set pieces – water lily pond, stones, grasses, bamboo, and ferns – Graf labelled most precisely in the plan. Otherwise, he gave only summary indications: "Cactus windows", "Potted plants", "Flower arrangements".

1955
Rothus Secondary School
▸ Wegmühlegässli 103, Bolligen

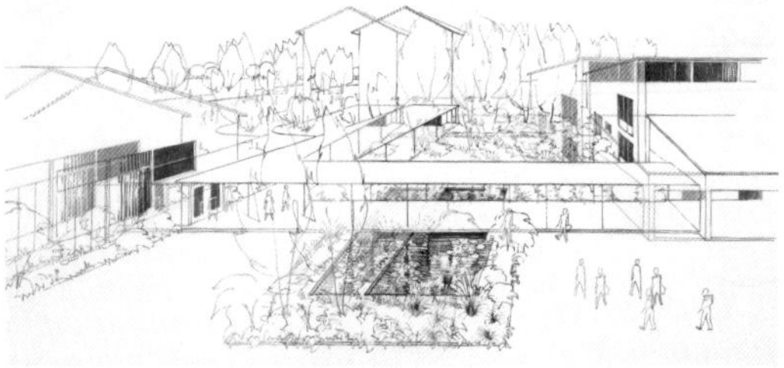

In the perspective, drawn probably by Hans Jakob Barth, the rectangular basin serves to create a transition in the Japanese manner between the interior space of the courtyard and the outlying areas.

1967
Rihs, H.
▹ n. s., Evilard / Leubringen
Two courtyard-like beds with stone settings, grasses, heather, pines, junipers, and rhododendrons are here augmented by East Asian plants such as forsythia, ornamental quince, Japanese kerria, and an ornamental beauty bush, which originates from China.

1970
Gerber, Th. → p. 192
▹ n. s., Ittigen

1971
Botanical Garden
▹ Altenbergrain 21, Bern

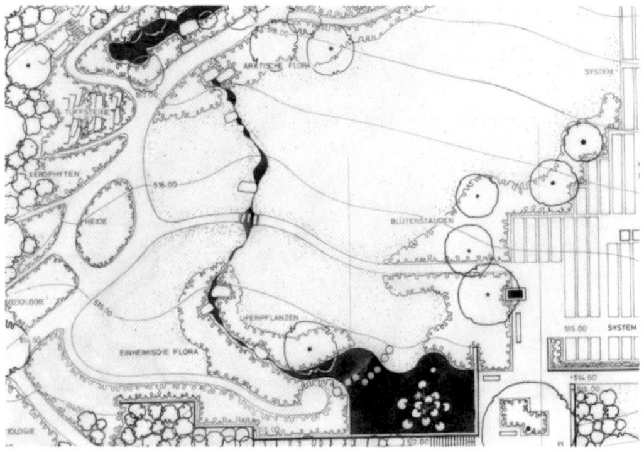

The naturalistic element is a stream that gushes from the Alpine garden towards the River Aare. Its widening and narrowing course is guided by strategically placed stones and short rapid sections before it feeds into two organically shaped lily ponds, one of which can be crossed via circular stepping stones – iconic since the G59.

1972
School
▸ Rosenweg 9, Rüfenacht
Courtyard-like green strips flank the two wings of the school building, one of which has a three-part fountain. The green areas use plantings of East Asian origin or of a corresponding character including European larch, Scots pine, Norway maple, silverleaf maple, Persian ironwood tree, Chinese juniper, carpet dwarf loquat, and Japanese rose. Set a little way apart is an enchanting pond.

1973
Föhrenweg building development
▹ n. s. [Föhrenweg 14, 16], Lyss

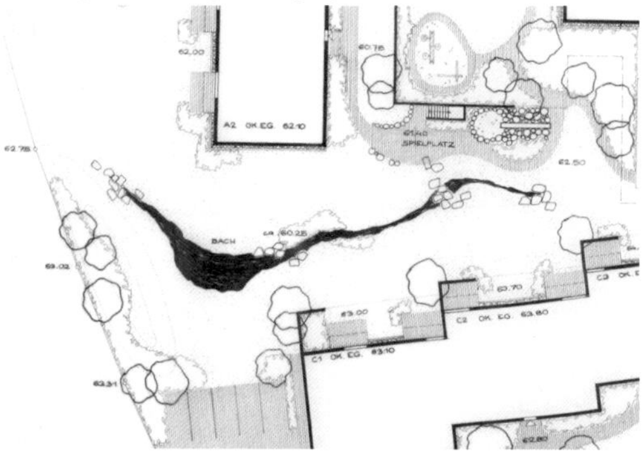

At the centre of the design is a stream with stone settings that widens into a small pond.

1974
Kiener, Architect
▹ n. s. [Hostalenweg], Herrenschwanden
Planting is used to give the rectangular water basin a near-natural form. A planted island, water lilies, stone settings, and pine trees as well as the backdrop of the adjacent forest heighten its Japanese impressions.

1974
Stüdeli, Hans
▸ n. s. [Brombeerweg 6], Bellach (SO)
Graf noted only the plantings of East Asian origin in the plan – a clear indication of its Japanese aspirations: pagoda dogwood, Japanese ornamental cherry, Japanese maple.

1976
Schweizer, Rudolf
▹ Bernastrasse 12, Bern
A dry stream bed "flows" between the stone slabs of a canopied seating area and "pours" into a lily pond adorned with boulders and rhododendrons.

1977
Diemant, H. → p. 187
▹ n. s. [Fischerstrasse 3], Zollikofen

1977
Von Dach, Nursery
▸ n. s. [Jägerweg 20], Lyss

1977
Büchler
Siedlung Halen, Herrenschwanden
Architecture: Atelier 5
Two inner courtyards – one as a dry garden with sunny perennials, the other as a wet garden with water plants. Both are furnished with boulders and designed as Japanese miniatures.

1966/1977
Graf, Hans
▸ Krauchtalstrasse 6, Bolligen
Lit. Anon. 1966b; Graf 1988.

1977
Breitfeld II building development
▹ n. s. [Bächimattstrasse 16, 18, 20, 22], Rüfenacht

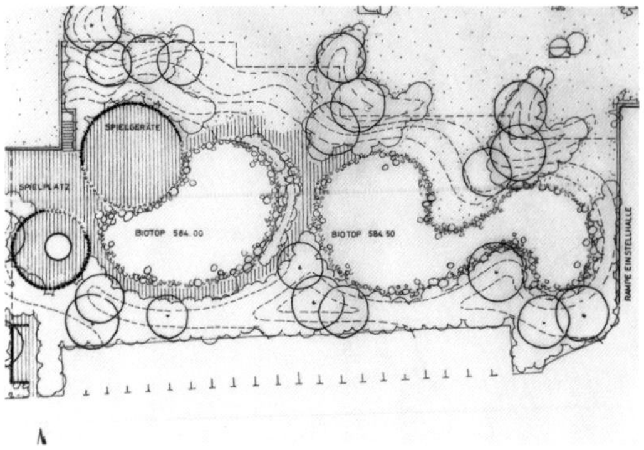

Japanese features inform the ecological interpretation of two biotopes bordered by stones. Undulations in the terrain indicate where the ponds (now drained) once were.

1978
Somm P[eter]
▸ k. A. [Hostalenweg 118], Herrenschwanden

1978
Tschan W.
▹ n. s. [Musterplatz 4], Bolligen

1978
Kohler, F.
▹ n. s., Burgdorf

1970–1979
Hübeli building development, house Loeb → p. 194
▹ n. s. [Hübeliweg 15], Muri
Architecture: Hans-Ruedi Ramseier, Anton Stucki, Gümligen

1979
Hehl
▹ n. s., Hinterkappelen

The perspective shows another example of the combination of baroque parterre and Japanese-influenced pond.

1979
Klossner
▹ n. s., Oberlindach

1979
Suter, O. → p. 193
▹ n. s., Stettlen

1980
Linder
▹ n. s., Langnau

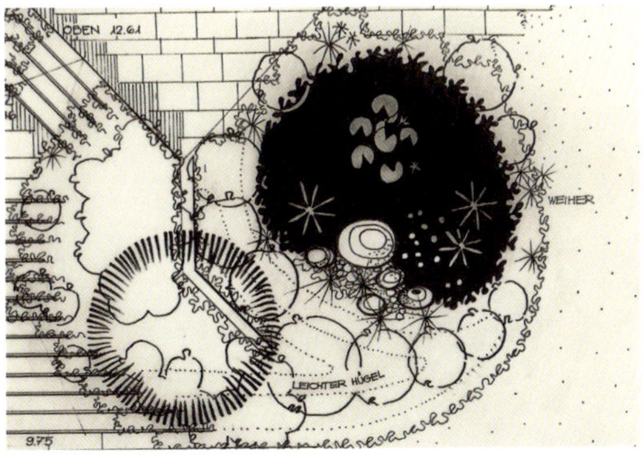

From the veranda, one has a view of the pond and lotus blossoms, which features strategically placed stone settings.

1980
Herrmann, U. → pp. 186, 197
▹ n. s., Biel

1981
Luder → p. 190
▸ n. s. [Höchhusweg 10], Steffisburg

1982
Eicher, W. → p. 195
▹ n. s., Oberlindach

1982
Bäggli Walter
▹ n. s., Frauenkappelen

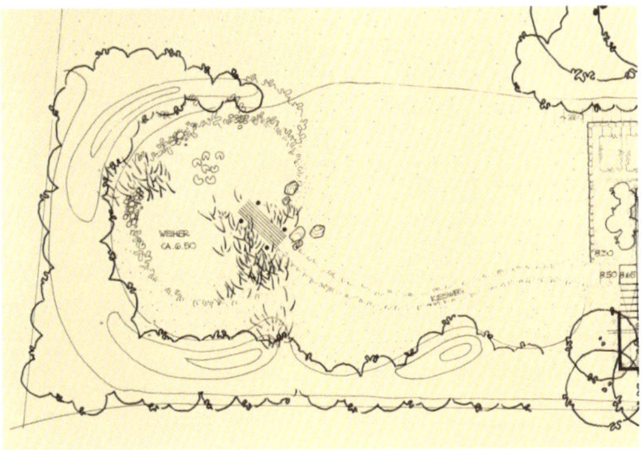

A footbridge over the lily pond doubles as a moon-viewing platform.

1982
Wüthrich, R. → p. 189
▹ n. s., Bremgarten

1986
Nurses' Garden, Health resort
▸ Im Ländli 16, Oberägeri
The garden with its organic pond form fed by a nearby stream is conceived as an extended living area.
Lit. Anon. 1986.

1987
Schindler, Kurt → p. 191
▹ n. s., Gümligen

1987
Häni
▸ n. s. [Im Gäuer 34], Zuzwil

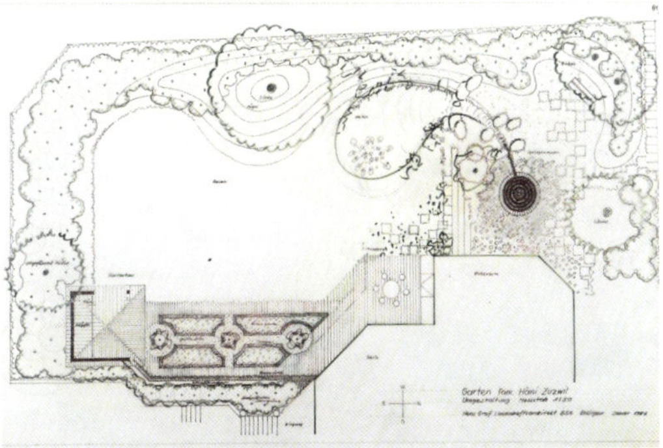

Much like his project for Luder in Steffisburg [→ p. 190], Graf combined a geometric but ornamental rose garden with an organically shaped pond and a stream, stone settings, and a hill.

1988
Retirement home
▸ Sägestrasse 10, Schüpfen
From a spring concealed by stones in the Japanese style gushes a stream that feeds a pond. A footbridge recalls the moon-viewing decks of Japanese gardens.

1990
Schneider, H.
▹ n. s., Münchenbuchsee

1991
Steiner
▹ n. s., Spiegel nr. Bern
A Japanese maple signifies the garden's Japanese character, setting up a dialogue through an "old wall" with several "bonsai". Further Japanese dwarf varieties and bamboo and iris plantings mark the orthogonal boundaries of a Japanese miniature landscape.

1995
Hospice in the Park
▸ Stollenrain 12, Arlesheim
A water lily pond with stone settings introduces a Japanese feature to a wildflower meadow. In the first version, Graf planned a terrace extending over the water that recalls traditional Japanese moon-viewing platforms.

1995
Seewer, J. E. → p. 196
▹ n. s., Sinneringen

1997
Schmid, P. & E. → p. 193
▹ Kapellenweg [50], Steffisburg

undated
Leu, A.
▹ n. s., Boll

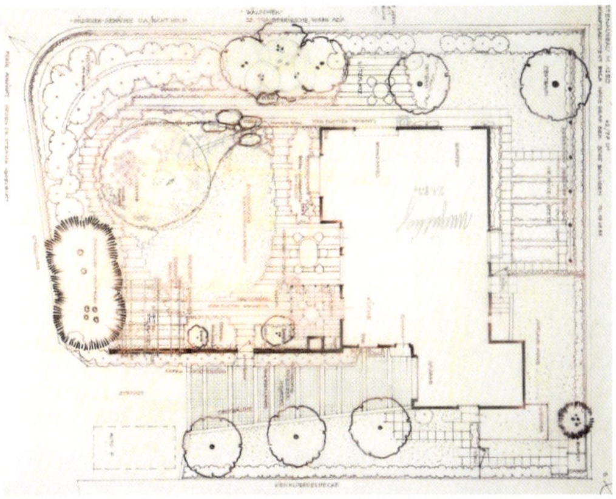

The garden plan drawn in 1990 shows a fish pond with water lilies fed by a waterfall splashing over stones.

undated
Schuetz Ernst
▸ n. s. [Moosgasse 10], Münchenbuchsee

Fritz Klauser and Fredy Klauser

Fritz Klauser

1919
“Garden in Ticino” → p. 201
▸ n. s., unknown location

1924
Stierlin-Naumann, Paul [Joachim]
→ p. 203
n. s., Wängi

1928–1930
Boller → p. 208
▸ n. s., Turbenthal

1930
Kappeler, Fritz
▷ n. s., Frauenfeld

The precisely elliptical outline of the water basin is “softened” by planting with water plants.

1930
Tobler, W. → p. 203
▷ n. s., Thal

1935
Häuptli → p. 202
▷ Tellstrasse, Aarau
Architect of the house: Hans Loepfe
The first draft even features a tea pavilion.

1938
Gartenbauausstellung
(Horticultural Exhibition, Gasga) → pp. 198, 199
St. Gallen

1944
Schmucki-Niederberger
▷ n. s., Will

Paul Zülli featured a similar bird bath on pages 2 and 3 of his 1932 article entitled “Wasserfreuden im Garten” in *Das ideale Heim*.

1946
Müller, H. → p. 204
▷ n. s., Rheinfelden

1948
Henz, J. → p. 205
▷ Jurastrasse, Aarau

1948
Hürlimann, E.
▷ n. s., Wil

A geometric version of the spring and water lily pool that recalls similar combinations by Gustav Ammann.

Fritz and Fredy Klauser

1950
Griesser
▷ n. s., Aadorf

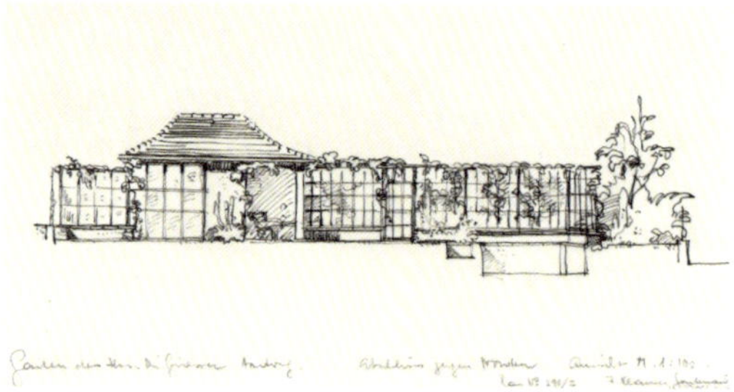

The garden house as Japanese citation. The project marks the handover from Fritz to Fredy Klauser. Some plans show the typical drawing style of the father, others the more modern “signature” of the son.

1950
Keller, Karl → p. 207
▷ Weinhalde, Rorschach

Fredy Klauser

1951
Bütler → p. 209
▸ n. s., Kradolf

1952
Loew, Hans
n. s., Oberaach

As with the Griesser house, the garden house has a very Japanese character.

1954
Leutenegger, O.
▷ n. s., Sirnach
A small pool with marsh plants conceived as a charming Japanese miniature.

1955
Kopp-Ploy, Werner → p. 208
▷ Seeacker [Neuseeland 24], Rorschacherberg
The house and garden still exist in their general form; however, the stream has been routed underground.

1955
Rusch-Hälg, Carl
▷ n. s., St. Gallen

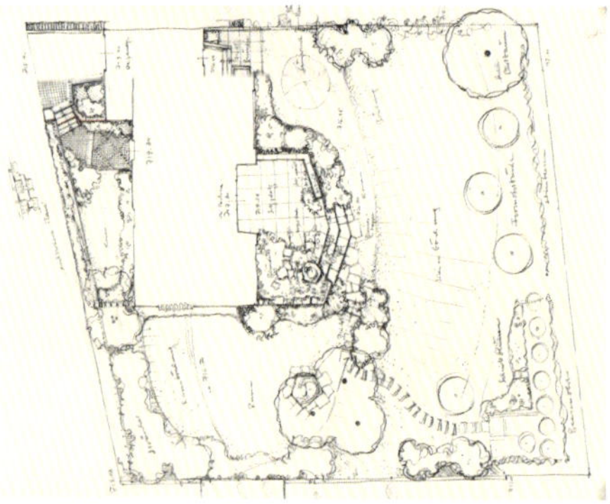

The bird bath is modelled after a washing basin from a Japanese tea ceremony. An opening in the wall allows the adjacent part of the garden to spill through.

1956
Cantonal Teacher Training Seminar
(today: Pedagogical University of St. Gallen, PHSG)
▷ Seminarstrasse 27, Rorschach
Klauser borrowed the “seminar pond” and mirrored it as the focal point of the “place for open-air theatre” in the northern part of the park.

1956–1957
Tugendhat, Fritz → p. 206
▸ Ludwigstrasse 22, St. Gallen
The garden was remodelled as part of major alterations to the house in the 1980s. The house, but not the garden, has since been restored to its original state.

1957
Schneider
▹ n. s., St. Gallen

Fredy Klauser incorporated the Schneider House a square, planted water basin that serves as a miniature landscape reduced to its essential Japanese features.

1957
Sloman, Ricardo
▹ n. s., Altnau

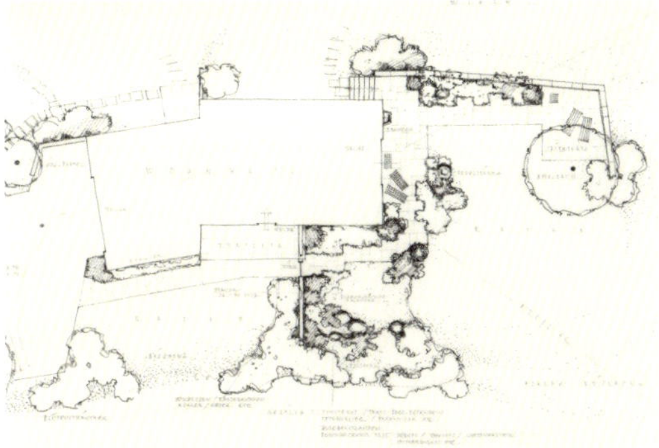

Fredy's garden rooms were typically an integral part of the green space but adjoining or growing out of the house. In the design for Ricardo Sloman, he equipped the intimate seating area with a bird bath reminiscent of washing basins from a Japanese tea ceremony. A wall, in front of which he placed an arrangement of stones, served both to screen and facilitate interaction with the realm beyond, allowing the rhododendrons to reach into the garden interior. Finally, Klauser marked out vistas using wild shrubs as a privacy screen with openings revealing strategic views.

Walter Leder

1932
Salvisberg, Otto Rudolf
▸ Restelbergstrasse 97, Zurich

1939
Meyer-Kaiser, F. → p. 225
▹ Sursee

1939
Special garden, National exhibition
▹ Zurich

1941
Michel, O.
▹ Im Appisberg, Männedorf

1943
Hotz-Dufour, Heinrich Willy
▹ Dreilindenstrasse 67, Lucerne

1943
Kofler-Truniger, E.
▸ Gesegnetmattstrasse [14], Lucerne

1943
Wettstein-Westerheim, von
▹ Saleggi nr. Ascona

1944
Bruppacher, J. → p. 222
▹ n. s., Uitikon

1944
Häberlin, Ed.
▹ Müllheim (TG)

1944
Schnetzer, W.
▹ n. s., Niederuzwil

1944
Vogt, O. → p. 214
▹ Tunaustrasse 20, Reinach

1944
Weber → pp. 210, 212, 220
▹ Gossetstrasse, Wabern (BE)

1944–1947
Kyburz
▹ n. s., Oberentfelden

An organically curved pond and a meandering stream with a small bridge exude a Japanese atmosphere.

1945
Frey, M. → p. 229
▹ n. s., St. Niklausen

1945–1946
Geiser-Hofer, Willy
▹ n. s., Langenthal

1946
Baumann
▹ n. s., Belp

1946
Della Casa
▹ n. s., Burgdorf

1946
Meyerhans, Emil
▹ Magdenaustrasse/Rebenstrasse 1, Weinfelden

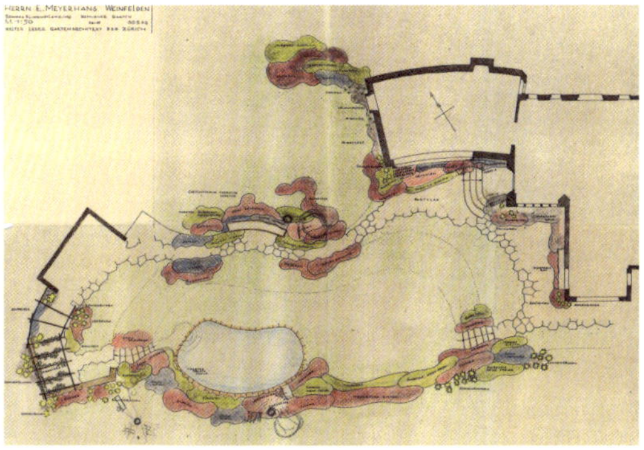

The planting plan for the Meyerhans Estate, 1949: red, blue, and yellow are reflected in the choice of cosmea, verbena, phlox, snapdragon, fire sage, amaranth, zinnia, and nasturtium; of aster, lobelia, sage, and trumpet tongue; and of marigolds, calendula, rudbeckia, and sunflower. Areas of white cover are achieved using flowering tobacco and ornamental maize.

1946–1948
Pulver, M. → p. 218
▹ n. s., Lucerne

1947
Rutz, D. H.
▹ n. s., Herisau

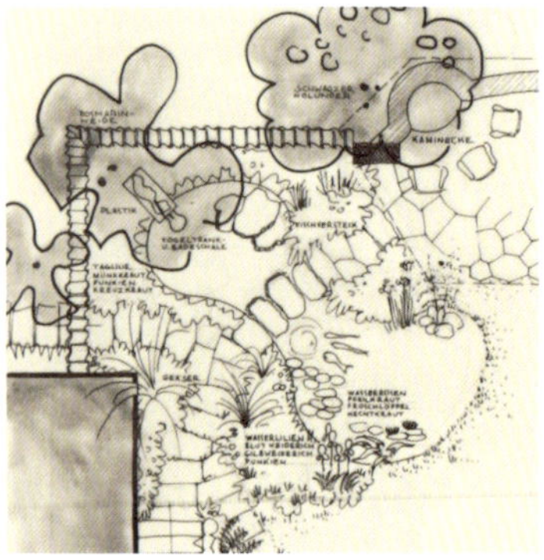

1947
Schwyter
▹ n. s., Zurich

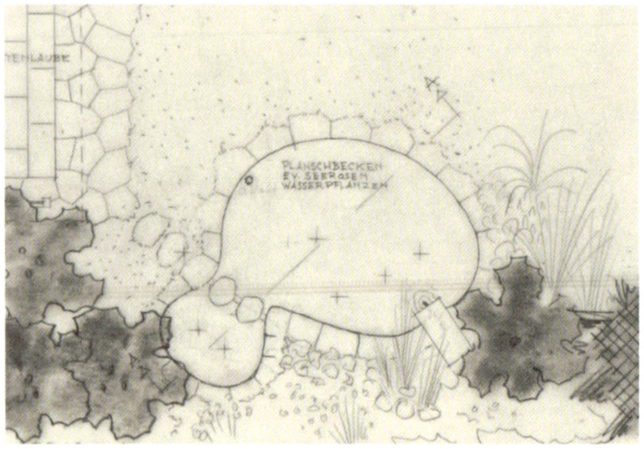

Here, the natural pond features as a paddling pool with stepping stones crossing it à la japonaise at the narrowest point.

1948
Gehrig, Fritz
▹ Rebwiesstrasse 30, Zurich-Rehalp

1948
Maeder, Theo
▹ n. s., ("House on the lake"), Biel

1948–1949
Luchsinger, E.
▹ Alte Landstrasse, Küsnacht

1951
Bühler, René / Bühler-Gredig, Adolf
▹ "Sonnenhügel", Uzwil

1951
Dierauer, Bruno → p. 224
▹ n. s., Oberuzwil

1951
Schulhaus Gönhard
▹ Weltistrasse 20, Aarau

1950–1951
Moser, P.
▹ n. s., Hasle-Rüegsau nr. Burgdorf

1952
Heusser, A.
▹ Grünenbodenweg, Köniz

1954
Handschuh-Custer, R.
▸ Im "Bindschädler" [Zollerstrasse], Herrliberg

1955
Hofmaier, G.
▹ n. s., Flums

1957
Brandestini, A.
▹ n. s., Küsnacht

1957
Leder, Walter
▹ Zurich

Circular stepping stones, stone settings, a water lily pool fed by a cascade, and Japanese maples were the Japanese ingredients of Leder's own garden.

1957–1959
Oberle, A.
▹ n. s., Döttingen

1958
Thyll, L.
▹ n. s., Reutte, Tirol (A)

1958
Weber, E.
▹ n. s., Gerlafingen

1959
G59 – 1. Schweizerische Gartenbau-Ausstellung
(1st Swiss Horticultural Exhibition) → p. 107
▸ Belvoirpark, Schneeligut, Zurich
with Johannes Schweizer

1959
Community Centre → p. 217
▹ Oberhauserstrasse 25, Opfikon

1959–1960
Treichler → p. 228
▹ n. s., Herrliberg

1960
Zollinger, E. → p. 216
▹ n. s., Rüti

1963
Evangelical parish centre
▸ Zürcherstrasse 14, Rapperswil

1963–1965
Kux, E.
▹ n. s., Küsnacht

1964
Brasch, Heinz → pp. 222, 223
▹ Neubruchstrasse [6], Forch

1964
Guggenheim
▸ Goldauerstrasse, Zurich

1964
Mukogaoka-Park → p. 110
Tōkyō

1965
Lakeshore design → p. 226
▹ Chamerstrasse, Gartenstrasse, Zug
Competition, Motto "ELDORADO"

1966
Töchterschule IV (Girls school)
▸ Schanzengasse 17, Zurich

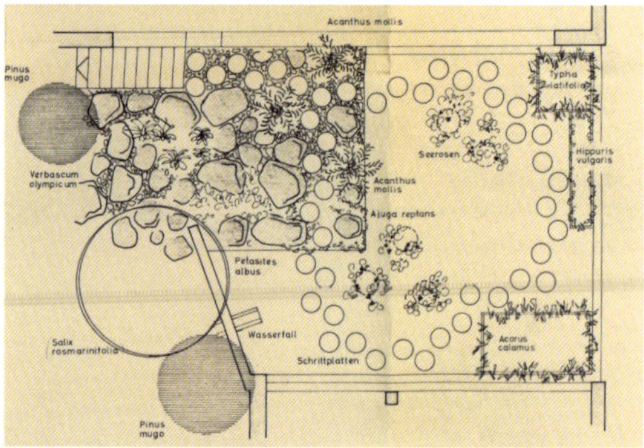

Leder gave two of the five atriums of the Töchterschule IV in Zurich a distinctly Japanese flair: the first with lily ponds, round stepping stones, and stone settings, and the third as a dry landscape.
Lit. Anon. 1969b.

1967
n. N. → p. 227
▹ n. s., St. Gallen
with Klaus Leder

undated
Hugenin und Allemand → p. 213
▹ n. s., location unknown

undated
Kaufmann → p. 221
▹ n. s., Thörishaus

undated
Lücker, C.
▹ Aux Crètes, Clarens, Montreux (VD)

undated
Lüdi-Zumstein, F.
▹ n. s., Grenchen

undated
Schönenberger → p. 213, 224
▹ n. s., Uznach

undated
Schnurrenberger-Kurth, Ernst
ACCUM AG
▹ Laufenbachstrasse 23, Gossau

undated
Wolf-Bender, H. → p. 218
▹ n. s., location unknown [Goldbach]

Evariste Mertens, Mertens Brothers, Mertens & Nussbaumer

Evariste Mertens

1871
Merian-Ziegler, Louis and [Maria Cecilia]
▹ Höllstein-Steinen (Lörrach)
with Arnold Neher

1891
Schwarzenbach-Zeuner, Robert
→ p. 233
▹ Seestrasse, Rüschlikon
For the villa of the silk manufacturer Wilhelmine and Robert Schwarzenbach-Zeuner, built by the opera house architects Fellner & Helmer, Mertens designed a "little stream with rocks". Emerging from a spring with an agglomeration of rocks, it ran in meandering curves along the north side of the plot, occasionally forming ponds with small islands before tumbling over a waterfall. The park itself was laid out according to a design by the Swedish landscape architect Carl Gustav Svensson (1891–1910).
Lit. Schwahn 2012, p. 78 ; F. K. 1904.

1892
Waldkirch von
▹ Meisenweg 4, Schaffhausen
The "rock section with waterfall & basin" that Mertens drew in the plan takes the form of a Japanese miniature and served him as a precursor for the project for Jacques Schuler-Ganzoni.

1895
Trümpy-Heer, Jacques → p. 230
▹ n. s. [Hauptstrasse 6], Mitlödi (GL)
Built in 1865 by Fridolin Trümpy, the property that today is known as the "White Villa" was acquired 20 years later by the factory owner Jacques Trümpy-Heer, who in 1886 extended and transformed it into a palatial residence. After Jacques Trümpy-Heer's death in 1906, the property remained in the family. In 1939 and 1947, his widow Emily Trümpy-Heer commissioned Gustav Ammann to partially replant the grounds (NSL 2-0584). In 1951, however, the property was sold, subdivided, and part of the park grounds were built upon. Mertens' design probably disappeared around this time. The villa was renovated in 2012.

1898–1901
Passavant-Fichter, G. [Georges]
▹ Gellertstrasse, Lehenmattstrasse, St. Albanteich, Basel
Georges Passavant (1862–1952), a banker and lawyer, may have explicitly requested a Japanese garden from Mertens, having ten years earlier travelled extensively through North America as well as Japan, China, Thailand, India, and the Indonesian island of Java.
Lit. Schneider/Lüthi 2007, p. 114.

1905
Schuler-Ganzoni, Jacques → p. 132, 235
▸ Gerichtshausstrasse 58, Glarus

1907
Gessner-Heusser, Emil
▹ Gessnerweg 5, Wädenswil
Mertens' archive contains two versions of the design for the park for the "Villa Gessner" (now "Rosenmatt" parish hall) built for the silk manufacturer Emil Gessner-Heusser in 1899. One contains a small pool with a hexagonal pavilion, the other a miniature of a wild, flowing stream with a pond and rapids. What was implemented, however, is a version based on a project by Ernst Klingelfuss.
Lit. Mertens O 1916, p. 68.

Mertens Brothers

1911
Jung, Carl Gustav and Emma
(today: Museum Haus C. G. Jung)
▹ Seestrasse [228], Küsnacht
The impressive avenue still exists, but the Japanese-inspired stream, pond, island, and stone settings do not.

1913
Bally-Gamper, Eduard
▹ Schönenwerd

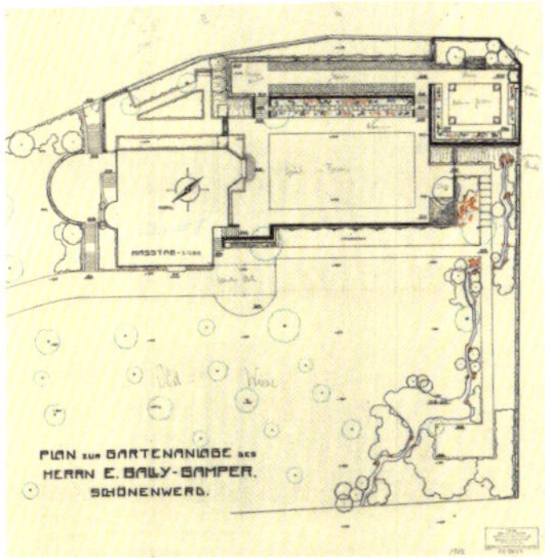

The brothers used an "existing spring" to create a Japanese-style garden area with a stream and broad pond.
Lit. Mertens O 1916, p. 71.

1913
Bucher-Guyer
▹ n. s., Niederweningen

1926
Glarner, P. / Lané
▹ n. s., Wildegg

1933
Reinhart, Alfred und Susanna, Villa Tannhof
(today: Landgut Reinhart) → pp. 10–15, 236
▸ Husertalstrasse 15
Ebertswil, Hausen am Albis (ZH)

1935
Bunge, M.
▹ Susenbergstrasse, Zurich

1933–1936
Cemetery Enzenbühl → p. 9
▸ Forchstrasse 384, Zurich
with Konrad Hippenmeier
Lit. Freytag 1940.

1937
Bührle, Emil Georg → p. 238, 239
▸ Zollikerstrasse 178, Zurich

1938
Schlimme-Burkhard, A.
▹ Bergstrasse 23, Zurich
The brothers' plan dated 24 December 1938 explicitly designates an area as a "Japan garden", but records show that they had already planned a water course streaming over rocks and flowing into an organically shaped "pond with water lilies" as early as 11 October. To add Japanese flair, they specified a magnolia and ornamental quince as solitary trees, as well as a Japanese maple on the terrace in front of the house as a Japanese accent.

1938
Staub, Emil und Alma, Villa Alma
▸ Seestrasse 80, Männedorf
Among the Japanese motifs still visible today in the garden are a small waterfall gushing out between stones, a pond, the planting of pines, and Japanese ornamental cherry and maple trees.

1938–39
Zurich Congress Hall
Architects: Haefeli, Moser, Steiger (HMS)
with Paul Speck
"In the richness of its accents and in its delicacy, this little garden has something Japanese about it […]," wrote Peter Meyer about the small inner courtyard between the vestibule and the garden hall. The lines written by sculptor Paul Speck about the garden breathe the spirit of Japan: the plants within the "silent chamber" of the garden were not merely "decorative elements" but rather populated the garden "as beings", which one comprehends precisely at the point "where their incomprehensibility touches us". The extent of spatial relationships in this chamber lies "in the immeasurable". Speck's intention was to elevate the garden "into the infiniteness of paradise". The motif of paradise is echoed by Walter Mertens, who described the garden courtyard as being "as quiet as a small island in surging seas" and an "enchanting fairy-tale island".
Lit. Meyer 1943, p. 313; Speck 1943; Mertens W 1943.

1941
Scheer, Karl
▹ Enzenbühlstrasse, Zurich

1943
Keiser, Rudolf
▹ Brummelstrasse, Buchs (AG)
The brothers designed a strolling garden in the style of Chisen-kaiyū-shiki with a "footpath by the water" around the existing pond, which they made the focal point of the park landscape.

undated
Bircher-Benner
▹ n. s. [Keltenstrasse 48], Zurich

Mertens & Nussbaumer

1942–1954
Tiefenbrunnen Lido
▸ Bellerivestrasse 200, Zurich

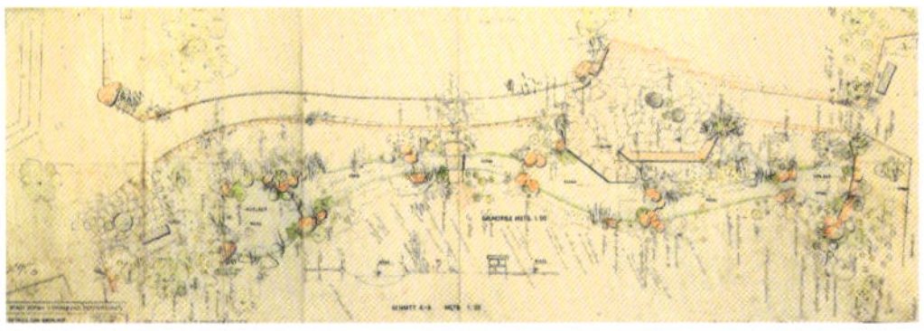

Because the planning and construction phase for the Tiefenbrunnen lido extended over a period of more than ten years, the project passed through the hands of the Mertens brothers, then Mertens & Nussbaumer together with the architects Josef Schütz and Otto Dürr, before Hans Nussbaumer completed it on his own. The plans credit only Mertens and Nussbaumer as the authors, while the magazine *Anthos* mentions only Hans Nussbaumer. Mertens & Nussbaumer used a Japanese notation in the plan for the course of a stream in the Tiefenbrunnen Lido a year before it opened in 1954.

Lit. Anon. 1962c

1946
Della-Balda, G.
▹ n. s., Grenchen

1946
Hussnigg, Aug[ust]. C[arl].
▹ Eisernen Zeit [Im Eisernen Zeit], Zurich
The influence of Japanese inspirations is signalled by the circular stepping stones crossing the pond and a stream spanned by a small bridge.

1947
Bolli, A.
▹ Seefeldquai 41, Zurich
Architect: Julius Kunkler
The villa, built in the 1890s, was demolished between 1970 and 1973 along with the Japanese elements of the garden including the pond, stream, gingko, Japanese maple, and apricot.

Lit. Zürcher Denkmalpflege 1975, p. 287.

1947
Rieben, J.
▹ n. s., Burgdorf

1947
Käferholz School
▸ Käferholzstrasse 271, Zurich

1947–1948
Liebenfels Cemetery
▸ Zürcherstrasse 108, Baden

Lit. Mertens O 1951, p. 148.

1948
Schollenberger, H. W.
▹ n. s., Wülflingen-Winterthur

Willi Neukom

1951
Hunsperger, J. H.
▹ n. s., Uitikon
For J. H. Hunsperger, vice-director of the Zurich Insurance Company, Neukom sketched a design for a garden with a curved pond lined with individual constellations of rock and boulders. Sections of concrete tube placed upright in the pond and planted with water lilies served as islands.

1952
Straumann-Schmidlin
▹ n. s., Waldenburg

As if inspired by Japanese rules for the design of rivulets, the stream is guided and channelled into rapid sections by strategically placed stones before flowing into a naturalistic pool.

1958
Schulhaus Hirzenbach
▸ Luchswiesenstrasse 186, Zurich

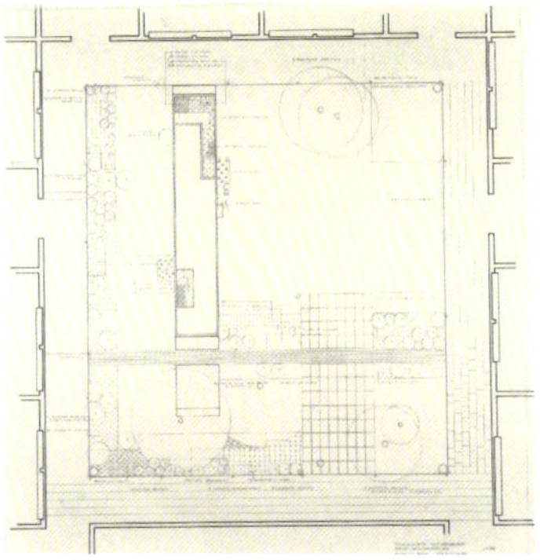

Neukom gave the inner courtyard an East Asian character: the long rectangular water basin, populated with islands planted with Japanese swamp iris and Japanese sweet flag, was nestled between empress trees (*Paulownia imperialis*), Japanese maple (*Acer japonicum*), rose gold pussy willow (*Salix gracilistyla*), Japanese spirea (*Spiraea japonica Bullata*), and variegated-leaf hardy kiwi (*Actinidia kolomikta*).

1958–1959
Staehelin-Peyer, Willy → p. 248
▸ Im Hausacher 35, Feldmeilen

A shrub, probably a magnolia, graces the terrace in the Japanese tradition of front planting (senzai).

1959
G59 – 1. Schweizerische Gartenbau-Ausstellung
(1st Swiss Horticultural Exhibition 1959)
→ p. 97, 100–101, 104–106
▸ Bellerivestrasse / Tiefenbrunnen Lido / Höschgasse, Zurich
with Ernst Baumann

1960
Meyer, Michael
▸ Schlimbergstrasse 38, Kilchberg
Architect: Helmut Rauber

Lit. Anon. 1962.

Erratically placed stone blocks and various plants of Japanese origin – azalea, East Asian sedge, yellow wax bell, and false bamboo – along with the courtyard and its specially grown pine and thread-leaved maples contribute to the Japanese aura.

1960–1961
Halpern, E. → p. 261
▹ n. s., Zurich-Witikon
Architect: Helmut Rauber

1960–1964/1973–1975
Nordheim Crematorium
▸ Käferholzstrasse 101, Zurich
Architect: Albert Heinrich Steiner

The City of Zurich, which lists Nordheim Crematorium as a protected site, considers it one of Neukom's key works, reflecting "his trip to Japan in 1964 and his interest in Scandinavian cemeteries". The centrepiece of the Japanese-inspired design is the courtyard furnished with a water basin, pine trees, and boulders.

Inventarergänzung 2013, p. 132–133; Anon. 1969; Schlappner 1981.

1961
Weinberg-Bär, Rolf
▹ Im Sonnenberg [?], Zurich

Design study for the Weinberg-Bär Garden in Zurich, 1961: even the drawing conveys a Japanese feel through the rendering of the plants with a feathery ink brush.

1961–1962
Lakeside resort
▸ Seestrasse 144, Uetikon am See
Architect: Oskar Stock
Neukom adapted his design for the lakeshore at the G59.

1962
IBM → pp. 134–136, 254, 256
▸ Säumerstrasse 4, Rüschlikon

1962
[Francesco und Elena Maria, formerly], Villa Saroli → p. 260
▸ Lugano

1962
Wirz, Rolf → p. 261
▹ n. s., Münchenstein

1963
District hospital → p. 244
▸ Spitalstrasse 24, Bülach

1963
St. Josef Monastery → p. 262
▹ Baselstrasse 27, Solothurn
Architects: Studer+Staeuble

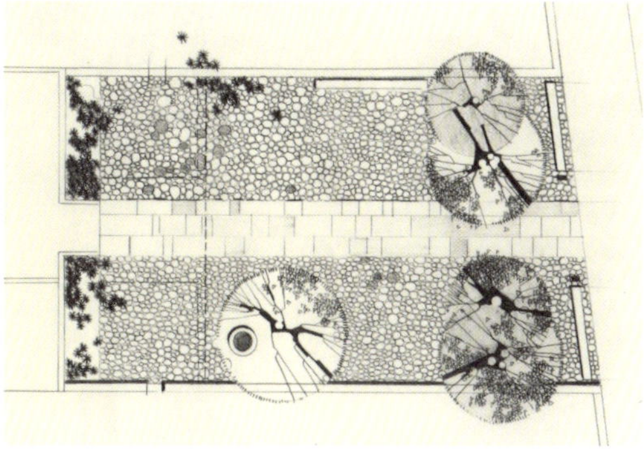

Neukom planned the creation of an organically shaped pond outside the monastery walls as a borrowed landscape for a water basin in the inner courtyard of the cloister. He drew numerous variants, including a sketch expressly labelled as "Japanese". For the area between the church and the cloister, he sketched a water garden that bore a striking resemblance to that at the G59. For the forecourt, he envisaged either a paving similar to that used by Ernst Cramer for the entrance to the teacher training college in Menzingen or two dry gardens furnished with river pebbles, bamboo, and ornamental shrubs. Neither his designs nor those of Cramer, who also competed to undertake the project for the church and state authorities, were carried out. The art historian Heinrich Weber has traced the lengthy process of building the new monastery, in which many were involved, among them Hermann Baur, Linus Birchler, Franz Füeg, and Alfred Roth.

Lit. Weber 2013.

1963
Schär, Max
▹ Untere Wart, Thun
The design employed circular stepping stones that led over a square water basin, giving it the impression of being a miniature of the Nymph Pond of the G59.

1963
Seeuferweg (lakeshore path)
→ p. 100–101
▸ Zürichhorn, Zurich

1963–1970
Nurses' School, Triemli Hospital
→ p. 263
▸ Paul Clairmont-Strasse 30, Zurich
Architects: Architektengemeinschaft Stadtspital Triemli Zurich (ASTZ)

The courtyard of the nurses' school exudes a contemplative Japanese atmosphere.

1964
Feller, Adolf → pp. 246, 247
▸ Stotzweid [Bergstrasse 77], Horgen

1964
Holiday homes Hurden
▹ House I, Brantschen, house II, Koella n. s., Hurden-Freienbach (SZ)
Neukom designed the relationship to the lake, the stepping stone boulders, and the bank reinforcements made of erratic blocks similar to those used at the G59.

1964
Swiss National Exhibition Expo 64 → p. 111
▹ Lausanne
Neukom sketched a water landscape that looks a lot like the precursor to that of the Grün 80.

1964
Vienna International Horticultural Exhibition (WIG64)
▹ Swiss Garden

1965
Retirement home
▹ Fällandenstrasse 22, Dübendorf
Neukom designed two courtyards, a garden courtyard with a geometrically delineated water basin planted with swamp irises, bulrushes, sedges, bamboo, willows as well as birch trees, and an inner courtyard planted with ornamental cherry, Japanese maple, and coastal pine on a modulated terrain that simulated an organic water surface.

1965
Vocational Training College Zurich
(today: Technical Skills College Zurich)
▹ [Ausstellungsstrasse 70, 72], Zurich
Neukom sketched five variants for the inner courtyards of the school. The most mature proposal for the longer of the two courtyards bore obvious resemblance to the Daisen-in. In place of gravel, however, Neukom intended to use actual water, planting the borders with water lilies and grasses and creating islands that emerge from the water's surface. Inspired by the Daisen-in, he structured the courtyard in a 2:3 ratio and placed two bald cypresses at the "threshold".

1965
Lins, Albert → p. 249
▸ Bergstrasse 27, Küsnacht

1962–1966
Fondli School → p. 244
▸ Bohnackerstrasse 5, Dietikon
Architects: Rudolf & Esther Guyer, Conradin Alder, Paul Dorer
The complex has been listed as a site of conservation interest of supra-municipal importance since 2019.

1967
Limmattal Hospital
▹ Urdorferstrasse 100, Schlieren

A striking dry landscape on the one hand contrasted with broad expanses of water on the other.

1967–1971
Thermal baths and health resort
▸ Dr. Martin Erb Strasse 11, Bad Zurzach

1967–1977
ETH Zurich Hönggerberg
→ p. 252, 253, 255
▸ Schafmattstrasse 16, 16a, 18–20, 30–36, Wolfgang-Pauli-Strasse 16, 16a, 18, Zurich
with Hans Epprecht
Architects: Albert Heinrich Steiner, Werner Gehry/ Max Ziegler, Erik Lanter

Design for the garden courtyard of the teaching building from 16 November 1972: the water landscape that floods into the "Central Square" can be read as nature assimilated, while its "replica" in the garden courtyard as scenery borrowed from it.

1968
ELVA Agroscope Reckenholz
▸ Reckenholzstrasse 191, Affoltern

1968–1970
Cemetery
▸ Riedmattstrasse 14, Ebikon

1969–1972
Rosengarten retirement home
▹ Aathalstrasse 21, Oberuster

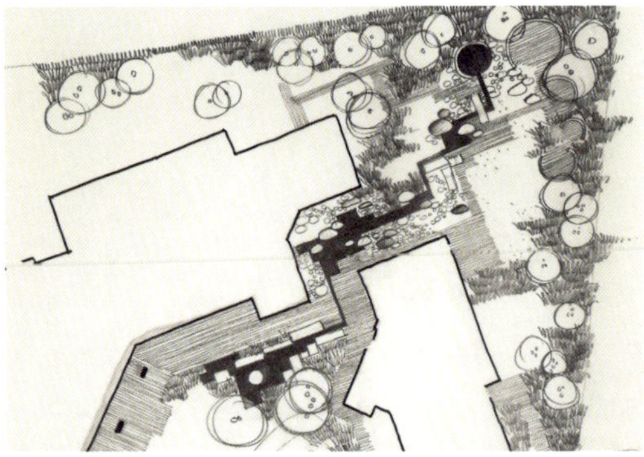

Willi Neukom designed various versions of a geometrically curved stream course. To counter right-angled edges, he employed Japanese-style stone settings on the banks, which in places is formed into a dry landscape.

1973
Joggenrain Estate
▸ Joggenrainweg 17, 19, Männedorf

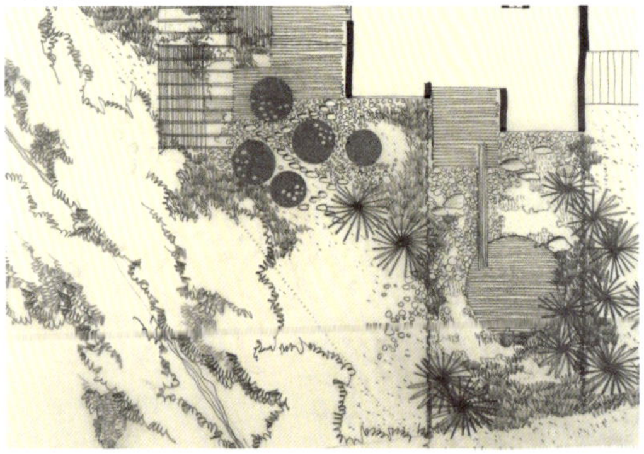

The garden of house no. 17 is designed as a dry landscape of scattered rocks and stone settings. It simulates a pond bridged by a narrow footbridge. In the garden of house no. 19, five circular, connected, and planted water basins are encased in a border of pebbles to act as a borrowed landscape of the nearby stream.

1973
SUVA Swiss Accident Insurance Institution
▹ Bodenächer, Mutschellenstrasse 2, Bellikon
For SUVA in Bellikon, Aargau, he availed himself of various Japanese elements, distributing them across three atriums: a green courtyard, water courtyard, and stone courtyard. For the green courtyard, he envisaged mountain pine, spindle tree, ferns, ivy, hydrangeas as well as yellow, orange, red, and purple rhododendrons. For the water courtyard, he reduced the palette to Scots pine, spindle tree, and various heather species. He reserved the richest assortment for the stone courtyard: Scots pine, mountain witch alder, foxglove, dwarf birch, alpine and blue fescue, feather bristle grass, Iceland poppy, black mullein, carpet phlox, wall bellflower, lavender, broom, baby's breath, thistle, mountain avens, field thyme, Missouri evening primrose, cotton lavender, sunflower, rock foils, and shrub veronica.

1974
Wolfsberg Training Centre
→ p. 73, 254
▸ Wolfsbergstrasse 15, Ermatingen (TG)
Architects: Rudolf & Esther Guyer
Lit. Anon., 1976.

1976
Guyer, Esther and Rudolf → p. 128, 256
Seldwyla Estate
▸ Tobelmühlistrasse 16, Zumikon

1976
University Zurich → p. 258, 259
Strickhof site development, phase 1
Interior courtyard and foyer hall
▸ Winterthurerstrasse 190, Zurich
Architect: Max Ziegler

1976–1977
Botnar, O. O.
▹ n. s., Villars-sur-Ollon
Neukom borrowed the adjacent stream and channelled it into the garden, guided by stones and bridged by footbridges, to feed a pond.

1977
Scheiwiler, Bruno → p. 243
▹ Immenweg [35]

1978
Wettstein, R. → p. 257
▸ Rebhusstrasse [51, 51a], Zumikon

1978–1980
Contraves building development Ettenfeld
▸ Schaffhauserstrasse 550, 550b, Zurich (at the time of construction Nr. 580)

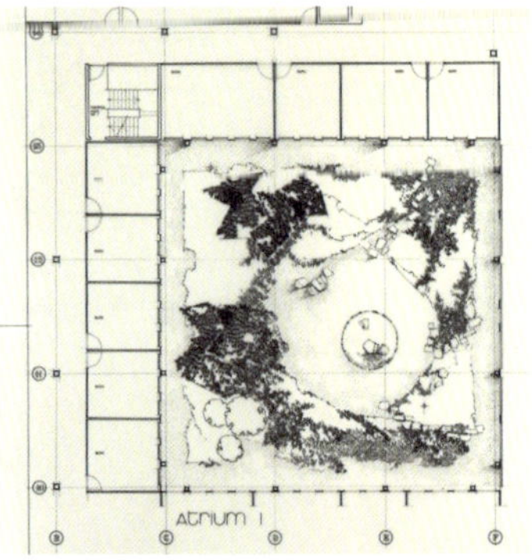

Analogous to the concept for the SUVA, Neukom designed three atriums as a green, stone, and water courtyard along with a fourth as a bamboo courtyard – reminiscent of Grün 80.

1978–1980
Grün 80, "Beautiful Gardens"
→ pp. 114–121, 125–127
▸ Rainstrasse 10, Münchenstein
Project partnership with Adolf Zürcher

1979
Pestalozzi Park Competition, Motto "Öpfel" [apple]
▹ Herrenbergli/Dachsleren, Zurich

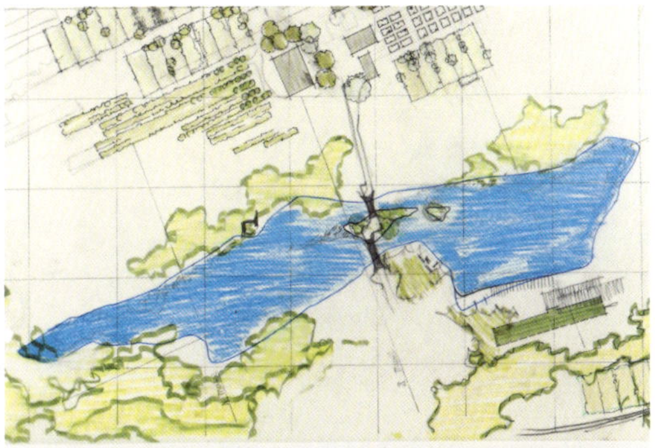

The site was originally earmarked for the Grün 80, which Zurich had applied to host. After the Grün 80 was awarded to Basel/Münchenstein, the city held a competition for the site to create a park landscape as a local recreation area. Neukom was awarded 5th place while the competition was won by Jens Lüpke. The project was never realised. Neukom's design of a large pond, an island, and staggered bridges is reminiscent of a Japanese paradise garden.

Lit. Sierts 1979.

1980
Small indoor swimming pool Bläsi
→ p. 240
▸ Limmattalstrasse 154, Zurich

1980
Kreuz building development
▹ St. Gallerstrasse 30, Jona
The central element was called the "bamboo courtyard" – probably in reference to the area of the same name at the Grün 80. Two of the four bamboo species chosen there – green powder bamboo and Japanese arrow bamboo – appear again here. They accentuate the Japanese flair of the water basin, whose edges were lined with stone blocks and planted with bulrushes, club rushes, irises, and water lilies.

1982
Baumann
▹ n. s., Schöftland

1982
Gepag (Corporation for development and planning of construction projects)
Allmend/Gisler building development
▹ Küsnacht
Japanese maple, wild apple, and ornamental quince as well as pine trees surround an organic pond with stone settings.

1982
Ungricht, Elsbeth → p. 245
▹ Brunnenmattstrasse, Dietikon
▸ Schürlistrasse 36, Dietikon
Neukom drew two variants – one for an unspecified location on Brunnenmattstrasse and one for the site at Schürlistrasse 36, which was ultimately realised. In both cases, a pond is the focal point of the garden. In one case, it was fed by a stream whose course was determined by placed stones and was bridged in typical Japanese manner by a footbridge made of staggered planks. In the second executed version, he placed a similar construction over the pond, which he also furnished with an island.

1983
PAX building development → p. 241
▸ Bifangstrasse 15, 17, 19, 21, Wohlen

1984
Gehrig, Louis
▹ Fuederholzstrasse 18, Herrliberg

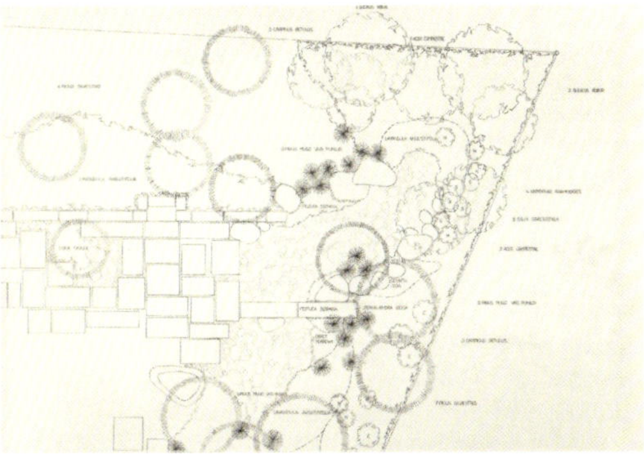

One of Neukom's most enchanting "inversions": as in the Japanese dry landscape, pebbles symbolise a body of water. True lavender lines the shore in places, as if it were both source and outlet. A hollowed-out stone echoes the ceremonial washing basin in a Japanese tea garden, and the planting of snow heather, Scots pine and dwarf mountain pine, Japanese sedge, and spiky fescue contributes to the general Japanese flair.

Adolf Zürcher

1962
Wymann, H.
▹ n. s., location unknown
Zürcher chose a Japanese flowering cherry (Prunus serrulata 'Sekiyama', syn. 'Kanzan') to plant as a Japanese accent in the otherwise "empty" garden area.

1964
Church Parish Centre
▹ Kirchweg 1, Buttikon
Competition project

1964
Troxler, Roger → p. 273
▹ n. s., Bellikon

1964–1969
"Alpenblick" building development
→ p. 264, 265
▸ Alpenblick 2–12, Cham
Architect: Joseph Stöckli, Zug
Zürcher used a number of East Asian plants, including Himalayan pine, winged spindle tree, beauty bush, Chinese gold rose, Japanese rose, and Mandarin rose.

1965
Fux, Christian
▹ n. s., Visp

A dry landscape with a dry stream bed of pebbles, its course directed by strategically placed stones and lined with bamboo.

1965–1966
Biland-Wyttenbach, F. → p. 272
▹ n. s., Holzhäusern

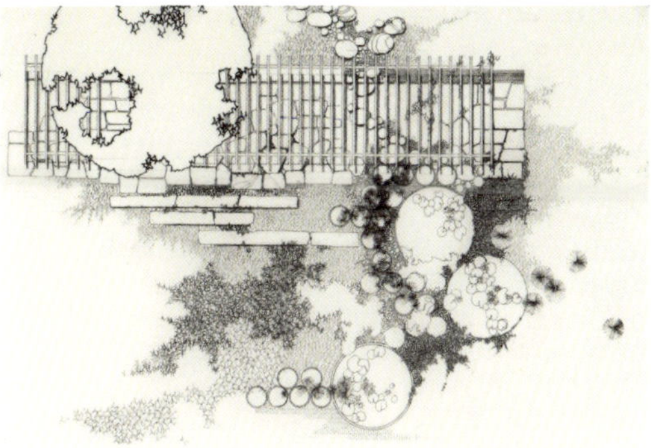

A variant of the sketch shown on page 272 showing rounded rather than sharp-edged stone blocks and circular planting areas echoing the form of the water basins.

1965–1966
Sisters of the Holy Cross Institute, St. Franziskus Hospital → pp. 80, 269
(today: Sisters of the Holy Cross Nursing Home)
▸ Seminarstrasse 11, Menzingen
Architect: Hanns A. Brütsch

1966
Hübeli School → p. 270, 271
▸ Waldeggstrasse 46, Emmenbrücke

1967–1968
Jung Garden Colony
▹ Hünenbergerstrasse 17, 17a–d, 19, 19a
Cham Architects: Gysin + Flueler, Zug
Zürcher placed miniatures of dry landscapes as common elements in front of the houses of the garden colony, which are grouped around a central green space. The design concept of the colony is still discernible, but the details have been lost.

1967
Hodel, Jul.
▹ Brüschrain, Zug

1967–1970
Loreto School → p. 270
▸ Loretostrasse 10, Zug

1968–1969
Bezzola → p. 278
▸ n. s. [Im Stümel 1], Aeugst am Albis

1969
Federal Horticultural Exhibition
→ p. 279
▹ Westfalenpark, Dortmund

1969
Wicander, H. J.
▹ n. s., Baar

1973
Retirement home in Sandbühl
▸ Färberhüslistrasse 9, Schlieren

1973
Shopping centre
▹ n. s., Emmen

1973–1975
Single-family house in Schönau
→ pp. 282, 283
Jürg and Renata Müller-Caviezel
▸ Oberwiler Kirchweg 11
Architect: Flueler, Zug

1976
EAWAG (Swiss Federal Institute of water supply, wastewater treatment and water protection)
(today: Swiss Federal Institute of Aquatic Science and Technology) → p. 276
▸ Seestrasse 79, Kastanienbaum
Architect: Roland Mozzatti
The pond still exists but is now an ecological biotope, and most of its original Japanese-inspired character has been lost.
Lit. G. R. 1973.

1977
Bruggenmatt building development
→ p. 284
▸ Bruggenmattweg, Bonstetten
Lit. Anon. 1983.

1978
Anderhub Trumpf AG
▹ Neuhofstrasse 12, Walterswil-Baar
Zürcher sketched a courtyard in which a meandering ribbon of blue-flowering creeping forget-me-not symbolises a body of water, while the yellow blossoms of three-leaved forest stonecrop form its shore. The blue-yellow colour scheme was complemented by a "multi-stemmed" Japanese red maple.

1978–1980
Grün 80 "Beautiful Gardens"
→ pp. 114–121, 125–127
▸ Rainstrasse 5, Münchenstein, Basel
with Willi Neukom

1978–1984
Mibelle AG
▹ Bolimattstrasse 1, Buchs

1979
Überbauung
▹ n. s., Lugano
Architect: Alex Huber
with Ivo Negrini

1979–1986
Dürrbach building development
▸ Am Dürrbach, Engelberg

Zürcher used river terracing to create a cascade that feeds a naturally shaped pond. Blocks of stone and rubble form islands within it.

1980–1981
Vocational training college in Herdschwand
▸ Oberhofstrasse 45, Emmenbrücke
Architects: Koch, Schmidig, Nosetti / Kunz, Gemperli

1980–1984
Girschweiler, Alfred → pp. 280–281
▸ n. s. [Rickenbach 6], Ottenbach

1981–1982
District hospital
▹ Seilerweg 29, Bad Hersfeld, Germany

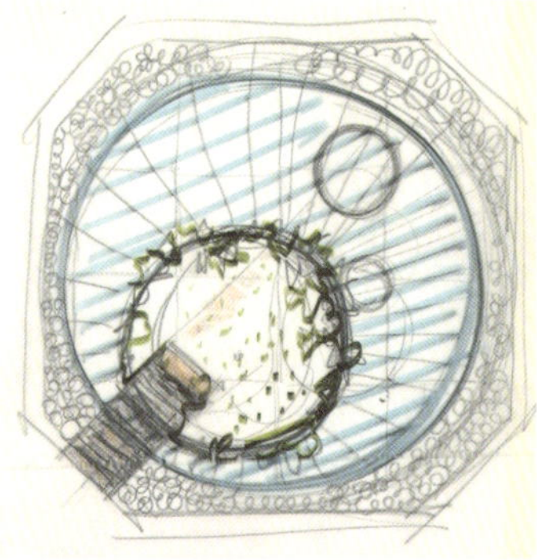

In 1981–1982, Zürcher worked intensively on the design of the inner courtyard of Bad Hersfeld hospital, sketching numerous variants. A platform cantilevers over an island within the circular pool, recalling traditional moon-viewing decks.

1983
Hofmatt retirement home
▹ Hofmatt 1, Weggis

1983
Birkenwäldli
Strandweg/Seehofstrasse, Unterägeri
Competition project

1984
Kauer, C.
▹ Hörnlirain 28, Walchwil

1986
Schnadt-Schuerch, Philippe → p. 276
▸ Hofstrasse 36, Zug

1988
Nicklaus, Karl
Two single-family houses
▹ n. s., Hünenberg

1988–1989
Building development
Single-family house VI, Bentom AG
▸ Waldheimstrasse 30, Zug

1989
Metro Park → p. 268
▹ Rochester, NY

undated
Mittler → p. 267
▹ n. s., Beckenried

undated [photographed in 1984]
Wiede → p. 282
▹ n. s., Buonas

undated [photographed in 1989]
Hermans → p. 280
▹ n. s., Oberägeri

[photographed in 1999]
Vasella → pp. 274, 275
▸ n. s. [Aabachweg 3], Risch

Plant catalogues

Sei Shōnagon, The Pillow Book – Makura no Sōshi

· Allium, Onion, Chives [*Allium tuberosum, Allium thunbergii, Allium japonicum*]
· Amaranth, Variegated [*Amaranthus tricolor*]
· Apricot, Japanese [*Prunus mume "Kobai"*]
· Arrowhead, Chinese [*Sagittaria trifolia*]
· Aster, Chinese [*Callistephus chinensis*]
· Ayafugusa [*unidentified rootless plant*]
· Azalea, "Satsuki", "Iwatsutsuji" [*Rhododendron indicum*]
· Bamboo, Chinese [*Phyllostachys "kuretake"*]
· Bayberry, Chinese [*Myrica rubra*]
· Bellflower, Chinese [*Platycodon grandiflorus*]
· Bottle Gourd [*Lagenaria siceraria*]
· Burnet, Great [*Sanguisorba officinalis*]
· Bur-reed, Branched [*Sparganium erectum*]
· Bushclover, Shrubby [*Lespedeza bicolor or thunbergii*]
· Camphor tree [*Cinnamomum camphora*]
· Chaff flower, Oriental, "Hikage" [*Achyranthes japonica*]
· Caramel tree, Japanese, "Katsura" tree [*Cercidiphyllum japonicum*]
· Carnation, Chinese pink [*Dianthus chinensis*]
· Carnation, Japanese large [*Dianthus superb*]
· Cedar, Hiba, Hiba arborvitae, "Hiba" [*Thujopsis dolabrata*]
· Cedar, Japanese cedar, "Sugi" [*Cryptomeria japonica*]
· Chinaberry tree, White cedar [*Melia azedarach*]
· Cherry, Japanese [*Prunus serrulata*]
· Chickweed, Clammy [*Cerastium glomeratum*]
· Chinquapin, Japanese [*Castanopsis cuspidata*]
· Chrysanthemum [*Chrysanthemum japonicum*]
· Cleyera, Japanese, "Sakaki" tree [*Cleyera ochnaceae*]
· Cypress, "Hinoki" Japanese cypress [*Chamaecyparis obtusa*]
· Daphne, False, "Yuzuri" [*Daphniphyllum macropodum*]
· Date, Chinese, "Jujube" [*Ziziphus jujuba*]
· Dayflower, Asiatic [*Commelina communis*]
· Fern, Bracken, "Devil fern" [*Pteridium aquiline*]
· Fern, Weeping, "Shinobugusa", "a grass that endures" [*Lepisorus thunbergianus*]
· "Gampi", "Ganpi", "Kami no ki" [*Daphne sikokiana, Wikstroemia sikokiana Franch*]
· Gentian, Japanese [*Gentiana scabra*]
· Grass, Lawn, "Shiba" [*Zoysia japonica*]
· Grass, Cogon [*Imperata cylindrica*]
· Grass, Red [*Themeda japonica*]
· Ground pine [*Lycopodium clavatum*]
· Hackberry, Chinese [*Celtis sinensis*]
· Hollyhock [*Alcea rosea*]
· Hops, Japanese [*Cumulus japonicus*]
· Hosta [*Hosta*]
· Indigo plant [*Indigofera tinctoria*]
· "Itsumade" [*unidentified creeper*]
· Iris, Japanese swamp iris [*Iris ensata, Iris kaempferi Sieb. ex Lem.*]
· Ivy, Japanese [*Hedera rhombea*]
· Kerria, Japanese, "Yamabuki", Japanese gold rose [*Kerria japonica*]
· Knotweed, Japanese [*Fallopia japonica*]
· Kotonashi [*unidentified wayside grass*]
· Kudzu vine, "Kudzu", [*Pueraria lobata*]
· Lantern, Chinese [*Physalis alkekengi*]
· Lotus [*Nelumbo*]
· Madder, Common [*Rubia tinctorum*]
· Magnolia, "Honoki" magnolia [*Magnolia obovata*]
· Mallow, Common [*Malva sylvestris*]
Maple:
· Maple, Japanese [*Acer japonicum, Acer palmatum*]
· Maple, Pointed-leaf [*Acer argutum*]
· Maple, Hawthorn [*Acer crataegifolium*]
· Maple, Red [*Acer rubrum*]
· Marlboro [*Ardisia japonica*]
· May flower shrub, Slender deutzia [*Deutzia gracilis*]
· Mercury [*Mercurialis leiocarpa*]
· Mistletoe, "Yadorigi" [*Viscum*]
· Moonseed, Chinese [*Sinomenium acutum*]
· Morning glory, Japanese [*Pharbitis nil, today Ipomoea nil*]
· Moss, "Koke" [*not specified*]
· Mugwort, Common, "Sashimo" [*Artemisia vulgaris*]
· Muku tree [*Aphananthe aspera*]
Oak:
· Oak, Japanese evergreen [*Quercus acuta*]
· Oak, Bamboo-leaf [*Quercus myrsinifolia*]
· Oak, Japanese emperor [*Quercus dentata*]
· Oak, Japanese chestnut [*Quercus acutissima*]
· Palm, True date [*Phoenix dactylifera*]
· Pear, Wild [*Pyrus pyraster*]
· Peach [*Prunus persica*]
· Peony, Chinese white [*Paeonia lactiflora*]
· Peony, Japanese [*Paeonia japonica*]
· Photinia, Japanese [*Photinia glabra*]
Pines:
· Pine, Japanese white [*Pinus pentaphylla*]
· Pine, European black [*Pinus nigra*]
· Pine, Scots [*Pinus sylvestris*]
· Plum [*Prunus domestica*]
· Plum, Red cherry [*Prunus cerasifera "Nigra"*]
· Poison bulb [*Crinum asiaticum var. japonicum*]
· Pondweed [Potamogeton differentus]
· Princess tree [*Paulownia tomentosa*]
· Privet, Japanese [*Ligustrum japonicum*]
· Reed, Chinese giant silver grass [*Miscanthus floridulus*] or Silver grass, Chinese silver grass [*Miscanthus sinensis*]
· Reed, Common [*Phragmites communis*]
· Rice, Wet rice, Manchurian wild rice [*Zizania latifolia*]
· Rush [*Juncaceae*]
· Safflower [*Carthamus tinctorius*]
· Shepherd's purse [*Capella bursa-pastoris*]
· Spindletree [*Euonymus sieboldianus*]
· Spirea, Japanese [*Spiraea japonica*]
· Strawberry [*Fragaria*]
· Strawberry, False [*Duchesne chrysantha*]
· Valerian, Golden [*Patron scabiosifolia*]
· Violet [*Viola arcuata*]
· Walnut [*Juglans regia*]
· Water chestnut [*Trapa bispinosa var. Iinumai, also Trapa natans*]
· Water lily, East Asian yellow [*Nuphar japonica*]
· Waterlily, Prickly [*Euryale ferox*]
· Willow [*Salix subopposita*]
· Windmill palm, Chusan palm [*Trachycarpus excelsus var. typicus*]
· Wisteria, Japanese [*Wisteria floribunda*]
· Wood, "Sappan" [*Caesalpinia sappan*]
· Wood sedge [*Carex sylvatica*]
· Woodsorrel, Creeping [*Oxalis corniculata*]
· Yam vine, Tokoro, "Devil vine" [*Dioscorea tokoro*]

Shōnagon 2006.

Gustav Ammann
Plant list [approx. 1934]

- *Lilium auratum* (Goldbandlilie) [Golden-rayed lily]
- *Lilium longiflorum* [Trumpet lily]
- *Lilium speciosum* [Japanese lily]
- *Iris kaempferi* [Japanese swamp Iris]
- *Chrysanthemum* [Chrysanthemum]
- *Ipomoea morning glory* [Purple morning glory]
- *Paeonia arborea* [Tree peony]
- *Azalea* [Azalea]
- *Acer* [Maple]
- *Prunus pseudocerasus* [Chinese sour cherry]
- *Prunus mume* [Japanese apricot]
- *Prunus persica* [Peach]
- *Prunus armeniaca* (Aprikose) [Apricot tree]
- *Camellia* [Camelia]
- *Magnolia* [Magnolia]
- *Actinidia* [Mini kiwi]
- *Elaeagnus* [Oleaster]
- *Diospyros*
- *Kaki* [Japanese persimmon]
- *Citrus* (Orangen + Zitronen) [Oranges Lemons]
- *Pyrus sinensis* [Japanese pear]
- *Vitis* [Vine]
- *Phyllostachys*
- *Fargesia* or *Sasa* [Bamboo]
- Evergreens: *Daphniphyllum* [False Daphne]
- *Viburnum odoratissimum* [Sweet viburnum]
- *Eurya japonica* [East Asian eurya]
- *Osmanthus* [Holly Leaf Osmanthus]
- *Mahonia japonica* [Japanese mahonia]
- *Ardisia* [Japanese marlberry]
- *Rhaphiolepis* [Hawthorn]
- *Aucuba japonica* [Japanese spotted laurel]
- *Aralia elata* [Japanese Angelica tree]
- *Coniferales* [Conifers]
- *Pinus* [Pine]
- *Juniperus* [Juniper]
- *Chamaecyparis* [Cypress]
- *Cryptomeria japonica* [Japanese cypress or Sugi]
- *Tsuga* [*Pseudotsuga japonica*]
- *Tsuga sieboldii* [Japanese Hemlock or Araragi Hemlock]
- *Abies* [Fir]
- *Thujopsis* [Hiba cedar]
- *Sciadopitys verticillata* [Japanese umbrella pine]
- *Ginkgo* [Ginkgo]
- Deciduous trees: *Acer* [Maple]
- *Rhus japonica* [Japanese sumac]
- *Albizia* [Persian silk tree]
- *Hydrangea* [Hortensia]
- *Hibiscus* [Rose mallow]
- *Lespedeza thunbergii* [Bushclover]
- *Berberis* [Barberry]
- *Wisteria* [Wisteria]
- *Kadsura japonica* [Sanekazura, Kadsura vine]
- Palms: *Rhapis* [Bamboo palm]
- *Cycas* [*revoluta*] [Japanese sago palm or Japanese palm fern]
- Ferns: *Davallia bullata* [Squirrel's-foot fern]
- *Adiantum pedatum* [Northern maidenhair fern]

Ammann n. d.a gtaA.

Willi Neukom, Adolf Zürcher: Grün 80: Dune, Grasses, and Heather Garden

Grasses

- *Pennisetum alopecuroides* 'Japonicum' [Foxtail fountain grass]
- *Festuca amethystina* [Tufted fescue]
- *Festuca scoparia* [Bearskin fescue]
- *Stipa barbata* [Feather grass]
- *Festuca ovina* [Sheep's fescue]
- *Bouteloua gracilis* [Blue grama grass]
- *Leymus arenarius* [Lyme grass]
- *Molinia caerulea* [Purple moor-grass]
- *Calamagrostis epigejos* [Bushgrass]
- *Avena sempervirens* [*Helictotrichon sempervirens*, Blue oat grass]
- *Carex Grayi* [Mace sedge]
- *Pennisetum macrourum* [African feather grass]

Small shrubs and woody plants

- *Hippophae rhamnoides* [Sea buckthorn]
- *Elaeagnus commutata* [Silverberry, Silver oleaster]
- *Nothofagus antarctica* [Antarctic beech]
- *Caragana pygmaea* [Pygmy peashrub]
- *Hebe armstrongii* [Shrubby veronica]
- *Cotoneaster adpressus* [Creeping cotoneaster]
- *Cotoneaster conspicuus* [Tibetan cotoneaster]
- *Perovskia atriplicifolia* [Russian sage]
- *Indigofera potaninii* [Chinese indigo]
- *Corokia cotoneaster* [wire-netting bush]
- *Senecio greyi* [*Brachyglottis greyi*, Daisy bush]
- *Ephedra distachya* [Joint fir]
- *Salvia rosmarinus* [Rosmary]

Heather

- *Pinus sylvestris* [Scots pine]

Undergrowth

- *Saxifraga umbrosa* [Pyrenean saxifrage]
- *Cotula squalida* [*Leptinella squalida*, Brass buttons]

Partially shaded locations

- *Carex morrowii* [Japanese grass sedge]
- *Luzula nivea* [Snowy wood-rush]
- *Astilbe chinensis* [Chinese astilbe]
- *Epimedium pinnatum* [Barrenwort]
- *Corydalis lutea* [Yellow fumitory]
- *Muehlenbeckia axillaris* [Creeping wire vine]
- *Sedum spurium* [Caucasian stonecrop]

Sunny locations

- *Achillea tomentosa* [Woolly yarrow]
- *Campanula* [Bellflowers]
- *Helianthemum* [Rock rose]
- *Dryas x suendermannii* [Dyras]
- *Gypsophila repens* [Creeping chalk-lover]
- *Ceratostigma plumbaginoides* [Hardy plumbago]
- *Lavandula angustifolia* "Hidcote Blue" [Common lavender]

Moorland soil

- *Betula pubescens* [Downy birch]
- *Salix helvetica* [Swiss Willow]
- *Fothergilla major* [Mountain witch alder]
- *Juniperus communis* "Hornibrookii" [Hornibrook juniper]
- *Juniperus sabina* "Tamariscifolia" [Tam juniper]
- *Juniperus communis* "Hibernica" [Irish juniper]
- *Juniperus virginiana* "Tripartita" [Eastern red cedar]
- *Juniperus horizontalis* [Creeping juniper]
- *Erica vagans* "Mrs. D. F. Maxwell" [Cornish heather]
- *Erica cinerea* "Rosabella" [Bell heather]
- *Erica carnea* "Winter Beauty", "Vivellii" [Snow heather]
- *Calluna vulgaris* "Alportii", "Foxii", "H.E. Beale", "Mullion" [Common heather]
- *Genista pilosa* [Creeping broom]
- *Genista hispanica* [Spanish broom]
- *Genista lydia* [Dwarf broom]
- *Cytisus praecox* [Broom plants]
- *Cytisus scoparius* [Scotch broom]
- *Cardamine trifolia* [Trefoil cress]
- *Luzula nivea* [Snowy wood-rush]
- *Luzula sylvatica* [Great wood-rush]
- *Rhododendron repens* [Dwarf azalea]
- *Rhododendron Luteum /Azalea pontica* [Yellow azalea, also Pontic azalea]
- *Rhododendron Luteum* "Coccinea Speciosa" [Deciduous azalea]
- *Rhododendron* Knap Hill [Knap Hill Azalea]
- *Gaultheria* [Wintergreen]
- *Gaultheria procumbens* [American wintergreen]
- *Vaccinium vitis-idaea* [Lingonberry]
- *Astilbe chinensis* [Chinese astilbe]
- *Blechnum spicant* [Hard fern]
- *Polystichum setiferum* "Plumosum Densum" [Soft shield fern]
- *Polystichum setiferum* "Proliferum" [Soft shield fern]
- *Adiantum pedatum* [Northern maidenhair fern]
- *Carex morrowii* [Japanese grass sedge]
- *Carex morrowii variegata* [Variegated Japanese sedge]

Archive: wne.144.1.2-Bepflanzungsplan-26

Walter Leder: Plant list for "Swiss Garden", Mukogaoka Park, Tōkyō, 17 March 1964

1. In front of the main entrance of the garden
· *Cercidiphyllum japonica*, KATSURA [Caramel tree]
· *Mahonia japonica*, HIIRAGINANTEN [Japanese mahonia]
· *Osmanthus ilicifolius*, HI-I-RAGI [Holly Leaf Osmanthus]
· *Juniperus chinensis pfitzeriana*, BYAKUSHIN, IBUKI, KAIZUKAIBUKI [Blue pfitzer juniper]
· *Aspidistra elatior*, HARAN [Cast iron plant]
· *Cercidiphyllum japonica*, KATSURA [Caramel tree]
· *Ophiopogon japonicus*, RYUNOHIGE [Mondo grass]

2. Inside the main entrance
· *Nandina domestica*, NANTEN [Heavenly bamboo]
· *Hamamelis japonica*, MANSAKU [Japanese witch hazel]
· *Trachelospermum asiaticum*, TEIKAKAZORA [Yellow star jasmine]
· *Pachysandra*, KICHIJISO [Japanese spurge]
· Eriobotrya japonica [Japanese loquat]
· *Osmanthus ilicifolius*, HI-I-RAGI [Holly Leaf Osmanthus]
· *Cornus florida dogwood* [Flowering dogwood]
· *Polystichum falcatum*, ONISHIDA [Fishtail fern]
· *Pinus densiflora*, AKAMATSU [Japanese red pine "Akamatsu"]
· *Pinus densiflora "Umbraculifera"*, TAGYOSHO [Japanese red pine "Tagyosho"]
· *Pinus parviflora "Glauca"*, HIMEKOMATSU [Japanese blue maiden pine]
· *Pinus mugo var. Mughus*, CHISAI [Mountain pine]
· *Mattenccia struthiopteris*, KOSASOTETSO [Ostrich fern]

3. Section of the shore near the main entrance
· *Ligularia tussilaginea*, TSUWABUKI [now Farfugium japonicum] [Leopard plant]
· *Acorus gramineus*, SEKISHO [Japanese sweet flag]
· *Hosta "Undulata"*, GIBOSHI [Variegated wavy hosta]
· *Equisetum arvense*, SUGINA [Field horsetail]
· *Equisetum hyemale*, TOKUSE [Rough horsetail]

4. Under cherry tree near the smaller pond
· *Taxus cuspidata* "Nana", KYARA [Japanese yew]
· *Lespedeza penduliflora*, MIYAGINOHAGI [Shrub bushclover]

5. North of the long pergola-shelter
· *Pittosporum tobira*, TOBERA [Japanese pittosporum]
· *Kerria japonica*, YAMABUKI [Japanese Kerria]
· *Weigelia hortensis*, TANIUTSUGI [Japanese weigela]
· *Camellia sansaqua*, SAZANKA [Sasanqua camelia]
· *Camellia*, TSUBAKI [Japanese camelia]
· *Thea sinensis*, CHA [Chinese tea plant]
· *Laurus nobilis*, GEKKEI JU [Bay tree]
· *Rhododendron obtusum*, KIRISHIMATSUTSUJI [Japanese Azalea]
· *Rhododendron obtusum sakamotoi*, KURUMETSUTSUJI [Japanese Azalea]
· *Rhododendron japonicum*, HINODIGIRI, HINOMAYO, HATSUGIRI [Japanese Azalea]
· *Selaginella involvens*, IWAHIBA [Involute spike-moss]
· *Selaginella japonica*, KURUMAGOKE [Tarry spike-moss]

6. South of the long pergola-shelter
· *Magnolia grandiflora*, TAIZANBOKU [Evergreen magnolia]
· *Cryptomeria elegans*, SUGI [Japanese cedar "Elegans"]
· *Sciadopitys verticillata*, KOYAMAKI [Japanese umbrella pine]
· *Chamaecyparis obtusa* "Nana", HINOKI CHIISAI [Hinoki Japanese cypress]
· *Juniperus procumbens*, HAIBYAKUSHIN [Creeping juniper]
· *Acer palmatum sanguineum*, NOMURA M. [Nomura-momiji] [Blood-red Japanese maple]
· *Lagerstroemia indica*, SARUSUBERI [Crape myrtle]
· *Rhododendron indicum*, SATSUKI [Satsuki Azalea]
· *Spiraea japonica*, SHIMOTSUGE [Shimotsuke] [Japanese spirea]
· *Hypericum Henryi* [St. John's wort]

7. Sitting place south of the fountain
· *Phyllostachys nigra* "Punctata", HACHIKU [Black bamboo]
· *Phyllostachys edulis*, MOSOCHIKU [Mōsō-chiku] [Moso bamboo]
· *Trachycarpus excelsa* [Trachycarpus fortunei], SHURO [Chusan palm]
· *Pittosporum tobira*, TOBERA [Japanese pittosporum]
· *Cinnamomum camphora*, KUSUNOKI [Camphor tree]
· *Fatsia japonica*, YATSUDE [Japanese aralia]
· *Euonymus radicans*, TSURUMASAKI [Common wintercreeper]
· *Hosta "Undulata"*, GIBOSHI [Variegated wavy hosta]
· *Aspidistra elatior*, HARAN [Cast iron plant]
· *Taxodium distichum*, RAKU-U-SHO [Bald cypress]

8. At both sides of the terrace path
· *Acer palmatum*, MOMIJI [Japanese maple]
· *Acer palmatum dissectum chisio*, CHISIO [Threadleaf Japanese maple Chisio]
· *Paeonia suffruticosa*, BOTAN [Tree peony]
· *Hamamelis japonica*, MANSAKU [Japanese witch hazel]
· *Cornus controversa*, MIZUKI [Wedding-cake tree]
· *Cornus kousa*, YAMABOSHI [Japanese flowering dogwood]
· *Daphne laureola odora*, JINCHOGE [Spurge laurel]
· *Podocarpus macrophyllus maki*, RAKANMAKI [Yew plum pine]
· *Ruscus aculeatus* [Butcher's broom]
· *Ruscus hypoglossum* [Spineless butcher's broom]
· *Hydrangea macrophylla*, GAKUAJISAI [Big leaf hydrangea]
· *Hydrangea m. otaksa*, AJISAI [Hortensia "Otaksa"]
· *Ilex crenata*, INUTSUGE [Japanese holly]
· *Mahonia repens* [Creeping mahonia]
· *Pachysandra terminalis*, KICHIJISO [Japanese spurge / Green sheen]
· *Ophiopogon japonica*, RYUNOHIGE [Mondo grass]
· *Bambusa* [Fargesia] *nitida*, CHISAI [Blue fountain bamboo]
· *Miscanthus sinensis*, SUSUKI [Chinese silver grass]
· *Iris orientalis sibirica* [Siberian iris]
· *Selaginella involvens*, IWAHIBA [Involute spike-moss]
· *Selaginella japonica*, KURUMAGOKE [Tarry spike-moss]

9. Water and marsh vegetation
· *Typha latifolia* [Bulrush]
· *Iris pseudacorus*, KISHOBU [Yellow iris]
· *Iris japonica* [kaempferi] [Japanese clematis Iris]
· *Nymphaea* [Water lily]
· *Lotos* [*Nelumbo*, Lotus flower]

Leder 1964c ASLA

Bibliography

Adam 2019
Adam, Hubertus, "Behutsam weiterentwickelt", in: *Deutsche Bauzeitung*, 153 (2019), No. 3, pp. 107–111.

Allaer 2007
Allaer, Robin, *De Les van Japan. Perceptie van Japan in Teksten van westerse Architeceten*, Faculty of Engineering and Architecture, Department of Architecture and urban planning, Master's thesis, University Gent, 2007.

Altherr 1927
Altherr, Alfred, "Von Künstlern und Gelehrten. Prof. Ernst Grosse", in: *Das Werk*, 14 (1927), No. 2, pp. XXVIII, XXXI.

Ammann 1914/15
Ammann, Gustav, "Über den Garten", in: *Wissen und Leben*, 15 (1914–1915), pp. 579–583.

Ammann 1926
"Sollen wir die Form ganz zertrümmern?", in: *Die Gartenkunst*, 39 (1926), No. 6, pp. 81–85.

Ammann 1931
"Garten am Monte Brè und Garten in Altstätten im Rheintal: Gartenarchitekten O. Froebels Erben, Zürich", in: *Das Werk*, 18 (1931), No. 3, pp. 69–82.

Ammann 1933a
"Die Gartenanlagen" in: H. J. K., "'Züga' die Zürcher Gartenbau-Ausstellung 1933: Architekten Karl Egender, Wilhelm Müller, Gartenarchitekt: Gustav Ammann, Zürich", in: *SBZ*, 101/102 (1933), No. 10, pp. 115–125 (sgd. "G. A.").

Ammann 1933b
"Otto Valentien, Zeitgemässe Wohngärten", in: Meyer, Peter, "Gartenbücher", in: idem, "Zürcher Gartenbauausstellung ZÜGA 24. Juni bis 17. September 1933", in: *Das Werk*, 20 (1933), No. 7, pp. 206–207.

Ammann 1933c
"Rückschau auf die 'Züga'", in: *Die Gartenkunst*, 46 (1933), No. 11, pp. 167–172.

Ammann 1934
"Gartenkultur", in: *SBZ*, 103/104 (1934), No. 10, pp. 114–118.

Ammann 1936
"Der Garten", in: *NZZ* 157 [15 March 1936], No. 446, Sheet 10.

Ammann 1938
Inserat, in: *Das ideale Heim*, 12 (1938), No. 3 [3 March 1938], n. pag. [Documents III, p. 30, NSL 2-S-8-3].

Ammann 1939
"Die Gärten an der Schweizer. Landesausstellung 1939", in: *SBZ*, 113/114 (1939), No. 18, pp. 213–217.

Ammann 1943a
"Der Garten des Wohlfahrtshauses", in: *Werkmitteilungen – Werkzeugmaschinenfabrik Oerlikon Bührle & Co.*, 3 (1943), No. 4, pp. 49–51.

Ammann 1943b
"Garten und Landschaft – Organische Gartengestaltung und landschaftsfremde Gärten", in: *Werk*, 30 (1943), No. 9, pp. 285–288.

Ammann 1946
"Paul Schädlich", in: *Schweizerisches Gartenbau-Blatt*, 67 (1946), No. 34, p. 184.

Ammann 1950
"Die Entwicklung der Gartengestaltung von 1900–1950", in: *Der Gärtnermeister*, 53 (1950), No. 41, pp. 1–11.

Ammann 1952
"Wie bepflanze ich meinen Teich?", in: *Schweizer Garten*, 22 (1952), No. 10, pp. 301–307.

Ammann 1953a
"Wie wir Wasser im Garten am besten zur Geltung bringen", in: *Schweizer Garten*, 22 (1953), No. 8, pp. 226–228.

Ammann 1953b
"Die persönliche Note im Garten", in: *Das ideale Heim*, 27 (1953), No. 10, pp. 405–412.

Ammann 1955
Blühende Gärten, Erlenbach-Zurich 1955.

Andermatt 2001
Andermatt, Erich, "Zum Gedenken an Dölf Zürcher, Landschaftsarchitekt BSLA", in: *Anthos*, 40 (2001), No. 1, pp. 74–75.

Anliker 1954
[Anliker, Johann], "Dendrologische und gärtnerische Sehenswürdigkeiten in St. Gallen und Umgebung. (Exkursion der Schweizerischen Dendrologischen Gesellschaft, 10. Juli 1954.)", in: *Schweizerisches Gartenbau- Blatt*, 75 (1954), No. 30 [29 July 1954], pp. 667–668 (sgd. "J. A.").

Anon. 1869
Anon. "Books received", in: *The Canadian Entomologist*, 1 (1869), No. 10 [15 May 1869], pp. 90–92.

Anon. 1885
"A Japanese Village in London", in: *The British Architect*, 11 (1885), [9 Jan 1885], p. 15.

Anon. 1903
"Das japanische Haus", in: *SBZ*, 41/42 (1903), No. 10, p. 120.

Anon. 1904a
"Bernische Kunstgesellschaft", in: *Der Bund*, 55 (1904), No. 97 [6 April 1904], p. 4.

Anon. 1904b
"Das Schweizerfest in Yokohama", in: *NZZ*, 125 (1904), No. 256, [14. September], p. 5.

Anon. 1907a
"† Evariste F. R. Mertens", in: *Chronik der Stadt Zürich*, 9 (1907), No. 15 [13 April 1907], p. 115.

Anon. 1907b
"Mannheimer Ausstellungstage", in: *NZZ*, 128 (1907), No. 138 [19 May 1907], pp. 1–2.

Anon. 1907c
"Die Architektur der Kultbauten Japans", in: *SBZ*, 49/50 (1907), 8 June, p. 287.

Anon. 1907d
"Bau- und Gartenkunst auf der Mannheimer Jubiläums-Ausstellung 1907", in: *SBZ*, 49/50 (1907), No. 10, pp. 117–123.

Anon. 1907e
"Country Homes Old & New – Hinchingbrooke, Huntingdonshire, The Seat of the Earl of Sandwich", in: *Country Life*, 22 (1907), No. 565 [2 Nov 1907], pp. 630–636.

Anon. 1910a
"Die japanisch-britische Ausstellung", in: *NZN*, 6 (1910), No. 128, [13 May 1910] edition 2, sheet 2, p. 1.

Anon. 1910b
"Personalnachrichten. Kleine Mitteilungen", in: *Die Gartenkunst*, 12 (1910), No. 5, p. 5.

Anon. 1910c
"Gardening at the Japan-British Exhibition", in: *The Gardeners Chronicle*, 48/3 (1910), No. 1239 [24. September 1910], pp. 227–228.

Anon. 1933
"Die Sondergärten in der 'Züga'", in: *NZZ*, 154 (1933), No. 1493, [20 August 1933], sheet 5.

Anon. 1935
"Wohnhaus Dr. K. und Wohnhaus Dr. B. in Küsnacht (Kanton Zürich), Architekt Max Ernst Haefeli BSA, Zürich", in: *Das Werk*, 22 (1935), No. 1, pp. 1–14.

Anon. 1936
"Brunnen", in: *Das Werk*, 23 (1936), No. 2, pp. 37–43.

Anon. 1937
"Die Pavillons der fremden Staaten", in: *Das Werk*, 24 (1937), No. 11, *Sonderheft Weltausstellung Paris 1937*, pp. 322–341.

Anon. 1938
"Ueber konkrete Kunst", in: *Das Werk*, 25 (1938), No. 8, pp. 250–256.

Anon. 1958
"Der Schweizer Pavillon an der Internationalen Weltausstellung 1958 in Brüssel", in: *Werk*, 45 (1958), No. 10, pp. 345–348.

Anon. 1962a
"Garden for an Atriumhouse at Kilchberg, Zurich", in: *Anthos*, 1 (1962), No. 1, pp. 20–23.

Anon. 1962b
"Die Seite des Redaktors", in: *Anthos*, 1 (1962), No. 2, pp. 1–3.

Anon. 1962c
"Parkbad Tiefenbrunnen in Zürich", in: *Anthos*, 1 (1962), No. 2, pp. 8–11.

Anon. 1962d
"An ideal garden", in: *Anthos* 1 (1962), No. 2, pp. 12–16.

Anon. 1962e
"Einfamilienhaus am Sonnenberg in Zürich", in: *SBZ*, 49 (1962), No. 1, pp. 10–15.

Anon. 1964a
"ifla news", in: *Anthos*, 3 (1964), No. 3, pp. 1–6.

Anon. 1964b
"The gardens of the Mayor's Mansion, Osaka, Japan", in: *Anthos* 3 (1964), No. 3, pp. 28–31.

Anon. 1964c
"The green spaces of the Expo 1964", in: *Anthos*, 3 (1964), No. 3, pp. 34–39.

Anon. 1965a
"A garden at Zurich", in: *Anthos* 4 (1965), No. 4, pp. 18–23.

Anon. 1965b
"The residential quarter 'Alpenblick' at Cham", Switzerland *Anthos*, 4 (1965), No. 4, pp. 24–27.

Anon. 1966a
"She just loves shrubs and bushes", in: *Anthos*, 5 (1966), No. 2, pp. 10–12.

Anon. 1966b
"The garden of the garden architect's house", in: *Anthos*, 5 (1966), No. 3, pp. 28–33.

Anon. 1967a
"Planting of a pond garden", in: *Anthos*, 6 (1967), No. 1, pp. 31–33.

Anon. 1967b
"Überbauung Alpenblick in Cham", in: *Werk*, 54 (1967), No. 5, pp. 2–4.

Anon. 1968
"'Alpenblick' Development at Cham", Switzerland, *Anthos*, 7 (1968), No. 3, pp. 7–9.

Anon. 1969a
"Park with water-course of 'Hübeli' Emmenbrücke school", in: *Anthos*, 8 (1969), No. 3, pp. 15–19.

Anon. 1969b
"The garden design of Nordheim Crematorium Zurich", in: *Anthos*, 8 (1969), No. 4, pp. 30–31.

Anon. 1969c
"Erweiterungsbauten der Töchterschule IV auf der Hohen Promenade, Zürich: Architekt Paul W. Tittel", in: *Werk*, 56 (1969), No. 7, pp. 469–471.

Anon. 1970a
"Park of the Institute for Apprentice Nurses of Heiligen Kreuz, Menzingen, and the New Building of the St. Franziskus Hospital", in: *Anthos*, 9 (1970), No. 3, pp. 6–7.

Anon. 1970b
Anon. [Zürcher, Adolf], "Die Landschaft beginnt beim Kinderspielplatz", in: *Werk*, (1970), No. 4, SWB-Kommentare 15, pp. 261–272.

Anon. 1971
"Ein japanischer Garten für Genf", in: *NZN* 67 (1971) [19 Oct 1971], No. 244, n. pag.

Anon. 1976
"Ausbildungszentrum Wolfsberg, Ermatingen: Architekten Rudolf + Esther Guyer", in: *Werk*, 63 (1976), No. 6, pp. 415–418.

Anon. 1977a
"Garden on a Slope at Küsnacht", in: *Anthos*, 16 (1977), No. 1, pp. 20–22.

Anon. 1977b
"Results of the Competition for the 'Grün 80' Exhibition: 2nd Swiss Exhibition for Garden and Landscape Design in Basle, 1980", in: *Anthos*, 16 (1977), No. 4, pp. 28–30.

Anon. 1977c
"Results of the Competition for 'Grün 80' – 2nd Swiss Exhibition of Garden and Landscape Design 1980 in Basle", in: *Anthos*, 16 (1977), No. 3, pp. 32–37.

Anon. 1978
"'Grün 80' 2nd Swiss Exhibition for Garden and Landscape Design Basle 1980: Sector Competition: Prize-winning Projects for the 'Beautiful Gardens' Sector", in: *Anthos*, 17 (1978), No. 1, pp. 22–23.

Anon. 1980
"Grün 80", in: *Wohnen*, 55 (1980), No. 6, p. 129.

Anon. 1983a
"Bruggenmatt Development, Bonstetten ZH", in: *Anthos*, 22 (1983), No. 2, pp. 17–19.

Anon. 1983b
"Abschied von Willi Neukom", in: *Anthos*, 22 (1983), No. 4, p. 35.

Anon. 1986
"Sisters' Garden, 'Ländli' sanatorium in Oberägeri ZG" in: *Anthos*, 25 (1986), No. 4, p. 12.

Anon. 1987
"Ernst Cramer: biography", in: *Anthos*, 26 (1987), No. 2, pp. 2–4.

Anon. 1991
"Naturnahe Schul- und Familiengärten"/"Formale Gärten", in: *Anthos*, 30 (1991), No. 4, p. 56.

Anon. 1998
"Nachrufe', in: *Thurgauer Jahrbuch* 73 (1998), pp. 197–199.

Anon. 2002
'Raumgreifend: Villa Staehelin, Feldmeilen ZH" in: Heimatschutz, 97 (2002), No. 3, p. 26.

Anon. n. d.
"Das Sonnenburggut", https://sonnenburg-gaertnerei.ch/aktuelles/das- sonnenburggut (retrieved: 5 Dec 2022)

Arioli 1963a
Arioli, Richard, "The International Garden Exhibition at Hamburg", in: *Anthos*, 2 (1963), No. 3, pp. 38–41.

Arioli 1963b
"IFLA news", in: *Anthos*, 2 (1963), No. 4, pp. 38–40 (sgd. "R. A.").

Arioli 1964a
"ifla news", in: *Anthos*, 3 (1964), No. 2, pp. 39–41.

Arioli 1964b
"The Japanese people and their gardens", in: *Anthos*, 3 (1964), No. 3, pp. 1–6.

Arioli 1964c
"Editorial", in: *Anthos*, 3 (1964), No. 3, pp. 11–12.

Arioli 1964d
"Japanese Gardens", in: *Anthos*, 3 (1964), No. 3, pp. 13–27.

Arioli 1964e
"Garden lighting", in: *Anthos*, 3 (1964), No. 3, pp. 40–41.

Arioli 1965
"National gardens", in: *Anthos*, 4 (1965), No. 1, pp. 29–31.

Arioli 1984
"Obituary Hans Jakob Barth, Riehen", in: *Anthos*, 23 (1984), No. 3, p. 42.

Árokay 2003
Árokay, Judit, "Der gelenkte Blick: umi in der klassischen japanischen Poetik" in: *Asiatische Studien*, 57 (2003), No. 3, *Meer und Berge in der japanischen Kultur: europäische Japan-Diskurse III und IV*, pp. 471–484.

a. v. 1932
a.v., "Kantonale Schule für Obst-, Gemüse- und Gartenbau in Oeschberg", in: *Der Bund*, 83 (1932) [20 Sept 1932], No. 439, n. pag. [p. 5].

Badeja 1980
Badeja, Edmund "'Land and water' sector", in: *Anthos*, 19 (1980), No. 1, *Grün 80*, pp. 27–36.

Baiemi 2003
Baiemi, Silke K., *Reisephotographie in Ostasien und europäische Reisebilder in Wissenschaft, Kunst und Alltag von ca. 1860 bis 1914/18*, Locarno 2003.

Baltzer 1903
Baltzer, Franz, *Das Japanische Haus. Eine Bautechnische Studie*, Berlin 1903.

Baltzer 1907
Die Architektur der Kultbauten Japans, Berlin 1907.

Bargen 2015
Bargen, Doris G., *Mapping Courtship and Kinship in Classical Japan: The Tale of Genji and Its Predecessors*, Honolulu 2015.

Barlow Rogers 2008
Barlow Rogers, Elizabeth, "Letter from the Editor", in: *site LINES, A Journal of Place, A Publication of the Foundation for Landscape Studies*, 4 (2008), No. 1, p. 2.

Bärtels 1987
Bärtels, Andreas, *Kostbarkeiten aus ostasiatischen Gärten*, Stuttgart 1987.

Baumann A 1931/1932
Baumann, Albert, "Der japanische Garten", in: idem, "Neuanlagen", in: *Jahresbericht der Kantonalen Schule für Obst-, Gemüse- und Gartenbau sowie der Gutswirtschaft in Oeschberg- Koppigen*, published by Kantonale Schule für Obst-, Gemüse- und Gartenbau Oeschberg, Selbstverlag der Gartenbauschule XI (1931/1932), pp. 31–33.

Beglinger 1995
Beglinger, Fridolin, "Zum Gedenken an Hans Nussbaumer Landschaftsarchitekt BSLA, Feldmeilen", in: *Anthos*, 34 (1995), No. 3, p. 52.

Behrendt 1933
Behrendt, Walter Curt, *Japanische Architektur und Gärten*, (Wegleitungen Kunstgewerbemuseum Zürich 115), Zurich 1933.

Benesch/Benesch 1938
Benesch, Otto Heinrich Karl and Otto Benesch, *Die Spätmeister des japanischen Holzschnitts: Sharaku, Hokusai, Hiroshige*, Vienna 1938.

Bernard 1977
Bernard, Frank, "Histoire et philologie japonaises", in: *École pratique des hautes études. 4e section, Sciences historiques et philologiques*, Annuaire 1976–1977, 1977, pp. 1035–1058. https://do.org/ 10.3406/ephe.1977.6332, https://www.persee.fr/doc/ephe_0000-0001_1976_num_ 1_1_6332 (retrieved: 27 Sept 2022).

Berger 2014
Berger, Douglas L., "Editor's Preface", in: Heine, Steven, *Zen-Koans*, Dimensions of Asian Spirituality, edited by Douglas L. Berger, Honolulu 2014.

Berthier/Parkes 2000
Berthier, François and Graham Parkes, *Reading Zen in the Rocks. The Japanese Dry Landscape Garden*, Chicago 2000.

Beuchert 1983
Beuchert, Marianne, *Die Gärten Chinas*, Cologne 1983.

Bill 1949
Bill, Max, "Schönheit aus Funktion und als Funktion", in: *Werk*, 36 (1949), No. 8, pp. 272–274.

Bill 1955
"Tetsuro Yoshida. Das Japanische Wohnhaus [und] Tetsuro Yoshida. Japanische Architektur", in: *Werk*, 42 (1955), No. 11, p. 235.

Bischof/Guillain 1958
Bischof, Werner and Robert Guillain, *Japan*, Zurich 1954.

Blaser 1958
Blaser, Werner, *Wohnen und Bauen in Japan*, Teufen, 1958.

Blaser 1985
Atrium: Lichthöfe seit fünf Jahrtausenden, Basel 1985.

Blaser 1988
Tempel und Teehaus in Japan, Basel 1988.

Bloesch 1912/13
Bloesch, Hans, "Aus fernen Ländern", in: *Die Alpen*, 7 (1912–1913), No. 7–8, pp. 458–460.

Blundell-Jones 2008
Blundell- Jones, Peter, "The lure of the Orient: Scharoun and Häring's East-West connections", in: *Architectural Research Quarterly*, 12 (2008), No. 1, pp. 29–42.

B. O. 1976
B. O., "Die Neubauten für Bauwissenschaften und Architektur der ETH auf dem Hönggerberg", in: *SBZ*, 94 (1976), No. 42, pp. 623–633.

Boehn 1910
Böhn [Boehn], Otto Ernst Leo von, "Die Lösung der Zweckfragen bei der Anlage von Privatgärten", in: *Die Gartenkunst*, 12 (1910), No. 9, pp. 1–24.

Boerschmann 1923
Boerschmann, Ernst, *Baukunst und Landschaft in China: eine Reise durch zwölf Provinzen*, Berlin 1923.

Boesiger 1951
Boesiger, Willi (ed.), *Richard Neutra – Buildings and Projects*, Zurich 1951.

Boesiger 1957
Le Corbusier – Œuvre complète 1950–57, Zurich 1957.

Boller/Hokusai 1955
Boller, Willy, Katsushika Hokusai, *Hokusai: ein Meister des japanischen Holzschnittes*, Zurich, 1955.

Brasch 1965
Brasch, Heinz, *Kyoto. Die Seele Japans*, Olten 1965.

Brauchlin/Wehrli 2018
Brauchlin, Emil and Hans Peter Wehrli, *Strategisches Management*, Berlin, 2018.

Braun 1954
Braun, R. A., "Kalifornischer Lebensstil", in: *Tages-Anzeiger für Stadt und Kanton Zürich*, 61 (1954), No. 19 [22 Jan 1954], sheet 12.

Brinker 1974
Brinker, Helmut, *Das Gold in der Kunst Ostasiens*, Sonderausstellung des Museums Rietberg, Zurich 1974.

Brinker 2008
"Schriftbau-werke", in: *TEC21* 134 (2008) No. 42–43, pp. 34–39.

Brown E 1910
Brown, Eric, untitled ["Studio Talk"], in: *The International Studio An Illustrated Magazine of Fine and Applied Art*, 40 (1910), No. 159, pp. 220–246.

Brown K 2008
Brown, Kendall H., "Fair Japan: Japanese Gardens at American World's Fairs, 1876–1940", in: *site LINES, A Journal of Place, A Publication of the Foundation for Landscape Studies*, 4 (2008), No. 1, pp. 13–16.

Bucher 1996
Bucher, Annemarie, *Vom Landschaftsgarten zur Gartenlandschaft. Gartenkunst zwischen 1880 und 1980 im Archiv für Schweizer Gartenarchitektur und Landschaftsplanung*, Zurich 1996.

Bucher 1998
"Der Garten – ein Spiegel seiner Zeit: Geschichte der abendländischen Gartenkunst", in: *Heimatschutz*, 93 (1998), No. 4, pp. 2–7.

Bucher 2009
Das kleine Forum – Plakatausstellung G/59 zum 50-jährigen Jubiläum, published by Das kleine Forum, Zollikerberg, 2009.

Bucher 2015
"Chinoiserien revisited: der Chinagarten in Zürich als Symbol der Freundschaft", in: *Topiaria helvetica* (2015), pp. 61–70.

Bucher Bauer/Nater 2004
Bucher Bauer, Miriam, Beatrice Nater, "Pioniere der Landschaftsarchitektur in der Schweiz – Interview mit Fredy Klauser, Rorschach, in: *Der Gartenbau*, 125 (2004), No. 27, pp. 15–17.

Buchwald 1964
Buchwald, Konrad, "Ballungsgebiete und Naturpärke im modernen Japan", in: *Garten und Landschaft*, 74 (1964), No. 10, pp. 320–324.

bw. 1928
bw., “Die Thronfeier in Kyoto”, in: *NZZ*, 149 (1928), No. 2029 [6 Nov 1928], edition 2, sheet 4.

Chambers 1757
Chambers, William, *Designs of Chinese buildings, furniture, dresses, machines, and utensils, engraved by the best hands, from the originals drawn in China by Mr. Chambers, architect, member of the Imperial Academy of Arts at Florence to which is annexed a description of their temples, houses, gardens*, London 1757.

Chambers 1772
A dissertation on oriental gardening, London 1772.

Chen 1992
Chen, Lixian, *Art and architecture in Suzhou gardens*, Nanjing 1992.

Colomina 1996
Colomina, Beatriz, *Privacy and Publicity. Modern Architecture as Mass Media*, 2nd ed. Cambridge (Mass.) 1996.

Conder 1886
Conder Josiah, “The art of landscape gardening in Japan”, in: *Transactions of the Asiatic Society of Japan*, 14 (1886), No. 2, pp. 119–175.

Conder 1893/1894
Landscape gardening in Japan “ in ancient and modern times, text with numerous illustrations in the text and 37 plates”, Tōkyō 1893. “‘Op’ supplement to it, containing 60 typical views on 40 plates, reproduced in collotype by K. [Kazumasa] Ogawa, with short descriptive text”, Tōkyō 1894.

Conder/Lancaster 1964
Conder, Josiah und Clay Lancaster, *Landscape gardening in Japan*, New York 1964.

Cramer 1946
Cramer, Ernst, ‘Wo stehen wir heute im Gartenbau?’, in: *Werk*, 33 (1946), No. 3, pp. 85–88.

Cramer 1964
“Neuer Dorffriedhof in der Schweiz”, in: *Garten und Landschaft*, 74 (1964), pp. 366–369.

Cramer 1972
Lecture manuscript, 19 April 1972, printed in: Bucher Annemarie, *Vom Landschaftsgarten zur Gartenlandschaft. Gartenkunst zwischen 1880 und 1980 im Archiv für Schweizer Gartenarchitektur und Landschaftsplanung*, Zurich 1996, pp. 106–107.

Cramer 1987
“Das heutige Gestalten”, [‘(Richtlinien für den Bauherrn) – Text von E. Cramer, etwa 1955’], in: *Anthos*, 26 (1987), No. 2, p. 10.

Davidson 1982
Davidson, A. Keir, *Zen Gardening*, London 1982.

Deane n. d.
Deane, Andrew R., “Designing Men”, in: *A Japanese Garden Handbook, Part One, Design and Craft*, North American Japanese Garden Association, n. d. pp. 1–24. https://najga.org/handbook/designing-men/ (retrieved: 9 Jan 2023).

Delank 1996
Delank, Claudia, *Das imaginäre Japan in der Kunst. Japanbilder vom Jugendstil bis zum Bauhaus*, Munich 1996.

Desarzens 1963
Desarzens, A. F., “Palais de Beaulieu, Lausanne: the Japanese Garden”, in: *Anthos*, 2 (1963), No. 4, pp. 30–32.

Domenig 1965
Domenig, Gaudenz, “Das Japanische Wohnhaus”, in: *NZZ*, 186 (1965). No. 1278/1279 [27 March 1965], sheet 5/6.

Du Cane/Du Cane 1908
Du Cane, Florence and Ella Du Cane, *The Flowers and Gardens of Japan*, London 1908.

Dumoulin 1959
Dumoulin, Heinrich, Zen – *Geschichte und Gestalt*, Bern 1959.

Dümpelmann 2001
Dümpelmann, Sonja, *Karl Foerster. Vom großen Welt- und Gartenspiel*, Berlin 2001.

Dutton 2010
Dutton, Jacqueline, “‘Nonwestern’ utopian traditions”, in: Claey, Gregory (ed.), *The Cambridge Companion to Utopian Literature*, Cambridge, Mass. 2010.

Eckbo 1964
Eckbo, Garrett, in: Jellicoe, Geoffrey Alan, Garrett Eckbo, Maria Teresa Parpagliolo et al., “Eindrücke beim IFLA-Kongreß in Tokio”, in: *Garten und Landschaft*, 74 (1964), pp. 342–344.

Egli 1939
Egli, Charles Adolf, “Präsidentenkonferenz vom 28. Januar 1939 in Zürich. Berichte der Sektionen (Schluss). Sektion St. Gallen”, in: *Schweizer Kunst* (1939), No. 8/9 [March–April], pp. [97]–98.

Elsesser 2021
Elsesser, Corinne, *Das japanische Vorbild. Raumkonzeptionen bei Josef Frank und Bruno Taut. Ein Beitrag zur Ideengeschichte der Modernen Architektur*, Berlin 2021.

Encke 1910
Encke, Friedrich August Ernst, “Die Lösung der Zweckfragen bei der Anlage von Privatgärten, in: *Die Gartenkunst*, 12 (1910), No. 9, pp. 1–24.

Engelhardt 1910
Engelhardt, Walter Freiherr von, “Kultur und Natur in der Gartenkunst”, in: *Kunst und Kultur*, Vol. 6, Stuttgart 1910.

Epprecht 1956
Epprecht, Hans, “Gustav Ammann und sein Wirken”, in: *Werk*, 43 (1956), No. 8, pp. 238–244.

Erdberg 1964
Erdberg, Eleanor von, *Grundsätze des Wohnens im westlichen und östlichen Raum: Baustil und Bautechnik in Amerika und Japan*, (Forschungsberichte des Landes Nordrhein-Westfalen No. 1449), Cologne 1964.

Etter 1974
Etter, Alfred, “Die Mühle Weinfelden und die Thur”, in: *Thurgauer Jahrbuch*, 49 (1974), pp. 90–97.

Farbholzschnitte 1965
Farbholzschnitte aus Japan. Schenkung Julius Mueller an das Museum Rietberg, collated by Heinz Brasch, Zurich 1965.

Farrer 1904
Farrer, Reginald, “The Gardens of Tokyo” in: *Macmillan's Magazine*, 89 (1904), pp. 378–384.

Farrer 1912
“On Japanese Gardens”, in: *The Gardeners Chronicle*, 52/3 (1912), No. 1349 [2 Nov 1912], pp. 330–331.

Farrer 1913
My Rock Garden, London 1913.

Fecht 1995
Fecht, Monika, “L'exposition internationale des arts décoratifs et industriels modernes, Paris 1925, Teil II: Die Pavillons”, in: *Form + Zweck*, 27 (1995), No. 12, p. 114.

Fenollosa 1913
Fenollosa, Ernest Francisco, *Ursprung und Entwicklung der chinesischen und japanischen Kunst*, Leipzig 1913.

Fiévé 2012a
Fiévé, Nicolas, “Les jardins paysagers des seigneurs domaniaux de l'époque d'Edo et les manuels relatifs à la composition des jardins. Première partie: Des manuels de paysage médiévaux aux parcs paysagers prémodernes”, in: *Projets de paysage* 8 (2012), ENSAP Bordeaux, ENSAP Lille, ENSP Versailles-Marseille, INSA Centre Val de Loire/École de la nature et du paysage Blois, Institut Agro Rennes-Angers, pp. 8–26. https://journals.openedition.org/paysage/14328, DOI: https://doi.org/10.4000/paysage.14328 (retrieved: 9 Jan 2023).

Fiévé 2012b
“Les jardins paysagers des seigneurs domaniaux de l'époque d'Edo et les manuels relatifs à la composition des jardins. Seconde partie: L'apparition de la figure de l'architecte-paysagiste et la transmission du savoir à travers les manuels imprimés”, in: *Projets de paysage* 8 (2012), ENSAP Bordeaux, ENSAP Lille, ENSP Versailles-Marseille, INSA Centre Val de Loire/École de la nature et du paysage Blois, Institut Agro Rennes- Angers, pp. 27–50. https://journals.openedition.org/ paysage/14405, DOI: https://doi.org/ 10.4000/paysage.14405 (retrieved: 9 Jan 2023).

Fintelmann/Weiß 1903
Fintelmann, V. and R.-F. Weiß, “Die Gartenkunst in Japan”, in: *Die Gartenkunst*, 5 (1903), No. 11, pp. 13–15.

F. K. 1904
F. K., “Robert Schwarzenbach-Zeuner”, in: *Mitteilungen über Textilindustrie*, 11 (1904), No. 14 [15 July 1904], pp. [229]–231.

Fleig-Harbauer 1981
Fleig- Harbauer, Gisela, “Der Japanische Garten – Wege zu moderner Gestaltung”, in: *Anthos*, 20 (1981), No. 2, Denkmalpflege im Freiraum, p. 44.

Foerster 1917
Foerster, Karl, *Vom Blütengarten der Zukunft. Erfahrungen und Bilder aus der neuzeitlichen Gartenentwicklung*, Berlin, 1917.

Foerster 1934
Garten als Zauberschlüssel, Berlin 1934. https://depositonce.tu-berlin.de/bitstream/11303/4570/3/dill_antje.pdf.txt (retrieved: 9 Jan 2023)

Foerster 1936
Der Steingarten der sieben Jahreszeiten in Sonne und Schatten, Berlin/Bern 1936.

Fogel 2013
Fogel, Joshua A., *Japanese Historiography and the Gold Seal of 57 C.E. – Relic, Text, Object, Fake*, Leiden 2013.

Fowler 2013
Fowler, Michael D., “Hearing a shakkei: The semiotics of the audible in a Japanese stroll garden”, in: *Semiotica*, 44 (2013), No. 197, pp. 101–117, DOI 10.1515/sem-2013-0083 (retrieved: 9 Jan 2023).

Fowler 2018
“Spatial confluences between the ink paintings and garden designs of Sesshū Tōyō”, in: *Studies in the History of Gardens & Designed Landscapes, An International Quarterly*, 38 (2018), No. 2, pp. 144–157. doi 10.1080/14601176. 2016.1269474 http://dx.doi.org/10.1080/14601176.2016.1269474 (retrieved: 9 Jan 2023).

Franz 1978
Franz, Heinrich Gerhard, *Pagode – Turmtempel – Stupa: Studien zum Kultbau des Buddhismus in Indien und Ostasien*, Graz 1978.

Freytag 1940
Freytag, J., “Konrad Hippenmeier”, in: *SBZ*, 115/116 (1940), No. 7, pp. 76–77.

Froebel 1884
Froebel, Otto, “Ueber Alpenpflanzen und deren Cultur”, Vortrag, gehalten in der Section Uto den 29 Feb 1884. https://www.sac-cas.ch/it/le-alpi/ueber-alpenpflanzen-und-deren-cultur-7675/ (retrieved: 7 Dec 2022)

Froebel 1884/85
“Ueber Alpenpflanzen und deren Cultur” in: *Jahrbuch des Schweizer Alpenclub*, 20 (1884/85), pp. 394–418.

Füeg 1959
[Füeg, Franz], “Einfamilienhaus in Feldmeilen”, in: *Bauen + Wohnen*, 13 (1959), No. 12, pp. 418–423 (sgd. “ü. e.”).

Fujioka/Tsunenari 1982
Fujioka, Michio and Kazunori Tsunenari, *Japanese residences and gardens: a tradition of integration*, Tōkyō 1982.

Gadient 2009
Gadient, Hansjörg, *LHA – Leo Hafner Archiv*, Steinhausen, 2009, pp. 148–157.

Gantenbein 1959
Gantenbein, Werner, “Der Schweizer Pavillon an der Weltausstellung 1958 in Brüssel”, in: *SBZ*, 77 (1959), No. 35, pp. 557–562.

Gantner 1928
[Gantner, Joseph], “Frankfurt am Main. Sonderschau und Vorträge ‘Das flache Dach’”, in: *Das neue Frankfurt*, 2 (1928), No. 4, pp. 75–76 (sgd. “Gtr.”).

ghn. 1958
ghn., “Die erste Schweizerische Gartenbauausstellung G 59 im Aufbau”, in: *Die Tat*, 23 (1958), No. 295, p. 6.

Gollwitzer 1959
Gollwitzer, Gerda, "Gärten am See. Zur 1. Schweizerischen Gartenbauausstellung Zürich 1959", in: *Garten und Landschaft*, 69 (1959), No. 8, pp. 225–235.

Gollwitzer 1964
"Japan 1964 – Eindrücke während einer Reise", in: *Garten und Landschaft*, 74 (1964), No. 10, pp. 313–315.

Gothein 1926
Gothein, Marie-Luise, *Indische Gärten*, Berlin 1926.

Goto/Naka 2016
Goto, Seiko and Takahiro Naka, *Japanese Gardens – Symbolism and Design*, London/New York 2016.

G. R. 1973
G. R., "Ausbau des Hydrologischen Laboratoriums Kastanienbaum zum Seenforschungsinstitut der EAWAG", in: *SBZ*, 91 (1973), No. 1, pp. 1–6.

Graf 1959
Graf, Hans, "Vom Wesen des Gartens", in: *Schweizerische Gärtnerzeitung*, 62 (1959), No. 29, pp. 129–130.

Graf 1988
"Hans Graf's Garden in Bolligen, Berne", in: *Anthos*, 27 (1988), No. 1, pp. 16–17.

Groeningen Van 1996
Groeningen Van, Isabelle, *The Development of Herbaceous Planting in Britain and Germany from the Nineteenth to Early Twentieth Century*, 2 Vols., PhD thesis, University of New York 1996, Vol. 2.

Gröning 1997
Gröning, Gerd, "Ideological Aspects of Nature Garden Concepts in Late Twentieth-Century Germany", in: Wolschke-Bulmahn, Joachim (ed.), *Nature and Ideology: Natural Garden Design in the Twentieth Century*, Washington, D.C. 1997. https://books.google.ch/books?id= wQ9KaBPtL-cC&printsec= frontcover&hl=de#v=onepage&q = 241&f=falsehttps://archive.org/stream/wolschke-bulmahn-nature-and-ideology/Wolschke-Bulmahn_Nature%20and%20Ideology_djvu.txt (retrieved: 9 Jan 2023)

Grütter 1976
Grütter, Jörg, "Traditionelle japanische Architektur: die Beziehung des architektonischen Innenraums zur Aussenwelt", in: *Bauen + Wohnen*, 30 (1976), No. 12, pp. 465–468.

Gu 1979
Gu, Jiegang, 顾颉刚, "'Zhuang Zi' he 'Chuci' zhong Kunlun he Penglai liangge shenhua xitong de ronghe《庄子》和《楚辞》中昆仑和蓬莱两个神话系统的融合 ("Die Verschmelzung der beiden mythologischen Systeme Kunlun und Penglai in 'Zhuangzi' und 'Chu Ci'"), in: *Zhonghua wenshi luncong*, 中华文史论丛 (Sammlung chinesischer Literatur und Geschichte) 2 (1979), No. 10, pp. 31–57.

Gyori/Fekete/Herczeg 2018
Gyori, Peter, Albert Fekete, and Agnes Herczeg, "Sight and insight – creating space for religious practices", in: *Technical Transactions*, 115 (2018), No. 4, *Architecture And Urban Planning*, pp. 45–54. DOI:10.4467/2353737XCT.18.053.8365 (retrieved: 9 Jan 2023).

Haggenmacher 1947
Haggenmacher, Fritz, "ZÜKA, die große Ausstellung des Jahres 1947", in: *Der Gärtnermeister*, 50 (1947), No. 35 [30 August 1947], pp. 305–307.

Hájek/Forman/Forman
Hájek, Lubor, Werner Forman, and Bedřich Forman, *Der frühe japanische Holzschnitt*, 2nd ed., Prague 1957.

Hall/Brown/Jansen 1988–1999
Hall, John Whitney, Delmer Brown, and Marius Jansen, *The Cambridge History of Japan*, Vol. 1, Cambridge 1988–1999.

Hammerbacher 1964
Hammerbacher, Herta, "Japanische Städte – heute und morgen" in: *Garten und Landschaft*, 74 (1964), No. 10, pp. 326–335.

Hanisch 1910
Hanisch, F. [Fritz], "Schlesische Parkanlagen", in: *Die Gartenkunst*, 12 (1910), No. 7, pp. 107–112.

Hansjakob/Hansjakob 1979
Hansjakob, Gottfried and Toni Hansjakob, "Rheinaue Bonn and Bonn Federal Garden Exhibition 1979", in: *Anthos*, 18 (1979), No. 2, *Wohnstrassen*, pp. 36–38.

Harada 1922
Harada, Jirō, "The Japanese Garden", in: *The Studio*, 84 (1922), No. 356 (November), pp. 264–270.

Harada 1928
The Gardens of Japan, edited by Geoffrey Holme, London 1928.

Harada 1956
Japanese Gardens, London/Boston 1956.

Harper/Shirane 2015
Harper, Thomas, Haruo Shirane (eds.), *Reading the tale of Genji*, New York/Chichester, West Sussex 2015.

Hartmann Schweizer 2013
Hartmann Schweizer, Rahel, *Otto Kolb – Architekt und Designer*, Zurich 2013.

Hashimoto/Shiga 1927–1928
橋本基編, 志賀直哉, Hashimoto, Motoi and Naoya Shiga, 聚樂, 東洋古建築庭園畫集, Juraku, Tōyō kokenchiku teien gashū [Juraku. Kunstbuch für den orientalischen antiken Architekturgarten oder Juraku. Orientalisches altes Architektur-Garten- kunstbuch], 丸善 座右宝刊行会 Zauhō Kankōkai, Maruzen 東京 Tōkyō 昭3 Shōwa 2–3 [1927–1928].

Hashimoto/Shiga 1930–1933
聚樂, 東洋古建築庭園畫集, Juraku Tōyō kokenchiku teien gashū [Juraku. Kunstbuch für den orientalischen antiken Architekturgarten oder Juraku. Orientalisches altes Architektur-Gartenkunstbuch], 丸善 座右宝刊行会 Zauhō Kankōkai, Maruzen, 東京 Tōkyō 昭5–8 Shōwa 5–8 [1930–1933].

Hayashi 2016
Hayashi, Michiko, "Japanese National Treasures at the 1910 Japan-British Exhibition, and the Japanese Temples and Their Treasures", in: *BIJUTSUSHI, Journal of Japan Art History Society*, 66 (2016), No. 181/1, pp. 36–54.

H. B. 1945a
H. B., "Gustav Ammann", in: *Der Gärtnermeister*, 48 (1945), [7 July 1945], No. 27, p. 216.

H. B. 1945b
"Fritz Klauser", in: *Der Gärtnermeister*, 48 (1945), [7 July 1945], No. 27, p. 216.

Hearn 1892
Hearn, Lafcadio, "In a Japanese Garden", in: *Atlantic Monthly*, 70 (1892), No. 417, pp. 14–33.

Hearn 1894–1895
Glimpses of Unfamiliar Japan, 2 Vols., Boston/New York 1894–1895.

Hearn 1911
Japan. Ein Deutungsversuch, Frankfurt am Main 1911.

Heisinger 1933a
Heisinger, Marta, "Japanische Gärten", in: *Die Form*, 8 (1933), No. 6, pp. 202–208.

Heisinger 1933b
"Japanische Gärten", in: *Das Werk*, 20 (1933), No. 2, pp. 33–40.

Helfrich/Raymond/Whitaker 2006
Helfrich, Kurt, Antonin Raymond, and William Whitaker, *Crafting a Modern World: The Architecture and Design of Antonin and Noémi Raymond*, New York 2006.

Henkel 1910
Henkel, Friedrich, "Gartenkunst in Japan", in: *Die Gartenkunst*, 12 (1910), No. 2, p. 32.

Hentrich 1933
Hentrich, Helmut, "Japanische Wohnkultur", in: *Die Form*, 8 (1933), No. 6, pp. 193–201.

h. er. 1933
h. er. "Aus Kunstzeitschriften", in: *NZZ*, 154 (1933), No. 553, [28 March 1933], edition 2, sheet 4.

Herrigel 1964
Herrigel, Gusty L. *Der Blumenweg: eine Einführung in den Geist der japanischen Kunst des Blumenstellens*, 3rd new reviewed edition, Munich 1964.

Hesdörfer 1907
[Hesdörfer, Max], "Die grosse allgemeine Gartenbauausstellung in Mannheim", in: *Die Gartenwelt*, 11 (1907), No. 35, pp. 411–415 (sgd. "Vom Herausgeber").

Hesse-Wartegg 1901
Hesse-Wartegg von, Ernst, "Das japanische Kirschblütenfest", in: *NZZ*, 122 (1901), No. 147 [29 May 1901], pp. 1–2.

Hijiya-Kirschnereit 1988
Hijiya-Kirschnereit, Irmela, *Das Ende der Exotik: Zur japanischen Kultur und Gesellschaft der Gegenwart*, Frankfurt am Main 1988.

Histoire de l'art du Japon 1900
Commission impériale du Japon à l'Exposition universelle de Paris (ed.), *Histoire de l'art du Japon*, Paris 1900.

H. J. K. 1933
H. J. K., "'Züga' die Zürcher Gartenbau-Ausstellung 1933: Architekten Karl Egender, Wilhelm Müller, Gartenarchitekt: Gustav Ammann, Zürich", in: *SBZ*, 101/102 (1933), No. 10, pp. 115–125.

Hoemann 1908
Hoemann, Reinhold, "Die Einfachheit in der Gartenkunst", in: *Die Gartenkunst*, 10 (1908), No. 9, pp. 147–154.

Hoemann 1913
"Wiener Ausstellung", in: *Die Gartenkunst*, 15 (1913), No. 14, pp. 206–208.

Hofmann 1954
[Hofmann], "Japanische Gärten", in: *Tages-Anzeiger*, 61 (1954), No. 192, [18 August 1954], sheet 4 (sgd. "hf.").

Holme 1893
Holme, Charles Geoffrey, "Artistic Gardens in Japan", in: *The Studio*, 1 (1893), No. 4, pp. 147–153.

Honda 1890–1926
Honda, Kinkichirō, *Zukai teien zōhō* (図解庭造法, Illustrierte Gartenbaumethoden), 団々社, Dandansha, Tōkyō 1890 [New edition: 団々社書店, Dandansha Shoten, Tōkyō 1896, 六合館, Rokugōkan, Tōkyō 1909, 六合館, Rokugōkan, Tōkyō 1926].

Hopstock 2015
Hopstock, Lars, *Building landscapes to live in: Hermann Mattern* (1902–1971), PhD thesis, 2 Vols., University of Sheffield 2015, Vol. 1.

Huber 1957
Huber, Benedikt, "Le Corbusier – Œuvre complète 1950–1957", in: *Werk*, 44 (1957), No. 10, pp. 197–198.

Humbert 1870
Humbert, Aimé, *Le Japon illustré*, 2 Vols., Paris 1870.

Hunziker 1980
Hunziker, Wolf, "The Planning of the 'Grün 80': a Joint Venture", in: *Anthos*, 19 (1980), No. 1, pp. 2–4.

Hunziker /Voss/Meier 2000
Hunziker, Wolf, Jürgen Voss, and Hans Meier, "Grün 80 – nach 20 Jahren", in: *Anthos*, 39 (2000), No. 1, pp. 14–19.

Imbert 2009
Imbert, Dorothee, *Between Garden and City. Jean Canneel Claes and Landscape Modernism*, Pittsburgh 2009.

Inoue/Watt 2014
Inoue Hiroshi and Paul B. Watt, "Japan's Ritsuryō System and Shintō Shrines Arose as Twins", in: *Journal of Religion in Japan*, 3 (2014), pp. 36–46.

Internationale Gartenbau-Ausstellung 1904
Internationale Kunst-, kunsthistorische und grosse Gartenbau-Ausstellung, Vertriebsstelle der amtlichen Drucksachen der Ausstellung Schmitz & Olbertz, Düsseldorf 1904.

Inventarergänzung 2013
Inventarergänzung – Bauten, Gärten und Anlagen 1960 bis 1980, published by Stadt Zürich, Hochbaudepartement, Amt für Städtebau, Zurich, 2013.

Ishimoto 1956
Ishimoto, Tatsuo, *The art of growing miniature trees, plants and landscapes: Japanese bonsai and bonkei adapted to western conditions*, Cambridge 1956.

Ishimoto 1958
The art of the japanese garden, New York 1958.

Ishimoto 1959
A treasury of japanese flower arrangements, New York 1959.

Ishimoto 1960
Japanische Blumenkunst, Munich 1960.

Ishimoto/Ishimoto 1963
Ishimoto, Tatsuo and Kiyoko Ishimoto, *The japanese house: its interior and exterior*, New York 1963.

Itoh/Futagawa 1963
Itoh, Teiji, and Yukio Futagawa, *Architecture japonaise: espaces, formes et matériaux*, Fribourg 1963.

Itoh 1972
Itoh, Teiji, *The Classic Tradition in Japanese Architecture – Modern Versions of the Sukiya Style*. New York/Tōkyō/Kyōto 1972.

Itoh 1973
Nihon No Toshikukan [Urbaner Raum in Japan], Tōkyō 1973.

Itoh 1985
Die Gärten Japans, Cologne 1985.

Itoh 1998
The Gardens of Japan, 2nd Edition, Tōkyō 1998.

Itoh 1999
Die Gärten Japans [translated from the American by Dieter Kuhaupt], Cologne 1999, [Original edition: Tōkyō 1984].

Jacquet 2013
Jacquet, Benoît, "La villa Katsura et ses jardins. L'invention d'une modernité japonaise dans les années 1930", in: Jacquet, Benoît and Nicolas Fiévé, *Vers une moder-nité architecturale et paysagère*, 2013 pp. 99–139, halshs-0232 8911, https:// halshs.a chives-ouvertes.fr/ halshs-02328911 (retrieved: 31 Jan 2023).

Jäger 1988
Jäger, Hermann, *Lehrbuch der Gartenkunst*: Jäger, Hermann, *Lehrbuch der Gartenkunst oder Lehre von der Anlage Ausschmückung und künstlerischen Unterhaltung der Gärten und freien Anlagen*, Berlin and Leipzig 1877.

Japanese Temples 1910
Japanese temples and their treasures, Architectural monuments from pre-buddhistic to Tokugawa period 1, [volume of plates], Tōkyō 1910.

Japanische Tuschmalerei 1962
Japanische Tuschmalerei Nanga und Haiga aus der Sammlung Heinz Brasch, published by Kunstgewerbemuseum Zürich, (Wegleitungen des Kunstgewerbemuseums Zürich), Vol. 248, Zurich 1962.

Jelllcoe 1964
Jellicoe, Geoffrey Alan, "Eindrücke beim IFLA-Kongreß in Tokio", in: *Garten und Landschaft* 74 (1964), pp. 340–341.

Joedicke 1987
Joedicke, Jürgen, "Weissenhof 1927–1987. Die Weissenhofsiedlung in Stuttgart – Rückblick und Ausblick", in: *WBW*, 74 (1987), No. 11, pp. 100–198.

Johnson/Hart 2019
Johnson, Mark Dean and Dakin Hart (eds.), *The Saburo Hasegawa Reader*, Oakland 2019.

Jost 2015
Jost, Kilian Matthias, *Felsenlandschaften – Eine Bauaufgabe des 19. Jahrhunderts Grotten, Wasserfälle und Felsen in landschaftlichen Gartenanlagen*, Diss., ETH Zurich 2015.

Jubiläums-Ausstellung Mannheim 1907
"Sondergarten von Heinrich Henkel, Darmstadt (D. E. 13.)", in: *Jubiläums-Ausstellung Mannheim 1907, Führer durch die Internationale Kunst- und grosse Gartenbau-Ausstellung*, published by der Ausstellungs-Leitung, Mannheim 1907, pp. 44–46.

Kaempfer 1712
Kaempfer, Engelberto, *Amoenitatum exoticarum, Fasciculus* V, Lemgoviae 1712.

Kaempfer 1727
The History of Japan, edited by Hans Sloane, translated by Johann Caspar Scheuchzer, 2 Vols., London 1727, Vol. 2, Buch V.

Kaempfer 1853
An Account of Japan by Engelbert Kaempfer, M. D. […], abridged and arranged from the translation by J. G. Scheuchzer, 2 Vols., Vol. 1, Part 2, London 1853.

Kaempfer 2013
Engelbert Kaempfers Geschichte und Beschreibung von Japan: Beiträge und Kommentar, published by Deutsche Gesellschaft für Natur- und Völkerkunde Ostasiens (OAG), Berlin 2013.

Kashikie 1964
Kashikie, Isamu, *The ABC of japanese gardening*, Tōkyō 1964.

Katahira 2010
Katahira Manabe, Miyuki, "The History of the Perception of Japanese Gardens: Analysis of how their Authenticity was defined through Discourse", in: *Kokusai ronbunshū*, 43 (2010), No. 12, pp. 111–141.

Keswick 1978
Keswick, Maggie, *The Chinese Garden – history, art and architecture*, London 1978.

Keswick 2004
Der chinesische Garten, 2nd expanded and revised edition, Stuttgart 2004.

Kiener 1968a
[Kiener, Alfred], "Terrain Design with Quarry- stones – Ernst Baumann, Garden Architect BSG Thalwil/Zurich", in: *Anthos*, 7 (1968), No. 1, pp. 1–3 (sgd. "Red.").

Kiener 1968b
"The editor's page", in: *Anthos*, 7 (1968), No. 1, pp. 21–23.

King 1984
King, F. H., *4000 Jahre Landbau in China, Korea und Japan*, Munich 1984.

Kleine/Kohn 1999
Kleine, Christoph and Livia Kohn, "Daoist Immortality and Buddhist Holiness: A Study and Translation of the Honcho shinsen-den", in: *Japanese Religions*, 24 (1999), pp. 119–196.

Kleine 2004
Kleine, Christoph, "Rebirth and Immortality, Paradise and Hell – Conflicting Views of the Afterlife in Ancient Japan", in: Formanek, Susanne and William R. LaFleur (eds.), *Practicing the Afterlife: Perspectives from Japan*, Vienna 2004, pp. 63–98.

Klemmer 1993
Klemmer, Clemens, "Meister der Moderne – Architektur als Medizin: Richard Neutra (1892–1970)", in: *WBW*, 80 (1993), No. 9, pp. 74–78.

Knippschild 2020
Knippschild, Judith, "*Da wurde der Wunsch zur Begierde". Über Japansehnsucht und Künstlerreisen im 19. und frühen 20. Jahrhundert*, Dissertation, Faculty of Philosophy, Ruprecht-Karls-Universität Heidelberg, 2020.

Koella 2008
Koella, Rudolf, "Bill vor Bill: Zum Frühwerk des Künstlers", in: *Max Bill. Aspekte seines Werks*, published by Kunstmuseum Winterthur and Gewerbemuseum Winterthur, Sulgen/Zurich 2008, pp. 26–39.

Kosaburō 1910
Kosaburō, Usui, *Hanochi rinsen chō.* 花洛林泉帖 *Glimpses of the Historical Gardens in Kyoto and its Outskirts*, 2 Bde., Vol. 1, Kyōto 1910.

Kreiner 1993
Kreiner, Josef, "Vom paradiesischen Zipangu zum zurückgebliebenen Schwellenland – das europäische Japanbild vom 16. bis zum 19. Jahrhundert", in: Berliner Festspiele GmbH (ed.), *Japan und Europa 1543–1929*, exhibition catalogue accompanying the exhibition at the Martin-Gropius-Bau Berlin, Berlin 1993, pp. 18–26.

Kuck 1940/2016
Kuck, Loraine E., *The Art of Japanese Gardens*, New York 1940; New edition: London/ New York 2016.

Kühn 2012
Kühn, Norbert, "Landschaftserleben, Pflanzenschönheit und Gartenkunst: Karl Foersters Steingarten als Weltentwurf" in: *Topiaria helvetica* (2012), pp. 19–27.

Kuitert 2002
Kuitert, Wybe, "Japonaiserie in London and the Hague: A History of the Japanese Gardens at Shepherd's Bush (1910) and Clingendael (c. 1915)", in: *Garden History*, 30 (2002), No. 2, pp. 221–238.

Kuitert 2010
"The Stone and the Moon, Jôju-in", in: *Shakkei, Journal of the Japanese garden society*, 16 (2010), No. 4, pp. 2–7.

Kuitert 2015
"Borrowing scenery and the landscape that lends – the final chapter of Yuanye", in: *Journal of Landscape Architecture*, 10 (2015), No. 2, 32–43, doi:10.1080 /18626033.2015.1058570 (retrieved: 12 Dec 2022).

Kume 2202
Kume, Kunitake, *Die Iwakura-Mission: das Logbuch des Kume Kunitake über den Besuch der japanischen Sondergesandtschaft in Deutschland, Österreich und der Schweiz im Jahre 1873*, translated and edited by Peter Pantzer, Munich 2002.

Kümmel 1911
Kümmel, Otto, *Das Kunstgewerbe in Japan*, Berlin 1911.

Lamprecht 2000
Lamprecht, Barbara Mac, *Richard Neutra – Complete Works*, Goessel, Peter (ed.), Cologne 2000.

Lancaster 1989
Lancaster, Roy, *Travels in China: A Plantsman's Paradise*, Woodbridge 1989.

Lasius 1881
Lasius, Georg, "Ueber Gartenbaukunst, VIII", in: *Zeitschrift des Schweizerischen Gartenbauvereins*, 1 (1881), No. 11, pp. 357–362.

Le Corbusier 1923
Le Corbusier, *Vers une architecture*, Paris 1923.

Le Corbusier 1926
Kommende Baukunst, Stuttgart 1926.

Le Corbusier 1928
Une maison – un palais, Paris 1928.

Lechner 1963
Lechner, Max, "Kathedrale in Tokio", in: *Werk*, 50 (1963), No. 8, pp. 175–177.

Lechner-Laur 1967
Lechner-Laur, Sophie, "Japanische Töpferei", in: *Werk*, 54 (1967), No. 6, pp. 346–349.

Leder 1947
Leder, Walter, "Das Lebendige im Garten", in: *Schweizer Garten*, 17 (1947), No. 1, pp. 1–8.

Leder/Mathys 1985
Leder, Klaus and Heini Mathys, "Abschied von Walter Leder", in: *Anthos*, 24 (1985), No. 1, pp. 44–46.

Lehmann 1984
Lehmann, Jean-Pierre, "Old and New Japonisme: The Tokugawa Legacy and Modern. European Images of Japan", in: *Modern Asian Studies*, 18 (1984), No. 4, pp. 757–768.

Leitl 1942
Leitl, Alfons, and Otto Valentien, *Ein Buch vom Schönen Garten*, Gütersloh 1942.

Lerch 2006
Lerch, Gabi, "Evariste Mertens und die bürgerliche Konfektionierung des Landschaftsgartens", in: Burbulla, Julia et al. (ed.), *Stadtlandschaften: Schweizer Gartenkunst im Zeitalter der Industrialisierung*, Institut für Geschichte und Theorie der Landschaftsarchitektur GTLA/ HSR, Rapperswil, Zurich 2006, pp. 66–75.

Lerch 2009
"'Un métier, un art une passion': Porträt einer eigenwilligen, unprätentiösen und verantwortungsbewussten Gestalterin", in: *Anthos*, 48 (2009), No. 3, pp. 54–57.

Lerch 2012
ASLA-Nachlässe-MERTENS Evariste-Lebens-lauf-Lerch 2012, www.asla.ch › Mertens_Evariste › Nachlass_ Mertens_Evariste_ 170530 (retrieved: 27. March 2023)

Leuzinger 1967
Leuzinger, Elsy, "Orientalia helvetica. Neues aus dem Museum Rietberg in Zürich", in: *Asiatische Studien*, 21 (1967), No. 1–4, pp. 21–23.

Liä Dsi 1921
Liä Dsi, *Das wahre Buch vom quellenden Urgrund, Tschung Hü Dschen Ging, Die Lehren der Philosophen Liä Yü Kou und Yang Dschu*. translated by Richard Wilhelm, Jena 1921.

Lods 1947
Lods, Marcel, "Visite à Neutra", in: *Werk*, 34 (1947), No. 10, pp. 314–[322].

Loewenstein 1922
Loewenstein, Fritz E., *Die Handzeichnungen der Japanischen Holzschnittmeister*, Plauen im Vogtland 1922.

Lotz 1932
Lotz, Wilhelm, "Sonne, Luft und Haus für Alle!", in: *Die Form*, 7 (1932), No. 6, pp. 179–188.

Luchsinger 1997
Luchsinger, Christoph, "Szenische Architektur: zur Architektur von Hans Hofmann (1897–1957)", in: *WBW*, 84 (1997), No. 4, pp. 56–57.

Lüpke 1980
Lüpke, Jens, "Sektor 'Säen und ernten'", in: *Anthos*, 19 (1980), No. 1, pp. 66–72.

Lüthardt 1955
Lüthardt, Gertrud, "Cécile von Rodt – eine originelle Bernerin", in: *Der Bund*, 106 (1955), No. 531 [13 Nov 1955], n. pag., [p. 19].

Ma. 1942
Ma., "Ausstellung im Gewerbemuseum Basel. Neuerwerbungen der Sammlung", in: *Das Werk*, 29 (1942), No. 10, pp. XX, XXII.

Mappes 1936
Mappes, Michael, "Die Kunst des landschaftlichen Gartens in Japan", in: *Die Gartenkunst*, 49 (1936), No. 2, pp. 28–32.

Martin 1926
Martin, Henry, *L'art japonais*, Paris 1926.

Mathys 1959a
Mathys, Heini, "G/59 – Eine neue Gartenwelt am Zürichsee", in: *Schweizer Garten + Wohnkultur*, 29 (1959), No. 5, pp. 107–117.

Mathys 1959b
"Gedanken über einen Liebesgarten", in: *Schweizer Garten + Wohnkultur*, 29 (1959), No. 9, pp. 197–198.

Mathys 1959c
"Wege und Stege an der G/59", in: *Schweizer Garten + Wohnkultur*, 29 (1959), No. 9, pp. 199–201.

Mathys 1960
"Landsitz in Feldmeilen", in: *Schweizer Garten + Wohnkultur*, 30 (1960), No. 2, pp. 21–28.

Mathys 1972
"Swiss Garden Architecture in the 20th Century: A Review and Appraisal of Situation", in: *Anthos*, 11 (1972), No. 3, pp. 1–19.

Mathys 1978
"Ernst Cramer at 80", in: *Anthos*, 17 (1978), No. 4, pp. 32–33.

Mathys 1982
"Der Naturstein in der Gartengestaltung. Das Beispiel der Gärten von Ernst Baumann, BSG Thalwil", in: *Anthos*, 21 (1982), No. 2, pp. 23–36.

Mathys 1983
"Der Chinesische Garten an der IGA 83", in: *Der Gärtnermeister*, 86 (1983), No. 28, pp. 667–668 (sgd. "H. M.").

Mathys 1991
"Verena Dubach-Anderesen, Landschaftsarchitektin BSLA, Kolding/DK, zum 65. Geburtstag", in: *Anthos*, 30 (1991), No. 4, p. 55.

Mathys 1992
"Ernst Baumann, Oberrieden/ Thalwail, zum Gedenken", in: *Anthos* 31 (1992), No. 3, p. 50 (sgd. "HM").

Mattern 1953
Mattern, Hermann, "Das Wohnen und Gärtnern in Japan", in: *Pflanze und Garten* 3 (1953), No. 6, pp. 9–12.

Mattern/Mattern 1970a
Mattern, Hermann and Beate Mattern, "Aus den Gärten unserer Nachbarn: Japan. Gärten, die nicht jeder sieht", *Grün*, 1 (1970), No. 9, pp. 436–439.

Mattern/Mattern 1970b
"Aus den Gärten unserer Nachbarn: Japan II. Die Tradition lebt", *Grün*, 1 (1970), No. 10, pp. 488–493.

Matteini/Porcinai 1991
Matteini, Milena, Pietro Porcinai and Alessandro Giannini, *Pietro Porcinai – architetto del giardino e del paesaggio*, Milan 1991.

Maurer 2013
Maurer, Bruno, "American way of life am Zürichsee. Die Villa Staehelin in Feldmeilen", in: *Herrliberger Kalender* (2013), pp. 2–9.

Mawson 1926
Mawson, Thomas, *The Art and Craft of Garden Making*, [1st ed. 1900], 5th ed., London 1926.

Meili 1957
Meili, Armin, "Reiseskizzen aus Japan", in: *NZZ*, 178 (1957), No. 2158 [27 July 1957], sheets 3–4.

Mertens E 1881a
Mertens, E. [Evariste], "Aesthetik im Garten. Teil I", in: *Zeitschrift des Schweizerischen Gartenbauvereins*, 1 (1881), No. 2, pp. 39–40 (sgd. "Mertens, E.").

Mertens E 1881b
"Aesthetik im Garten. Teil II", in: *Zeitschrift des Schweizerischen Gartenbauvereins*, 1 (1881), No. 3, pp. 88–91.

Mertens 1881c
"Aesthetik im Garten. Teil V, Wasser- und Felsenanlagen", in: *Zeitschrift des Schweizerischen Gartenbauvereins*, 1 (1881), No. 6, pp. 201–205.

Mertens O 1916
Mertens, Oskar, "Über Gartenkunst" in: *Das Werk*, 3 (1916), No. 5, pp. 65–80 (sgd. "Mertens, O.").

Mertens O 1933
"Moderne Probleme des Hausgartens", in: *NZZ*, 154 (1933), No. 1145, [24 June 1933], *Zürcher Gartenbau-Ausstellung*, p. 13.

Mertens O 1951
"Das Friedhofproblem", in: *Werk*, 38 (1951), No. 5, pp. 145–148.

Mertens W 1913
Mertens, Walter, "Ein Ausstellungsgarten aus Breslau", in: *Die Gartenkunst*, 15 (1913), No. 16, pp. 239–241.

Mertens W 1934
"Ein Garten ohne Haus", in: *SBZ*, 103/104 (1934), No. 10, pp. 120–121.

Mertens W 1939
"Die Sondergärten" in: *NZZ*, 169 (1939) [16 July 1939], No. 1293, sheet 4.

Mertens W 1943
"Zum Gartenhof im Kongresshaus", in: *SBZ*, 121/122 (1943), No. 24, pp. 295–296.

Meyer 1926
Meyer, Peter, "Das japanische Haus. Zum Vortrag von Prof. D. A. Grosse", in: *SBZ*, 87 (1926), No. 20, pp. 258–260.

Meyer 1932a
"Wohnhaus Professor O. R. Salvisberg, Zürich, Restelbergstrasse", in: *Das Werk*, 19 (1932), No. 8, pp. 244–251 (sgd. "P. M.").

Meyer 1932b
"Eigenheim des Architekten Prof. O.R. Salvisberg, Zürich", in: SBZ, 99/100 (1932), No. 13, pp. 161–169 (sgd. "P. M.").

Meyer 1933
"Zürcher Gartenbauausstellung ZÜGA 24. Juni bis 17. September 1933", in: *Das Werk*, 20 (1933), No. 7, pp. 193–212.

Meyer 1936a
"Japanische Architekturzeitschriften", in: *Das Werk*, 23 (1936), No. 3, p. XXII.

Meyer 1936b
"Das japanische Wohnhaus (zum Buch von Tetsuro Yoshida)", in: *Das Werk*, 23 (1936), No. 3, pp. 78–84.

Meyer 1937a
"Ornamentfragen: umständliche Kunst, Ornament und Architektur" in: *Das Werk*, 24 (1937), No. 2, pp. 53–59.

Meyer 1937b
"Umfang und Verdienste des 'Jugendstils'", in: *Das Werk*, 24 (1937), No. 5, pp. 134–140.

Meyer 1939a
"Die Gärten der Schweizerischen Landesausstellung 1939", in: *Das Werk*, 26 (1939) No. 7, pp. 193–224.

Meyer 1939b
"Zur Erinnerung an die Gartenpracht der Schweizerischen Landesausstellung 1939 in Zürich. Garten, Landschaft und Architektur", in: *SBZ*, 113/114 (1939), No. 18, pp. 203–211.

Meyer 1939c
"Gärten und Bauten der LA" in: *NZZ*, 169 (1939), No. 1293, [16 July 1939], sheet 4.

Meyer 1940
"Situation der Architektur 1940", in: *Das Werk*, 27 (1940), No. 9. pp. 241–251.

Meyer 1943
"Das Kongresshaus in Zürich als Beispiel moderner Architektur", in: *SBZ*, 121/122 (1943), No. 26, pp. 310–313.

Meyer-Benteli 1944
Meyer- Benteli, Hans, "Konkrete Kunst", in: *Das Werk*, 31 (1944), p. 128–[p. 133].

Migge 1910
Migge, Leberecht, "Garten-Naturalismus", in: *Die Gartenkunst*, 12 (1910), No. 11, pp. 194–196.

Migge 1927a
"Die gute Garten- Wohnung", in: *Die Gartenschönheit*, 8 (1927), No. 1, pp. 10–13.

Migge 1927b
"Der kommende Garten", in: *Die Gartenschönheit*, 8 (1927), No. 3, pp. 64–65.

Migge 1932
Die wachsende Siedlung nach biologischen Gesetzen, Stuttgart 1932.

Migge 1936
"Brief eines Gartenarchitekten" in: *NZZ*, 157 (1936), No. 446 [15.3.1936], sheet 10.

Miyoshi/Makino 1907
Miyoshi, Manabu and Tomitaro Makino, *Pocket-atlas of Alpine plants of Japan*, 2 Vols., Tōkyō 1907.

Moll 2006
Moll, Claudia, "Ein persönliches Stück Gartengeschichte. Der Garten der Villa Schuler in Glarus", in: Sigel, Brigitt et al. (ed.), *Nutzen und Zierde*, Zurich 2006, pp. 258–263.

Moos von 2004
Moos, Stanislaus von, *Nicht Disneyland. Über Modernität und Nostalgie in der Schweiz*, Zurich 2004.

Morel 1987
Morel, Martine. "Reconstruire, dirent-ils. Discours et doctrines de l'urbanisme", in: *Les Cahiers de l'Institut d'Histoire du Temps Présent*,, 3 (1987), No. 5 [June], *Images, discours et enjeux de la reconstruction des villes françaises après 1945*, pp. 13–49.

Mori 1945
森蘊, Mori, Osamu, 平安時代庭園の研究, *Heian jidai teien no kenkyū* [Study on gardens of the Heian era], Tōkyō 1945.

Mori 1955
桂離宮の研究, *Katsura rikyū no kenkyū* [Study on the Villa Katsura], Tōkyō 1955.

Mori 1962
Typical Japanese Gardens, Tōkyō 1962.

Mori 1986
作庭記の世界, *Sakuteiki no sekai* [The universe of Sakuteiki], Tōkyō 1986.

Mori 1988
庭園, *Teien* [Gardens], Tōkyō 1988.

Moriya 1953
Moriya, Kenji, *Die japanische Malerei*, Wiesbaden 1953.

Morse 1972 (1885)
Morse, Edward Sylvester, *Japanese Homes and their Surroundings*, new edition, North Clarendon 1972 (First edition New York 1885).

Mottini 2014
Mottini, Roger, "Nicht immer Sonnenschein – 150 Jahre Beziehungen zwischen der Schweiz und Japan", in: *oag Notizen* (2014), No. 9, pp. 30–49.

M-r 1960
M-r, "Einflüsse japanischer Gartenkultur", in: *Das ideale Heim*, 34 (1960), No. 9, pp. 371–378.

Müller-Beeck 1926
Müller-Beeck, Fernando George, "Die japanische Gartenkunst unter dem Einfluss der Lehren des Buddhismus und Shintoismus", in: Buschan Georg (ed.), *Der Erdball. Illustrierte Zeitschrift für Menschen-, Länder- und Völkerkunde*, 1 (1926), No. 2, pp. 61–64. https://www.digi-hub.de/viewer/!toc/DE-11-00212 1697/1/-/ (retrieved: 21 Jan 2022).

Münsterberg 1904–1907
Münsterberg, Oskar, *Japanische Kunstgeschichte*, 3 Vols., Braunschweig 1904–1907.

Murray/Chamberlain/Mason 1891
Murray, John, Basil Hall Chamberlain and W. B. Mason, *A Handbook for Travellers in Japan*, London 1891.

Muthesius 1903
Muthesius Hermann, "Das japanische Haus", in: *Zentralblatt der Bauverwaltung*, 23 (1903), No. 49, pp. 306–307.

Muthesius 1907
Landhaus und Garten. Beispiele neuzeitlicher Landhäuser nebst Grundrissen, Innenräumen und Gärten, Munich 1907.

Näf 1946
Näf, Hans, "Alexander Calder, Ballettmeister der Materie", in: *Die Weltwoche*, 14 (1946), No. 681 [29 Nov.], p. 5.

Nakajima 1973
Nakajima, Ken, "Ein japanischer Garten in fremder Umgebung", in: *Garten und Landschaft*, 83 (1973), No. 8, pp. 397–398.

Nakamura M 1971
中村昌生, Nakamura, Masao, 茶室の研究 [A study on tea pavilions], 墨水書房, Bokusui Shobo, Tōkyō 1971.

Nakamura N 1958
Nakamura, Noriko, *Internationales Ikebana: die japanische Kunst des Blumenstellens*, Zurich 1958.

Natsumi 2008
Natsumi, Nonaka, "The Japanese Garden: The Art of Setting Stones", in: *site LINES, A Journal of Place, A Publication of the Foundation for Landscape Studies*, 4 (2008), No. 1, pp. 5–8.

Nedden 1964
Nedden zur, Beate, "Zwischen Tradition und Moderne", in: *Garten und Landschaft*, 74 (1964), No. 10, pp. 415–417.

Nerfin 2016
Nerfin, Pauline, "Jardin alpin, dites-vous?", in: *Kunst + Architektur in der Schweiz*, 67 (2016), No. 3, pp. 54–65.

Neukom 1959
Neukom Willi, "Wassergarten", in: Gollwitzer, Gerda, "Gärten am See. Zur 1. Schweizerischen Gartenbauausstellung Zurich 1959", in: *Garten und Landschaft*, 69 (1959), No. 8, pp. 225–235.

Neutra 1962
Neutra, Richard, *Auftrag für morgen*, Hamburg 1962.

Nishikawa 1936
Nishikawa, Issotei, *Floral Art of Japan*, (Tourist Library 1), Tōkyō 1936.

Nitschke 1993
Nitschke, Günter, *Japanische Gärten – rechter Winkel und natürliche Form*, Cologne 1993.

Noguchi 1964
Noguchi, Isamu, "Die Wirkung der Natur", in: *Garten und Landschaft*, 74 (1964), No. 10, pp. 318–319.

Nute 2004
Nute, Kevin, *Place, Time and Being in Japanese Architecture*, London/New York 2004.

Nute 2019
"Crossing Boundaries: the Transfer of Spatial Ideas across Culture and Medium in the Work of Two Modern American Architects", in: *The International Journal of Interdisciplinary Cultural Studies*, 14 (2019), No. 2, pp. 13–26.

Nyffenegger 2004
Nyffenegger, Brigitte, "Gestalter mit Sinn für Qualität: erste Annäherung an den Landschaftsarchitekten Willi Neukom", in: *Heimatschutz*, 99 (2004), No. 1, pp. 26–27.

Nyffenegger/Gonzalez/Sanchez 2016
Nyffenegger, Brigitte, Michaela Gonzalez, and Miguel Sanchez, *Landgut Tannhof, Ebertswil – Gartendenkmalpflegerisches Gutachten*, Zurich 2016.

Ortschy D' 1964
Ortschy D', Brigitte, "Wechselspiel: Japan und das Abendland", in: *Garten und Landschaft*, 74 (1964), No. 10, p. 316.

Osoegawa 2011
Osoegawa, Steffen, "Fernes Vorbild. Der Japangarten der Gartenbauschule Oeschberg", in: *Topiaria helvetica* (2011), pp. 73–76.

Osoegawa 2020a
'Der Gartenhistoriker Albert Baumann", in: *dergartenbau*, 141 (2020), No. 7, pp. 16–18.

Osoegawa 2020b
"Aus dem Leben von Albert Baumann", in: *dergartenbau*, 141 (2020), No. 21, pp. 22–24.

Osoegawa 2020c
"Der Gartenarchitekt Baumann", in: *dergartenbau*, 141 (2020), No. 22, pp. 22–24.

Ozawa 1889–1905
Ozawa, Keijirō, 小澤圭次郎, "En. en genryū-kō" "苑源流考" ["Consider ations on the sources of parks and gardens"], in: *Kokka* 國華, (1889), No. 5 – (1905), No. 181.

Ozawa 1894
"En.en genryū-kō", "園苑源流考" ["Considerations on the sources of parks and gardens"], in: *Kokka* 國華, (1894), No. 58 [Juli], pp. 195–196, No. 59 [August], pp. 210–213, No. 60 [September], pp. 230–232, No. 61 [October], p. 246.

p. 1945
p., "Wandlungen in der Gartengestaltung", in: *National-Zeitung*, 103 (1945), No. 542 [22 Nov 1945], n. pag., Gustav Ammann Documents IV, 1942–50, p. 43, NSL 2-S-8-4].

Papini 2021
Papini, Massimiliano, *Transcultural flows from Japan to the North East of England 1862–1913: visual and material culture in relation to the Anglo-Japanese interaction*, PhD thesis, Northumbria University, 2021 https://nrl.northumbria.ac.uk/id/eprint/50243/ (retrieved: 15 July 2023).

Pauly 2014
Pauly, Ulrich, "Die Jenseitsvorstellungen der Japaner", in: *OAG Notizen* (Deutsche Gesellschaft für Natur- und Völkerkunde Ostasiens), 05/2014, pp. 11–34. https://oag.jp/img/images/publications/oag_notizen/Feature_Pauly.pdf (retrieved: 9 Jan 2023).

Pekar 2000
Pekar, Thomas, "Der Japan-Diskurs um 1900. Ein Skizzierungsversuch", in: Gebhardt, Walter (ed.), *Ostasienrezeption zwischen Klischee und Innovation*, Munich 2000, pp. 227–254.

Porcinai/Mordini 1966
Porcinai, Pietro and Attilio Mordini, "Giardini del Giappone", in: *Giardini d'Occidente e d'Oriente*, Milan 1966.

Probst 1969/2012
Probst, R[aymond], "Bericht des Delegierten des Bundesrats für Handelsverträge. Reise nach Japan und Hong Kong. April 1967. Excerpt [No. 25, 31 May 1967]", in: *Diplomatische Dokumente der Schweiz*, 1.1.1967–31.12.1969, Vol. 24, Zurich 2012.

Qian 1982
Qian, Yun, *Classical Chinese Gardens*, Beijing/Hongkong 1982.

Red. 1982
Red., "Gärten und Gemälde von Otto Valentien", in: *Garten + Landschaft*, 34 (1982), No. 10, pp. 760–767.

Redesdale 1917
Redesdale, Lord [Freeman-Mitford, Algernon Bertram], Edmund Gosse, *Further Memories*, Hutchinson & Co, London 1917.

Reichert 1993
Reichert, Folker E., "Zipangu – Japans Entdeckung im Mittelalter", in: Croissant, Doris und Lothar Lederrose (eds.), *Japan und Europa 1543–1929*,.- Exhibition catalogue Martin-Gropius-Bau Berlin, Berlin 1993, pp. 25–36.

Reischauer 1955
Reischauer, Edwin O., *Ennin's Travels in T'ang China*, New York, 1955.

Remmele/Vegesack 2003
Remmele, Mathias, Alexander von Vegesack, (eds.), *Marcel Breuer – Design und Architektur*, exhibition catalogue. Vitra Design Museum, Weil am Rhein 2003, pp. 243–246.

Risler 1910
Risler, Georges, "Les citésjardins", Dec 1909, in: *Musée Social, Mémoires et Documents*, Paris 1910.

Robinson 1870
Robinson, William, *The Wild Garden*, London 1870.

Rodt 1903
Rodt, Caecilie von, *Reise einer Schweizerin um die Welt*, Verlag F. Zahn, Neuenburg 1903. Frz. Rodt, *Cécile de, Voyage d'une Suissesse autour du monde*. Préface de Mr A. Gobat, Neuchâtel 1904.

Roelly 2007
Roelly, Thom, "In Memoriam Fredy Klauser", in: *Anthos*, 46 (2007), No. 2, p. 60.

Rohrer 2009
Rohrer, Judith, "G|59 – ein herausforderndes Erbe", in: *Anthos*, 48 (2009), No. 2, *G|59 und 50 Jahre danach*, pp. 12–17.

Roth A 1927
Roth, Alfred, *Zwei Wohnhäuser von Le Corbusier und Pierre Jeanneret. Fünf Punkte zu einer neuen Architektur von Le Corbusier und Pierre Jeanneret*, Stuttgart 1927.

Roth A 1938
"Raum-Bild-Farbe in der heutigen Architektur", in: *SBZ*, 111/112 (1938), No. 2, pp. 18–20.

Roth G 1977
Roth, G., "Zum Hinschied von Gartenarchitekt Oskar Mertens, Zürich", in: *Anthos*, 16 (1977), No. 1, p. 39.

Röthlisberger 1917
Röthlisberger, Hermann, "Von schweizerischen Gärten", in: *Die Gartenkunst*, 19 (1917), No. 6, pp. 81–87.

Rotzler 1950
Rotzler, Willy, "Wohnbau und Grünfläche", in: *Werk*, 37 (1950), No. 3, pp. 65–72.

Rotzler 1955
"Baukunst in Japan", in: *NZZ*, 176 (1955), [17 Dec 1955], *Architektur*, n. pag. [2 pages]

Rotzler/Barth 1987a
Rotzler, Stefan and Hans Jakob Barth, "Gardens like bunches of flowers" (Excerpt from a recorded conversation with S. Rotzler on the 26 May 1984), in: *Anthos*, 26 (1987), No. 2, pp. 5–9.

Rotzler/Barth 1987b
"Restlessness and self-criticism" (Excerpt from a recorded conversation with S. Rotzler on the 26 May 1984), in: *Anthos*, 26 (1987), No. 2, p. 12.

Roy 1983
Roy, Louis Guillaume le, *Natuur uitschakelen – natuur inschakelen*, Uitgeverij Ankh-Hermes, Deventer 1973 (German title: *Natur einschalten, Natur ausschalten*, Stuttgart 1983).

Rüegg 2015
Rüegg, Jonas, "Aimé Humbert. Wertvorstellungen eines Bourgeois und das Japan der Bakumatsu-Zeit", in: *Asiatische Studien: Zeitschrift der Schweizerischen Asiengesellschaft*, 69 (2015), No. 1–2, pp. 47–71.

Ruoff 1993
Ruoff, Eeva, "Evariste Mertens, ein bedeutender Gartenarchitekt", in: *Unsere Kunstdenkmäler. Mitteilungsblatt für die Mitglieder der Gesellschaft für Schweizerische Kunstgeschichte*, 44 (1993), No. 1, pp. 87–97.

Ruoff n.d.
"Leopold Karl Theodor Fröbel – Kunstgärtner und Pflanzenzüchter", http://www.religio.de/froebel/genealo/garten.html (retrieved: 7 Dec 2022)

Rüttermann 1999
Rüttermann, Markus, "Pflege und Kritik der 'Tradierungen' (denju) – Zum Verhältnis zwischen Tradition und Strukturwandel der Öffentlichkeit im Japan der frühen Neuzeit", in: *NOAG*, 73 (1999), No. 165–166, pp. 45–144.

Saitō 1923
Saitō Katsuo, "Ryoanji no ishigumi 'toranoko watashi' no kaibo", in: *Teien*, 5 (1923), No. 8.

Saitō 1964
Magic Trees and Stones: Secrets of Japanese Gardening, (engl.), New York/Tōkyō 1964.

Saitō 1966
Zukai Sakuteiki, The Illustratet Sakuteiki, (jap.), Tōkyō 1966.

Saitō 1969
Japanese Gardening Hints: the romance of gleaming sand, rugged stones and shady trees in your garden, (engl.) Tōkyō 1969.

Satō 1933
佐藤昌, 外國人の見たる庭園, 園芸学会雑誌, 園芸学会, 第4巻第1号, 1933年4月, 88–106ページ, Satō, Akira, "Gardens seen by Foreigners", *Journal of the Japanese Society for Horticultural Science*, 4 (1933), No. 1 [25 April 1933], pp. 88–106.

Satō 1979
"Der japanische Garten", in: *Der Gartenbau*, 100 (1979), No. 27, pp. 1192–1196.

Satō 1991
"Satō, Akira", in: Matteini, Milena, *Pietro Porcinai and Alessandro Giannini, Pietro Porcinai – architetto del giardino e del paesaggio*, Milan 1991.

Schaarschmidt-Richter 1977
Schaarschmidt-Richter, Irmtraud, *Japanische Gärten*, Baden-Baden 1977.

Schaarschmidt-Richter/Mori 1979
Schaarschmidt-Richter, Irmtraud, and Osamu Mori, *Der japanische Garten – ein Kunstwerk*, Fribourg 1979 [Le jardin japonais, Fribourg 1979, Japanese Gardens, New York, 1979.]

Schelling von 1859
Schelling, Friedrich Wilhelm Joseph von, Philosophie der Kunst, in: *Sämtliche Werke*, I. Abt., Vol. 5, Stuttgart 1859.

Schlappner 1981
Schlappner, Martin, "Das Krematorium Nordheim in Zürich: Architekt: A.H. Steiner, Zürich", in: *Schweizer Ingenieur und Architekt*, 99 (1981), No. 45, pp. 999–1005.

Schmid 1947
Schmid, Willy, "Der Gartenbau an der 'Züka'", in: *NZZ* 168 (1947), No. 1907 [30. September 1947], sheet 7.

Schmocker-Willi 2002
Schmocker-Willi, Ursula, "Verena Dubach", in: *Anthos*, 41 (2002), No. 2, pp. 57–58.

Schneider 1910a
Schneider, Camillo Carl, "Über japanische Gartenkunst", in: *Die Gartenkunst*, 12 (1910), No. 1, pp. 1–8.

Schneider 1910b
"Über japanische Gartenkunst", in: *Die Gartenkunst*, 12 (1910), No. 2, pp. 19–22.

Schneider 1924
"Japanische Gartengestaltung", in: *Die Gartenschönheit*, 5 (1924), No. 7, pp. 129–136.

Schneider/Lüthi 2007
Schneider, Jürg and Barbara Lüthi, "Carl Passavant (1854–1887): eine Welt in Bildern", in: *Traverse*, 14 (2007), No. 3, pp. 113–122.

Schnieper 1980
Schnieper Walter, "'Grüne Zimmer' im urbanen Raum", in: *Luzerner Neuste Nachrichten*, 91 (1980), No. 16 (21/22 April), Magazine, p. 5.

Schöppler 2020
Schöppler, Luke, *The Meiji Legacy: Gardens and Parks of Japan and Britain*, 1850–1914, PhD thesis, University of Derby 2020. http://hdl.handle.net/10545/625069 (retrieved: 3 Jan 2022).

Schubert B 2001
Schubert, Bernd, "Zum Tod von Heini Mathys", in: *Anthos*, 40 (2001), No. 1, p. 74.

Schubert B 2009
"Zum 90. Geburtstag von Hans Graf", in: *Anthos*, 48 (2009), No. 3, p. 63.

Schubert M 1959
Schubert, Margot, "An beiden Ufern des Zürich-Sees … – Notizen zur 1. Schweizerischen Gartenbau-Ausstellung", in: *Pflanze und Garten*, 9 (1959), No. 8, pp. 152–153.

Schulze 1907
Schulze, Otto, "Von der Mannheimer Gartenbauausstellung 1907. II. Der Sonderkarten von Fr. Henkel, Darmstadt.", in: *Die Gartenkunst*, 9 (1907), Nr. 11, pp. 215–221.

Schwahn 2012
Schwahn, Michael, "Swensson (Svensson), Carl Gustav (1861–1910) Gartenarchitekt", in: *Österreichisches Biographisches Lexikon (ÖBL) 1815–1950*, published by Österreichische Akademie der Wissenschaften, 2003–2022, Vol. 14 (Lfg. 63), Vienna 2012, p. 78f. https://www.biographien.ac.at/oebl/oebl_S_43/Swensson_Carl-Gustav_1861_1910.xml (retrieved: 3 June 2023).

Schwarz 1957
Schwarz, Margot, "Der Japanische Garten", in: *NZZ*, 78 (1957), No. 3704, [14 Dec 1957], sheet 3.

Schwarz 1961
Armes reiches Ostasien. Reiseskizzen aus Bangkok, Tokio und Hongkong, Gute Schriften, Basel 1961.

Schwarz/Graf/Mathys 1974
Schwarz, Urs, Hans Graf, and Heini Mathys, "Up for discussion: can gardens assume a function in nature conservation?", in: *Anthos*, 13 (1974), No. 2, pp. 29–32.

Schwerin von 2014
Schwerin von, Sophie, "Ansichtssache – Emil Steiner", in: *Anthos*, 43 (2014), No. 3, pp. 70–73.

Schwerin von 2015
"Ansichtssache – Christian Stern", in: *Anthos*, 54 (2015), No. 1, pp. 70–73.

sda 1978
(sda) [Schweizerische Depeschenagentur], "Spross Gartenbau übernimmt Mertens und Nussbaumer", in: *NZZ* 199 [12 Jan 1978], No. 9, p. 46.

Seike/Kudō/Schmidt 1986
Seike, Kiyoshi, Masanobu Kudō and Walter Schmidt, *Japanische Gärten und Gartenteile*, Stuttgart 1983.

Senda 1997
Senda, Minoru, "Ideal Spaces beyond the Sea" ("海の向こうの理想的な場所"), in: Toru, Haga (芳賀徹) (ed.), *Ideal Places in History: East and West* (歴史における理想郷—東と西), International Research Center for Japanese Studies (国際日本文化研究センター), 10 (1997), engl. pp. 37–44, (jap. pp. 45–48).

Sfakiotaki
Sfakiotaki, Despina, *Analysis of Movement in Sequential Space. Perceiving The Traditional Japanese Tea And Stroll Garden*, Diss. Faculty of Technology, University of Oulu, Oulu 2005.

Shiga/Hashimoto 1935
Shiga, Naoya, and Motoi Hashimoto (eds.) Gardens of Japan, *A Pictorial Record of the Famous Palaces, Gardens and Tea-Gardens with Explanations in English* [Explanations by Matsunosuke Tatsui], Tōkyō 1935.

Shigemori 1936–1939
重森三玲, Shigemori, Mirei, 日本庭園史圖鑑, *Nihon teienshi zukan* [Japanisches Gartengeschichtsbuch], 24 Bde., 有光社, Yūkōsha, 東京 Tōkyō 1936–1939.

Shigemori 1971
– 日本庭園史大系, *Nihon teienshi taikei*, [History of Japanese Gardens], 35 Bde., 社会思想社, Shakai Shisōsha, 東京 Tōkyō 1971.

Shikibu 1966
Shikibu, Murasaki, *Die Geschichte vom Prinzen Genji*, translated by Oscar Benl, complete, unabridged edition, 2 Vols., Zurich 1966.

Shiota 1916
Shiota, Takeo, *Japanese Gardens and Houses*, New York 1916.

Shōnagon 2017
Shōnagon, Sei, *Kopfkissenbuch – Makura no Sōshi*, first complete [German] translation, reissued by Michael Stein, Zurich/Munich 2017. Shōnagon, Sei, *The Pillow Book*, English translation with notes by Meredith McKinney, London 2006 [Title of Original: *Makura no Sōshi*].

Siebold/Zuccarini 1835
Siebold, Philipp Franz von, Joseph Gerhard Zuccarini, *Flora Japonica sive Plantae quas in imperio Japanico coll., descripsit, ex parte in ipsis locis pingendas curavit: [Lugduni Batavorum apud auctorem 1835–1870]*, Leiden 1835.

Sierts 1979
Sierts, Hans, "Projekt-Wettbewerb Pestalozzipark", in: *Anthos*, 18 (1979), No. 3, pp. 30–34.

Sigel/Jong de 2010
Sigel, Brigitt, and Erik A. de Jong, *Der Seeuferweg in Zürich. Eine Spazierlandschaft der Moderne von 1963*, Scheidegger & Spiess, Zurich 2010.

Slawson 1991
Slawson, David A., *Secret Teachings in the Art of Japanese Gardens: Design Principles, Aesthetic Values*, Kodansha International Ltd., Tōkyō 1991.

Smits 2022
Smits, Ivo, "Riverside Mansion Mythologies. Retextualising the Past in Poetic Commentary", in: Negri, Carolina, Pier Carlo Tommasi (eds.), *Images from the Past: Intertextuality in Japanese Premodern Literature*, Ca' Foscari Japanese Studies 19, Arts and Literature 6, 2022, pp. 3–16.

Sommer 1977
Sommer, Helge, "Terrace Garden at Küsnacht", in: *Anthos*, 16 (1977), No. 1, pp. 17–19.

Speck 1943
Speck, Paul, "Zum Gartenhof im Kongresshaus", in: *SBZ*, 121/122 (1943), No. 24, p. 295.

Speidel 2007
Speidel, Manfred, "Stolze Leere. Die langsame Entdeckung der Architektur Japans durch die Moderne", in: *Japan und der Westen. Die erfüllte Leere*, published by Kunstmuseum Wolfsburg, Cologne 2007, pp. 253–264.

Speidel 2008
"Träume vom Anderen. Japanische Architektur mit europäischen Augen gesehen. Einige Aspekte zur Rezeption zwischen 1900 und 1950", in: *archimaera* 1 (Januar 2008), Special issue *FremdSehen*, pp. 79–96, https://www.archimaera.de/2007/fremd sehen/1198/archimaera001Speidel.pdf (retrieved: 21 Jan 2022).

Spezialinventar 2006
Reformierte Kirchen der Stadt Zürich – Spezialinventar, published by Hochbaudepartement der Stadt Zürich, Amt für Städtebau, Archäologie und Denkmalpflege, September 2006, pp. 92–95.

Spezialinventar 2008
Schulhäuser der Stadt Zürich – Spezialinventar, published by Amt für Städtebau der Stadt Zürich, Archäologie und Denkmalpflege, September 2008, p. 223.

St. 1979
St., "Der japanische Garten in Bonn, ein östliches Geschenk für westliche Betrachter", in: *Der Gartenbau*, 100 (1979), No. 27, pp. 1189–1192.

Steiner E. 1959
[Steiner, Emil], "Japanische Akzente an der G/95", in: *Schweizerisches Gartenbau-Blatt*, 80 (1959), No. 33 [20 August 1959], pp. 789–790 (sgd. "St").

Steiner A.-H. 1974
Steiner, Albert Heinrich, "Die Physikbauten und die Infrastruktur an der ETH Hönggerberg in Zürich", in: *SBZ*, 92 (1974) No. 18, pp. 450–454.

Stoffler 2006a
Stoffler, Johannes, *Es ist überall Erdbebenzeit. Gustav Ammann (1885–1955) und die Landschaften der Moderne in der Schweiz*, Diss. ETH Zurich 2006.

Stoffler 2006b
"Die Versöhnung mit der Natur – Der Garten Hauser-Studer in Zürich", in: Sigel, Brigitt et al. (ed.), *Nutzen und Zierde*, Zurich 2006, pp. 308–311.

Stoffler 2008
Gustav Ammann. Landschaften der Moderne, Zurich 2008.

Stoffler 2016a
Fliessendes Grün. Leitfaden zur Pflege und Wiederbepflanzung städtischer Freiflächen der Nachkriegsmoderne, Zurich 2016.

Stoffler 2016b
"Gesichter der Nachkriegsmoderne. Willi Neukoms Aussenanlagen der ETH Hönggerberg in Zürich", in: *Topiaria helvetica, Beton und Biotop. Gärten und Landschaften der Boomjahre*, Zurich 2016, pp. 21–29.

str. 1937
str., "Schweizerischer Werkbund SWB – Ausstellung Paris 1937", in: *Das Werk*, 24 (1937), No. 11, pp. XIV, XVI, XVIII.

Stucky 1959
[Stucky, Fritz] "Landhaus Gimmenen bei Zug: 1958, Architekten: Fritz Stucky und Rudolf Meuli, Zug", in: *Werk* 46 (1959), No. 6, pp. 3, 5 (sgd. "F. S.").

t., 1945
t., "25 Jahre Gartenbauschule Oeschberg – Aus der Tätigkeit der ehemaligen Schüler", in: *Schweizerisches Gartenbau-Blatt*, 66 (1945), No. 51 [20 Dec 1945], pp. 252–253.

t. 1964
t., "Weg zwischen Wasser und Bäumen", in: *NZZ*, 185 (1964), [1 June 1964], No. 2360, n. pag.

Tachibana/Daniels/Watkins 2004
Tachibana, Setsu, Stephen Daniels and Charles Watkins, "Japanese gardens in Edwardian Britain: landscape and transculturation", in: *Journal of Historical Geography*, 30 (2004), pp. 364–394.

Tagsold 2017
Tagsold, Christian, *Spaces in Translation. Japanese Gardens and the West*, Philadelphia 2017.

Tagsold 2019
Japanische Gärten in Deutschland, Museum für Gartenkunst, Stiftung Schloss und Park Düsseldorf 2019.

Takama 1983
Takama, Shinji, *Die wunderbare Welt des Bambus*, Cologne 1983.

Takashima 1937
Takashima, K., "Japanische Gartenideale II", in: *Die Gartenschönheit*, 18 (1937), No. 3, pp. 107–109.

Takei / Keane 2004
Takei, Jirō and Keane, Marc P., *Sakuteiki oder die Kunst des Japanischen Gartens. Die Regeln zur Anlage und Gestaltung aus den historischen Schriftrollen der Heian-Zeit*, Stuttgart 2004.
Takei, Jirō and Keane, Marc P., *Sakuteiki: Visions of the Japanese Garden: A Modern Translation of Japan's Gardening Classic*, Tōkyō/Vermont 2008.

Tamura 1925
Tamura Tsuyoshi, *Zoengaku gairon*, Tōkyō 1925.

Tamura 1937
Jardins japonais: ses origines et caractères dessins et plans, Tōkyō 1937.

Tamura 1964
"Landschaft und Garten in Japan", in: *Garten und Landschaft*, 74 (1964), No. 10, pp. 338–340 (fälschlicherweise sgd. "Tamura, Tsayoshi").

Tandy 1966
Tandy Cliff, *Landscape and Human Life. The Impact of Landscape Architecture Upon Human Activities*. A record of the 9th Congress of the International Federation of Landscape Architects, held in Japan in 1964, Jakarta 1966.

Tatsui 1936/1949
Tatsui, Matsunosuke, *Japanese Gardens*, Tourist Library 5, Japan Travel Bureau, 1934, 1936, 1938, 1947, 5th ed., Tōkyō 1949.

thw. 1958
thw., "Japanischer Bilderbogen", in: *NZZ*, 179 (1958), [29 Nov 1958], No. [illegible], n. pag. [between sheets 6 & 7].

Taut 1937
Taut, Bruno, *Houses and People of Japan*, Tōkyō 1937.

Taut 1997
Taut, Bruno, *Das japanische Haus und sein Leben*, edited by Manfred Speidel, Berlin 1997.

Taut 2004
"Das architektonische Weltwunder Japans – 12. Nov 1934", in: idem, *Ich liebe die japanische Kultur*, edited by Manfred Speidel, 2nd ed., Berlin 2004.

Theile 1957
Theile, Albert, "Tetsuro Yoshida. Der Japanische Garten", in: *Werk*, 44 (1957), No. 10, Exhibitions, pp. 198–199.

Thomas 2008
Thomas, Angela, *Mit subversivem Glanz. Max Bill und seine Zeit*, Vol. 1: 1908–1939, Zurich 2008.

Thunberg 1784
Thunberg, Carl Peter, *Flora japonica*, sistens Plantas Insularum Japonicarum secundum systema sexuale Emendatum Redactas ad XX Classes, Ordines, Genera, et Species, Leipzig 1784

Thunberg 1793–1795
Travels in Europe, Africa and Asia performed between the years 1770 and 1779, 4 Vols., London 1793–1795.

Trüdinger 1956
Trüdinger, Reni, "Tetsuro Yoshida †", in: *Werk*, 43 (1956), No. 11, p. 224*.

Tschichold 1940
Tschichold, Jan, *Der frühe chinesische Farbendruck*, Basel 1940.

Tunnard 2014 (1938)
Tunnard, Christopher, *Gardens in the Modern Landscape: A Facsimile of the Revised 1948 Edition*, University of Pennsylvania Press, Philadelphia 2014 (1st edition Architectural Press, London 1938).

Tschumi 2004
Tschumi, Christian, *Mirei Shigemori 1896–1975. Modernizing The Japanese Garden*, Dissertation ETH Zurich, 2004.

U. I. 1964
U. I., "Japanische Architektur", in: *NZZ*, 185 (1964), No. 66 [8 Jan 1964], sheet 3.

Unger 1909
Unger, Alfred, "Nochmals der Japangarten auf der Internationalen Luftschiffahrtausstellung in Frankfurt a. M.", in: *Die Gartenwelt*, 13 (1909), No. 50, pp. 594–595.

Unger, 1926
"Zur Geschichte der Einführung japanischer Pflanzen in Europa (Japan)", in: *Die Gartenwelt* 30 (1926), No. 38, pp. 604–605.

Unger 1929
– "Meine japanischen Azaleen nach dem strengen Winter (Hausgarten Unger, Heidelberg)", in: *Die Gartenwelt* 33 (1929), No. 32, pp. 445–446.

Unger 1929/1930
· "Japan-Betrachtungen und -Erinnerungen I", in: *Die Gartenwelt*, 33 (1929), No. 45, pp. 629–630.
· "Japan-Betrachtungen und -Erinnerungen II", in: *Die Gartenwelt*, 33 (1929), No. 47, p. 657.
· "Japan-Betrachtungen und -Erinnerungen III", in: *Die Gartenwelt*, 33 (1929), No. 49, pp. 684–685.
· "Japan-Betrachtungen und -Erinnerungen IV", in: *Die Gartenwelt*, 33 (1929), No. 51, pp. 712–713.
· "Japan-Betrachtungen und -Erinnerungen V", in: *Die Gartenwelt*, 34 (1930), No. 2, pp. 25–26.
· "Japan-Betrachtungen und -Erinnerungen VI", in: *Die Gartenwelt*, 34 (1930), No. 4, pp. 52–54.
· "Japan-Betrachtungen und -Erinnerungen VII", in: *Die Gartenwelt*, 34 (1930), No. 6, p. 80.
· "Japan-Betrachtungen und -Erinnerungen V", in: *Die Gartenwelt*, 34 (1930), No. 10, pp. 137–138.

Unger 1930
"Japanische Zier- Ahorn", in: *Die Gartenwelt*, 34 (1930), No. 6, pp. 76–77 (sgd. "Unger A.").

Valentien 1932
Valentien, Otto, *Zeitgemässe Wohngärten*, Munich 1932.

Valentien 1938
Gärten, Berlin 1938.

Valentien 1951
"Gärten", in: *Die Gartenkunst*, 51 (1951), No. 6, pp. 129–133.

Valentien 1954
Gärten, Tübingen 1954.

Valentien 1959
"Elemente des japanischen Gartens in Deutschland", in: *Garten und Landschaft*, 69 (1959), No. 8, pp. 236–239.

V. M. 1907
V. M., "Lafcadio Hearn. Blicke in das unbekannte Japan", in: *NZZ*,128 (1907), No. 58, [27 Feb 1907] edition 2, plts. 1–2.

Vogel 1968a
Vogel, Franz, "Der Garten Japan – 9. Internationaler Kongress der Garten- und Landschaftsarchitekten in Japan", in: *Der Bund* 115 (1964), No. 420 [1 Oct 1964] (evening edition), n. pag. [p. 21].

Vogel 1968b
"Japanische Gartenbilder", in: *Der Gartenbau*, 89 (1968), No. 30, pp. 1333–1340.

Vogt 1996
Vogt, Adolf Max, Le Corbusier – Der edle Wilde, Braunschweig/ Wiesbaden 1996.

Wagner 1932
Wagner, Martin, *Das wachsende Haus, ein Beitrag zur Lösung der städtischen Wohnungsfrage*, Berlin 1932.

Walker 2022
Walker, Sophie, "Zen-Challenge: The Unenterable Garden", in: dies., *The Japanese Garden*, New York 2022.

Wallace 1869
Wallace, Alexander, *Report on the Culture of the Japanese Silkworm, Bombyz Yama-maï, in 1867–68, in England*, Colchester 1869.

Wang 2015
Wang, Yi, *A Cultural History of Classical Chinese Gardens*, New York 2015.

Wasserfallen 1957
Wasserfallen, A. [Adolf], "Schulhausbau in Zürich / School building in Zurich", in: *Bauen + Wohnen / Building + home, 11* (1957), No. 11, pp. 405–410.

Watanabe 2016
Watanabe, Toshio, "A Kyoto Garden Renewal? From Meiji to Early Showa Period", in: Pitelka, Morgan, Alice Y. Tseng (eds.), *Kyoto Visual Culture in the Early Edo and Meiji Periods. The arts of reinvention*, London 2016, pp. 163–181. Excerpt pp. 1–30. https://ueaeprints.uea.ac.uk/id/eprint/66986/1/Chapter.pdf (retrieved: 31 Jan 2023)

Weber 2013
Weber, Heinrich, "Das Frauenkloster St. Josef in Solothurn: ein Klosterneubau in der Schweiz des 20. Jahrhunderts", in: *Jahrbuch für Solothurnische Geschichte*, 86 (2013), pp. 259–318.

Weese 1922
Weese, Artur, "Röthlisberger, Hermann", in: *Das Werk*, 9 (1922), No. 4, pp. 73–76.

Weh. 1908
Weh., "Japanische Kultur", in: *NZZ*, 129 (1908), No. 58, [9 Dec 1908], edition 5, sheets 1–2.

Weilacher 2001
Weilacher, Udo, *Visionäre Gärten. Die modernen Landschaften von Ernst Cramer*, Basel, Berlin, Boston 2001.

Weiss 1996
Weiss, Thomas (ed.), *Sir William Chambers und der Englisch-chinesische Garten in Europa*, Ostfildern-Ruit 1996.

Wettstein 1977
Wettstein, Hans, "Gartenarchitekt Albert Baumann, Oeschberg zum Gedenken – Albert Baumann als Lehrer und Gestalter einer Schule", in: *Anthos*, 16 (1977), pp. 37–38.

Wilhelm 1980
Wilhelm, Richard (ed.), *Liä Dsi [Liezi]. Das wahre Buch vom quellenden Urgrund*, Stuttgart 1980.

Winkler 1943
Winkler, Robert, "Wohlfahrtshaus der SWO, Werkzeugmaschinenfabrik Oerlikon Bührle & Co., Zürich-Oerlikon", in: SBZ, 121/122 (1943), No. 17, pp. 203–207.

Wolschke-Bulmahn 2009
Wolschke-Bulmahn, Joachim, "Gärten, Natur und völkische Ideologie", in: *Die Ordnung der Natur. Vorträge zu historischen Gärten und Parks in Schleswig-Holstein, edited by Rainer Hering, Published by Landesarchivs Schleswig-Holstein* Band 96, Hamburg 2009, pp. 143–187.

Wu 2021
Wu, Chunming, 吴春明, *The Prehistoric Maritime Frontier of Southeast China. Indigenous Bai Yue and Their Oceanic*, The Archaeology of Asia-Pacific Navigation, Vol. 4, Singapore, 2021.

Xiao/Xue 2014
Xiao, Jing, Charlie Q. L. Xue, "Architecture in Ji Cheng's The Craft of Gardens: a visual study of the role of representation in counteracting the influence of the pictorial idea in Chinese scholar gardens of the Ming period", in: *Studies in the History of Gardens & Designed Landscapes: An International Quarterly*, 35 (2015), pp. 218–234.

Yamada 2009
Yamada, Shoji, *Shots in the dark – Japan, Zen and the West*, International Research Center for Japanese Studies, Kyōto, Chicago and London 2009.

Yoshida 1935a
Yoshida, Tetsuro, *Das japanische Wohnhaus*, Berlin 1935.

Yoshida 1935b
"Das japanische Wohnhaus", in: *Deutsche Bauzeitung*, 69 (1935), No. 23, p. 461.

Yoshida 1952
Japanische Architektur, Tübingen 1952.

Yoshida 1954
Das japanische Wohnhaus, [2nd ed.], Berlin 1954.

Yoshida 1957
Der japanische Garten, Tübingen 1957.

Yoshimura/Halford 1957
Yoshimura, Yuji and Giovanna M. Halford, *Miniature trees and landscapes: the japanese art of miniature trees and landscapes: their creation, care, and enjoyment*, Rutland, Vermont 1957.

Zahner 1963
Zahner, Arnold, "Travelling in Japan", in: *Anthos*, 2 (1963), No. 4, pp. 20–29.

Zhaojun/Zhou 1994
Zhaojun, Liu, Yunfeng Zhou (eds.), *Zhuo Zheng Garden [Humble Administrator's Garden]*, Suzhou 1994.

Zhou 1994
Zhou Wen Xiang, *Suzhou Gardens*, Suzhou 1994.

Zhu 1992
Zhu, Junzhen, *Chinese Landscape Gardening*, Beijing 1992.

Zülli 1932
Zülli, Paul, "Wasserfreuden im Garten", in: *Das ideale Heim*, 6 (1932), No. 3, pp. 1–7.

Zulauf 1996
Zulauf, Rainer, "Über Abbilder hinaus", in: *Anthos. Zeitschrift für Landschaftsarchitektur* 35 (1996), No. 4, *Menschen und Gärten*, pp. 12–15.

Zürcher 1978
Zürcher, Adolf, "Garden yards: functionally designed", in: *Anthos*, 17 (1978), No. 1, pp. 14–16.

Zürcher 1980a
"The Garden as a 'Marginal Theme' of a Horticultural Exhibition?", in: *Anthos*, 19 (1980), No. 4, pp. 16–19.

Zürcher 1980b
"'Attractive Garden' sector", in: *Anthos*, 19 (1980), No. 1, *Grün 80*, pp. 39–48 (sgd. "Dölf Zürcher").

Zürcher Denkmalpflege 1975
Zürcher Denkmalpflege – 7. Bericht 1970–1974 (Stadt Zürich 1968–1973) – 1. Teil, published by Direktion der öffentlichen Bauten des Kantons Zürich, Zurich 1975.

Sources

Ammann, Gustav, NSL 2, gtaA, ETH Zurich

Ammann 1910 gtaA
Gustav Ammann, Letter to Vater Jakob Alfred Ammann und Mutter Carolina Ammann-Sauter, 17 July 1910, transcript Johannes Stoffler, 04.12.2014, gtaA, NSL, no rec.

Ammann 1934a gtaA
Peter Meyer, Letter to Gustav Ammann, 1 Jan 1934, 1 p., NSL 2-S-34 M3/3.

Ammann 1934b gtaA
"Japanische Gärten", draft/outline of an unpublished article for the magazine *Das Werk*, Ts., 1934, 3 pages, gtaA, NSL 2-T-3-10 M1/3.

Ammann n. d.a gtaA
"Japan. Kataloge", 1 page, supplement to: Ammann, Gustav, "Japanische Gärten", draft/outline of an unpublished article for the magazine *Das Werk*, Ms., 4 pages, NLS 2-S-34 M3/3.

Ammann n. d.b gtaA
"The Gardens of Japan. By Jiro Harada 1928 Edited by Geoffrey Holme, The Studio Ltd. 44 Leicester Square London W.C.2.", Ms., 6 pages, gtaA, NSL 2-S-34 M3/3.

Ammann n. d.c gtaA
"Buch v. Tetsuro Yoshida. Das japanische Wohnhaus. Verlag Ernst Wasmuth. Berlin 1935. Mit vielen masstäbl. Rissen + Konstruktionsdetails. *Das Werk* März 36", Ms., 1 page, gtaA, NSL 2-S-34 M3/3.

Ammann 1953 gtaA
"Wie wir heute Wasser im Garten am besten zur Geltung bringen", Ts. of the article "Wie wir Wasser im Garten am besten zur Geltung bringen" in: *Schweizer Garten*, 22 (1953), No. 8, pp. 226–228, 2 pages, NSL 2-T-3-5.

Baumann, Albert, ASLA

Baumann A n. d. [1930–32] ASLA
Literatur list of Maruzen Company, LTD., Tōkyō, untitled 3 pages, NL BAA, no rec.

Dubach, Verena, VD, ASLA

Dubach n. d. ASLA
Dubach, Verena, "Bambus", written for *Anthos*, unpublished Ms., pp. 1–3, ASLA, NL VD, n. d., no rec.

Dubach [1958] ASLA
"Der Schattenstauden-Garten oder der Garten für den Besinnlichen", Ts., 1 page, ASLA NL VD_10001_113-0001.

Dubach 1987 ASLA
"Raum-Erlebnis, Raum-Gestaltung, Grüner-Raum", Ts. to lecture given 26 March 1987 at the Association of Swiss Architects (BSA) in Zurich, pp. 1–13, ASLA NL VD, no rec.

Leder, Walter, ASLA

Leder 1960a ASLA
Akira Satō, Letter to Walter Leder, 17 May 1960, ASLA WL, no rec.

Leder 1960b ASLA
Akira Satō, Letter to Walter Leder, 15 August 1960, ASLA WL, no rec.

Leder 1963a ASLA
Satō, Akira, Letter to Walter Leder, 20 Dec 1963, ASLA NL WL, no rec. ["IFLA 1964"].

Leder 1963b ASLA
Letter to Akira Satō, 28 Dec 1963, ASLA NL WL, no rec. ["IFLA 1964"].

Leder 1964a ASLA
"I.F.L.A. Kongress Japan 1964", pp. 1–22, Ms., ASLA NL WL, no rec.

Leder 1964b ASLA
"Comments to the project for 'The Swiss Garden' in Mukogaoka Park in Tokyo", Ts., 22 Feb 1964, ASLA WL, no rec., pp. 1–6.

Leder 1964c ASLA
"Comments to the project for 'The Swiss Garden' in Mukogaoka Park in Tokyo – (continuation)", Ts., 22 Feb 1964, ASLA WL, no rec., pp. 7–14.

Leder 1972a ASLA
Satō, Akira, Letter to Walter Leder, 15 August 1972, ASLA NL WL, no rec. ["IFLA 1964"].

Leder 1972b ASLA
Letter to Pierre Zbinden, 15 August 1972, ASLA NL WL, no rec. ["IFLA 1964"].

Leder 1972c ASLA
Kitamura, Nobumasa, Letter to Walter Leder, 19 August 1972, ASLA NL WL, no rec. ["IFLA 1964"].

Leder 1972d ASLA
"List of Names of Japanese Visitors Group: Zurich 27–29 August 1972", ASLA NL WL, no rec. ["IFLA 1964"].

Leder 1972e ASLA
"Führung für japanische Gartenarchitekten vom Montag, den 28 August 1972", ASLA NL WL, no rec. ["IFLA 1964"].

Leder n. d. ASLA
untitled, n. d., Ms., ASLA WL, no rec., 2 pages

Leder n. d. ASLA, IFLA II
untitled, n. d., ASLA, WL, no rec. ["IFLA 1964 II"].

Mattern, Hermann, NHH, TU

Mattern 1926 NHH
Hermann Mattern, Letter to Herta Hammerbacher, 29 Oct 1926, NHH, TU.

Mattern 1927 NHH
Letters to Herta Hammerbacher, 18 Feb 1927 and 20 April 1927, NHH, TU.

Abbreviations

Archive entries
- ASLA, NL: Archiv für Schweizer Landschaftsarchitektur [Archive for Swiss Landscape Architecture], Fachhochschule OST [University of Applied Sciences OST], Rapperswil, Estate
- gtaA, NSL 2: Geschichte und Theorie der Architektur, Archiv, [Institute for the History and Theory of Architecture, Archive] ETH Zurich, Estate of Gustav Ammann
- HMB: Historisches Museum Baden [Historical Museum Baden]
- TU NHH: Technische Universität Berlin [Technical University of Berlin], Estate of Herta Hammerbacher

Journals
- SBZ: Schweizerische Bauzeitung
- WBW/wbw: Werk, Bauen + Wohnen (since 2000: werk, bauen + wohnen)
- Das Werk / Werk: Das Werk until 1942 / Werk since 1943

Newspapers
- NZZ: Neue Zürcher Zeitung
- NZN: Neue Zürcher Nachrichten

Sources
- Ms.: Manuscript
- Ts.: Typescript

- anon.: anonymous
- n. d.: undated
- No.: Issue number
- plt.: Plate (illustration)
- n. pag.: no page number
- no rec.: no record
- sgd.: signed
- Vol.: Volume

Subject index

Image Credits

Archives

gta Archives, ETH Zurich, Gustav Ammann

- Gustav Ammann: pp. 87 (Paul Schädlich), 89, 90, 139, 140, 142–146, 287

Archive of Swiss Landscape Architecture, ASLA

- Richard Arioli: pp. 52–54, 61, 62, 65, 66, 68–70
- Albert Baumann: pp. 85 (Paul Schädlich), 148 (bottom), 149, 150 (Kinkichirō Honda), 151, 288 (right)
- Ernst Baumann: pp. 98, 99 (bottom), 152, 154–163, 289, 290 (left, centre)
- Ernst Cramer: pp. 91, 92 (bottom), 93, 103, 164, 167, 168, 170–185, 290 (right), 291
- Hans Graf: pp. 186, 187, 189, 190, 191 (top), 192–197, 292, 293
- Fritz and Fredy Klauser: pp. 199, 201–209, 294, 295 (left)
- Walter Leder: pp. 71, 72, 94, 102 (bottom), 107, 109, 110, 210–229, 295 (centre, right), 296
- Heini Mathys: pp. 1, 5, 55, 59, 67
- Evariste Mertens: pp. 230, 233, 235
- Mertens Brothers: pp. 198, 236, 238, 239, 297
- Mertens & Nussbaumer: pp. 298 (left)
- Willi Neukom: pp. 60, 92 (top), 97 (top, Emil Steiner), 97 (bottom), 99 (top), 100–101, 104–106, 111, 114, 115, 117–119, 240, 243–263, 298 (centre, right), 299, 300, 301 (left, centre)
- Adolf Zürcher: pp. 113, 116, 120, 264 (bottom), 265, 267–284, 301 (right), 302

Photo essays

- Martin Linsi: pp. 9–15, 41–48, 74–79, 121–129, 134–136

Museums

- The Harry G. C. Packard Collection of Asian Art, Gift of Harry G. C. Packard, and Purchase, Fletcher, Rogers, Harris Brisbane Dick, and Louis V. Bell Funds, Joseph Pulitzer Bequest, and The Annenberg Fund Inc. Gift, 1975, 1975. 268.107, CC0: p. 21
- The Metropolitan Museum of Art, Gift of Mr. and Mrs. Earl Morse, in honor of Douglas Dillon, 1979, CC0: p. 22
- National Museum Kyōto, A甲228, www.kyohaku.go.jp/, https://de.wikipedia.org/wiki/Datei:Sesshu_-_View_of_Amano-Hashidate.jpg, CC0: p. 23
- Smithsonian's National Museum of Asian Art, Freer Gallery of Art Collection, The Mary and Cheney Cowles Collection, Gift of Mary and Cheney Cowles, F2021.4.12.1, https://asia.si.edu/exploreart- culture/collections/search/ edanmdm:fsg_F2021.4.12.1, CC0: p. 24
- Los Angeles County Museum of Art, Gift of Kochukyo in Honor of the Opening of the Pavilion for Japanese Art, M.88.186, https://collections.lacma.org/node/192098, CC0: p. 26
- The Metropolitan Museum of Art, Charles Stewart Smith Collection, Gift of Mrs. Charles Stewart Smith, Charles Stewart Smith Jr., and Howard Caswell Smith, in memory of Charles Stewart Smith, 1914, 14.76.15, CC0: p. 27
- Smithsonian's National Museum of Asian Art, Freer Gallery of Art Collection, of Charles Lang Freer, F1897.29, CC0: p. 29
- Smithsonian's National Museum of Asian Art, Freer Gallery of Art, Gift of Charles Lang Freer, F1901.166, CC0: p. 30
- Harvard Art Museums/Arthur M. Sackler Museum, Bequest of the Hofer Collection of the Arts of Asia, 1985.352.21.A: p. 32
- Harvard Art Museums/Arthur M. Sackler Museum, Bequest of the Hofer Collection of the Arts of Asia, 1985.352.43.A: p. 33
- Harvard Art Museums/Arthur M. Sackler Museum, Bequest of the Hofer Collection of the Arts of Asia, 1985.352.35.A: p. 34
- Harvard Art Museums/Arthur M. Sackler Museum, Bequest of the Hofer Collection of the Arts of Asia, 1985.352.50.A: p. 35
- Harvard Art Museums/Arthur M. Sackler Museum, Bequest of the Hofer Collection of the Arts of Asia, 1985.352.20.A: p. 36
- Harvard Art Museums/Arthur M. Sackler Museum, Bequest of the Hofer Collection of the Arts of Asia, 1985.352.52.A: p. 37
- National Museum of Modern Art, Tōkyō, https://commons.wikimedia.org/w/index.php?title=File:KitawakiI_Rio-anji.jpg&oldid=721201279, CC0: p. 108

Collections

ETH-Library

- Leo Wehrli: p. 2, ETH- Library, Dia_247-02146, http://doi.org/10.3932/ethza-000043905
- Walter Mittelholzer: p. 230, ETH- Library LBS_MH01-006441, http://doi.org/10.3932/ethza-000298312

Historisches Museum Baden

- Werner Nefflen: p. 102 (top), Historisches Museum Baden, Photo Werner Nefflen, Q.01.9761B, CC BY-SA 4.0

The author

- pp. 16, 73, 80, 130–133, 241, 288 (centre)

Web

- Troy McMullin: p. 7, https://lichenportal.org/portal/imagelib/imgdetails.php?imgid=1113352, CC0
- Unknow, CC0: p. 264 (top), Chamapedia / source unknown / licence: CC BY- NC-ND 3.0 CH /
- Link: https://www.chamapedia.ch/wiki/Datei:1503180_Alpenblick_Dachterrasse.jpg, CC0

Magazin

- Deutsche Gesellschaft für Gartenkunst und Landschaftskultur e.V. (DGGL): pp. 57, 191 (bottom)

Private sources

- Christof Wenger: p. 148 (top)

Biography

Rahel Hartmann Schweizer studied at the University of Zurich and "La Sapienza" Faculty of Architecture in Rome, earning a degree in East Asian Art History and a doctorate in Medieval and Early Modern Art History. As a publicist and curator, she is interested in the reciprocal influences of Western and East Asian culture, interdisciplinarity in architecture and engineering, and "cross-pollination" between the arts. She lives and works in Köniz near Bern.

Acknowledgements

I would like to thank Hansjörg Gadient, Professor of Design and Planning of Urban Open Spaces, Eastern Switzerland University of Applied Sciences (OST), for providing the initial impetus for this book. In his role as director of the Archive for Swiss Landscape Architecture (ASLA), he also connected me with the Swiss Foundation for Landscape Architecture (SLA). I am indebted to the Foundation, its president, Ingo Golz (landscape architect MLA BSLA/SIA), and its board members Brigitte Frei-Heitz, Christa Gebert, Ueli Graber, Christian Gubler, Ueli Leuthold, Margit Mönnecke, Roland Raderschall, Christian Roeder, Judith Rohrer, Johannes Stoffler, and Peter Wullschleger for agreeing to allow me to examine the bequests and reproduce images and plans from them. I would like to thank Sophie von Schwerin, research assistant at the Institute for Landscape and Open Space (ILF) at the OST and curator at the ASLA for her assistance, and likewise Simon Orga for the tireless help he gave me in the same capacity. Manuela Egerter and her team Gian Bamert, Frank Meier, and Nuria Orecchini are also to thank for their work in digitising material at the SLA.

I am thankful to Irina Davidovici, Head of the gta Archive at ETH Zurich, for the goodwill she showed to the project, and to her colleague Almut Grunewald for her untiring and invaluable support in helping me sift through Gustav Ammann's estate.

That I was able to visit and study several of the still existing private gardens and non-public parks of companies and institutions, as well as to arrange for some of them to be photographed, is thanks the generosity of the owners Gratian Anda, Esther and Rudolf Guyer, Annamarie and Paul Luder, Paul Mirani, Irene Gisela Prina, Karin Reinhart, Andrea Schuler and Karl Zünd as well as the company employees Hansueli Faust, Leonid Leiva Ariosa, Belinda Sosa Ivaldi and Katia Moskvitch at IBM Research – Zurich in Rüschlikon (ZH); Esther Gächter-Kobler, Cornelia Keller at the Sisters of the Holy Cross Institute in Menzingen (ZG); Max Weber at UBS Center for Education and Dialogue in Ermatingen (TG); David Wyss, Head of the Planning and Building Department of the Canton of Zug (Menzingen Cantonal School); and Domenico Quattrone at the Landgut Reinhart in Ebertswil (ZH). For this gem by the Mertens brothers, I am also grateful to Thomas Müller, Head of Documentation at Zurich Cantonal Monument Preservation Department for allowing me to consult the extensive documents stored in their archives. Franziska Wartmann took the trouble to search the company archives for photographic records of the courtyards that Ernst Cramer created for Wartmann & Cie AG in Brugg, and Martin U. Wieser (Arch. FH/SIA) provided me with numerous archive documents on the history of Menzingen Cantonal School and its building and gardens.

I am grateful to Jeff Steward, Director of Digital Infrastructure and Emerging Technology at Harvard Art Museums, and his colleague Katie Kujala, Post Production Developer, Digital Imaging and Visual Resources, for allowing me to reproduce their fantastic images of Tosa Mitsunobu's artworks. I would like to thank Stefan Schweizer, President of the Deutsche Gesellschaft für Gartenkunst und Landschaftskultur e.V. (DGGL) and Antje Popp, Head of the DGGL's Federal Office, for their permission to use images from the magazine *Die Gartenkunst*.

The only colour photograph of the Japanese garden created by Albert Baumann at the Oeschberg Cantonal School of Fruit, Vegetable and Horticulture in Koppigen was kindly made available to me by Christof Wenger (landscape architect FH BSLA).

I would like to thank Birkhäuser Verlag and its publishing director Ulrike Ruh for recognising the relevance of the topic and for their trust in my work. I very much appreciated Michael Wachholz's dedicated project management and stimulating collaboration in the development of the project. His astute editing of the text refined the coherence of the content, clarified complex trains of thought, and corrected the occasional linguistic faux pas.

To Katharina Kulke I am likewise extremely grateful for stepping into his shoes at the halfway point and guiding the project to completion with enthusiasm and vigilance, and commitment and competence in equal measure. I was moved by Julian Reisenberger's both creative and sensitive translation into English, which in places also resulted in valuable – literally illuminating – feedback for the German edition.

Claudia Pastors, Hilke Bemm (German) and Patricia Kot (English) were meticulous in spotting potential spelling and punctuation errors.

Martin Linsi, whose eye and camera expertly captured the Japanese nuances of the still existing gardens, has informed the visual tone of this book. His excellent photographs along with outstanding material from the archives provided the raw material for this book. The graphic designers Michael Heimann, Hendrik Schwantes and Haig Walta (Heimann + Schwantes) as well as the litography Pixelstorm have skilfully refined – also in the Japanese sense – and given form to this content. I am grateful to Angelika Gaal for her careful supervision of the production and Holzhausen, Wolkersdorf, for the accurate printing.

This project would not have come about in this form without corresponding financial support. For this I would again like to thank the SLA for their additional contribution in waiving fees for the image rights. The public sector – the cantons of Aargau, Solothurn, Thurgau, Zug and Zurich as well as the Burgergemeinde Bern and Grün Stadt Zürich – supported the publication, as did the Ernst Göhner Foundation, the Federation of Swiss Landscape Architects (BSLA), the company raderschallpartner ag as well as Sibylle Aubort Raderschall and Roland Raderschall.

Rahel Hartmann Schweizer
Borrowed Sceneries

Copy editing:
Patricia Kot

Project management:
Michael Wachholz,
Katharina Kulke

Translation
(German → English):
Julian Reisenberger

Production:
Angelika Gaal

Design:
Heimann + Schwantes, Berlin

Image Editing:
Pixelstorm Litho & Digital
Imaging, Vienna

Printing:
Holzhausen, the book-printing
brand of Gerin Druck GmbH

Paper:
Munken Lynx, 135 g
GardaPat, 130 g

Library of Congress Control
Number: 2024930654

Printed in Austria

FSC
www.fsc.org
MIX
Paper | Supporting responsible forestry
FSC® C108696

Bibliographic information published by the German National Library The German National Library lists this publication in the Deutsche Nationalbibliografie detailed bibliographic data are available on the Internet at http://dnb.dnb.de.

ISBN 978-3-0356-2647-6

e-ISBN (PDF)
978-3-0356-2648-3

German Print ISBN
978-3-0356-2645-2

Birkhäuser Verlag GmbH, Basel
Im Westfeld 8, 4055 Basel,
Switzerland
Part of Walter de Gruyter GmbH,
Berlin/Boston

9 8 7 6 5 4 3 2 1
www.birkhauser.com

Stadt Zürich
Grün Stadt Zürich

Kanton Zürich
Fachstelle Kultur

Burgergemeinde
Bern

Unterstützt vom
Kanton Zug

SWISSLOS
Kanton Aargau

SWISSLOS
KANTON THURGAU

kultur
SWISSLOS
Fonds des
Kantons Solothurn

ERNST GÖHNER
STIFTUNG

Schweizerische Stiftung für Landschaftsarchitektur SLA
Fondation suisse pour l'architecture paysagère FAP
Fondazione svizzera per l'architettura del paesaggio FAP

BSLA FSAP
Bund Schweizer Landschaftsarchitekten und Landschaftsarchitektinnen
Fédération Suisse des Architectes Paysagistes

raderschallpartner ag
landschaftsarchitekten bsla sia